TESTING AND TESTABLE DESIGN OF HIGH-DENSITY RANDOM-ACCESS MEMORIES

FRONTIERS IN ELECTRONIC TESTING

Consulting Editor
Vishwani D. Agrawal

Books in the series:

From Contamination to Defects, Faults and Yield Loss
    J.B. Khare, W. Maly
    ISBN: 0-7923-9714-2
Efficient Branch and Bound Search with Applications to Computer-Aided Design
    X.Chen, M.L. Bushnell
    ISBN: 0-7923-9673-1
Testability Concepts for Digital ICs: The Macro Test Approach
    F.P.M. Beenker, R.G. Bennetts, A.P. Thijssen
    ISBN: 0-7923-9658-8
Economics of Electronic Design, Manufacture and Test
    M. Abadir, A.P. Ambler
    ISBN: 0-7923-9471-2
I_{DDQ} Testing of VLSI Circuits
    R. Gulati, C. Hawkins
    ISBN: 0-7923-9315-5

TESTING AND TESTABLE DESIGN OF HIGH-DENSITY RANDOM-ACCESS MEMORIES

by

Pinaki Mazumder
The University of Michigan

and

Kanad Chakraborty
The University of Michigan

KLUWER ACADEMIC PUBLISHERS
Boston / Dordrecht / London

Distributors for North America:
Kluwer Academic Publishers
101 Philip Drive
Assinippi Park
Norwell, Massachusetts 02061 USA

Distributors for all other countries:
Kluwer Academic Publishers Group
Distribution Centre
Post Office Box 322
3300 AH Dordrecht, THE NETHERLANDS

Consulting Editor: Vishwani D. Agrawal, Lucent Technologies,
Bell Labs Innovations

Library of Congress Cataloging-in-Publication Data

A C.I.P. Catalogue record for this book is available
from the Library of Congress.

Copyright © 1996 by Kluwer Academic Publishers

All rights reserved. No part of this publication may be reproduced, stored in a retrieval system or transmitted in any form or by any means, mechanical, photo-copying, recording, or otherwise, without the prior written permission of the publisher, Kluwer Academic Publishers, 101 Philip Drive, Assinippi Park, Norwell, Massachusetts 02061

Printed on acid-free paper.

Printed in the United States of America

Dedicated to

All who have suffered from memory disorders, like Alzheimer's disease, amnesia, and autism. [1]

---

[1] This book deals with the design of computer memories with gargantuan storage capacity and high reliability. Medical researchers are constantly in pursuit of drugs that will alleviate ailments causing memory degeneration in human beings; hence the dedication.

CONTENTS

LIST OF FIGURES	xi
LIST OF TABLES	xix
ABOUT THE AUTHORS	xxiii
FOREWORD	xxv
PREFACE	xxvii
SYMBOLS AND NOTATION	xxxiii
ACKNOWLEDGEMENTS	xxxvii

1	**INTRODUCTION**	**1**
1.1	Introduction	2
1.2	Cell designs	3
1.3	Circuit implementations	4
1.4	Basic read/write mechanisms in RAMs	11
1.5	A brief history of RAM processing and circuit technology	13
1.6	Application-specific DRAMs and SRAMs	20
1.7	Circuit techniques for improved performance of RAMs	26
1.8	Testing of RAMs	36
1.9	Summary of the focus of the book	38
1.10	A primer on memory testing terminology	39
1.11	Problems	42

2	**ELECTRICAL TESTING OF FAULTS**		**45**
	2.1	Introduction	46
	2.2	Electrical fault modeling	46
	2.3	Description of the TMS4016	50
	2.4	Operation of the pins in TMS4016	51
	2.5	DC parametric testing	52
	2.6	AC parametric testing	58
	2.7	Electrical testing of dual-port SRAMs	65
	2.8	Conclusion	70
	2.9	Problems	72
3	**FUNCTIONAL FAULT MODELING AND TESTING**		**75**
	3.1	Introduction	76
	3.2	Layout-independent functional testing of RAMs	78
	3.3	Techniques dependent on layout and circuit design for functional test of RAMs	131
	3.4	Transparent test algorithms for RAMs	146
	3.5	Fault simulation of RAMs based on memory cell layout	147
	3.6	Conclusion	152
	3.7	Problems	154
4	**TECHNOLOGY AND LAYOUT-RELATED TESTING**		**157**
	4.1	Introduction	158
	4.2	Motivation for technology and layout-oriented techniques	158
	4.3	Faults in bipolar (multi-port) RAMs and their testing	160
	4.4	Faults in MOS SRAMs and their testing	165
	4.5	Gallium Arsenide SRAMs: design, fault modeling and testing	182
	4.6	Faults in MOS DRAMs	204
	4.7	Conclusion	216
	4.8	Problems	218
5	**BUILT-IN SELF-TESTING AND DESIGN FOR TESTABILITY**		**221**

Contents ix

	5.1 Introduction	222
	5.2 Concurrent, non-concurrent and transparent testing	223
	5.3 An overview of BIST approaches for RAMs	224
	5.4 Parallel signature analyzer (PSA): integrating test application with signature analysis	230
	5.5 DFT and BIST architectures based on deterministic testing	234
	5.6 DFT and BIST for parallel testing of DRAMs	249
	5.7 DFT for embedded dynamic content-addressable memories (CAMs)	282
	5.8 Transparent BIST algorithms and architecture	294
	5.9 BIST based on pseudorandom pattern testability of embedded memories	300
	5.10 Scan BIST implementation for embedded memories	317
	5.11 Board level scan-based DFT for embedded SRAMs	322
	5.12 Advantages of this approach	329
	5.13 Conclusion	330
	5.14 Problems	332
6	**CONCLUSION**	**339**
A	**GLOSSARY**	**343**
B	**COMMERCIAL RAM DATA**	**355**
C	**MARKET FOR RAMS**	**365**
	REFERENCES	**367**
	INDEX	**383**

LIST OF FIGURES

Chapter 1

1.1 Static RAM cell designs: (*a*) six-transistor full CMOS, (*b*) four transistor with resistor load NMOS, (*c*) dual-port SRAM with double-ended access, (*d*) static content-addressable memory cell (static CAM or SCAM), (*e*) five-transistor CMOS with a single access transistor, (*f*) dual-port SRAM with single-ended access ... 5

1.2 Dynamic RAM cell designs: (*a*) four-transistor cell, (*b*) three-transistor cell with two control and two I/O lines, (*c*) three-transistor cell with one control and two I/O lines, (*d*) three-transistor cell with two control and one I/O lines, (*e*) three-transistor cell with one control line and one I/O line, (*f*) one-transistor cell with one control line and one I/O line ... 6

1.3 Block diagram of basic internal organization of a 4 Kb SRAM; courtesy [136] ... 7

1.4 Block diagram of basic internal organization of a 16 Kb DRAM; courtesy [136] ... 8

1.5 A DRAM sense amplifier ... 9

1.6 Chip I/O signal timing: (*a*) read, (*b*) write; courtesy [174] ... 11

1.7 (*a*) Planar, (*b*) trench and (*c*) stacked capacitor cells; courtesy [136] ... 15

1.8 Memory cells using ferroelectric capacitors and their hysteresis curves; courtesy [38] ... 17

1.9 A DRAM cell structure gets simpler if a high dielectric constant material such as BST is used instead of standard oxinitride dielectric; courtesy [146] ... 18

1.10 Variations in DRAM demand and price/bit over a 3-year period; courtesy [123] ... 19

1.11 Evolution of multiplexed DRAM sense amplifiers ... 28

1.12 An ATD circuit; courtesy [174] ... 29

1.13 A flow chart of the organization of the book   38

Chapter 2

2.1 A Schmoo plot   48
2.2 The bathtub curve   49
2.3 The TMS4016 - an SRAM chip; courtesy [115]   50
2.4 Protection diodes for input and output pins   54
2.5 Read cycle timing waveforms for the TMS4016 SRAM chip; courtesy [115]   59
2.6 Determination of rise and fall times   60
2.7 Determination of setup, hold and release times   60
2.8 Block diagram of a dual-port SRAM; courtesy [137]   66
2.9 X/Y separation of the two ports in a dual-port SRAM system; courtesy [137]   68
2.10 Single timing combined with address formatting to achieve a time difference between X and Y addresses (the addresses applied to the two ports in an X/Y separation scheme); courtesy [137]   70

Chapter 3

3.1 A RAM system organization   78
3.2 Transition graph of Mealy machines modeling (i)&(ii) stuck-at-zero and stuck-at-one faults; (iii) a fault-free 1-cell memory, with 2 states 0 and 1; (iv) a fault-free 2-cell memory; (v) a state-coupling fault in which an 'up' ('down') i.e., $0 \to 1$ ($1 \to 0$) transition in cell i causes an up (down) transition in cell k; (vi) a transition fault in cell i that forces this cell to remain in its initial state (state upon switching on the power supply) always; courtesy [15, 16, 23]   80
3.3 Stuck-at fault (SAF) and transition fault (TF) model   88
3.4 Coupling fault (CF) model: (i) CFin – B changes state when A undergoes a transition; (ii) CFid – B is forced to 0 or 1 when A undergoes a transition; (iii) SCF – B is forced to a value opposite to that of A; (iv) BF – an 'OR' or 'AND' bridge is formed between A and B, so that both of them are forced, respectively, to the OR or AND of their fault-free states   89
3.5 Pattern-sensitive fault (PSF) model   89
3.6 A checkerboard pattern   93

List of Figures

3.7	State-transition graph for testing 3-coupling faults; courtesy [133]	98
3.8	An example of the ATS procedure	104
3.9	March sequences used in Algorithm B of Nair, et al.(1978); courtesy [122]	108
3.10	Algorithm for constructing an Eulerian sequence for a k-cube	117
3.11	Type-1 neighborhood	118
3.12	Type-2 neighborhood	118
3.13	The two-group method	120
3.14	Basic detection algorithm for NPSFs	121
3.15	Basic location algorithm for NPSFs	122
3.16	TLAPNPSF1G algorithm; courtesy [165]	123
3.17	TLAPNPSF2T algorithm; courtesy [165]	125
3.18	TLSNPSF1G (original version)	127
3.19	TLSNPSF1G (advanced version)[140]	128
3.20	TLSNPSF1T	129
3.21	An SRAM, a single-buffered memory (SBM), and a double-buffered memory (DBM); courtesy [169]	137
3.22	Pointer-addressed memory cell; courtesy [169]	138
3.23	A DRAM cell matrix array (4 Mbit); courtesy [131]	148

Chapter 4

4.1	Fault model used by the SDD technique; courtesy [78]	166
4.2	Circuit technique for (a) the SDD open-circuit test, and (b) the cell-array current detection test; courtesy [78]	168
4.3	Test patterns for Layout-Based Delay Time Testing; courtesy [70]	170
4.4	Bit-line model used in simulation; courtesy [70]	171
4.5	A defect causing a high-impedance output for a 2-input NOR gate; courtesy [152]	173
4.6	An I_{DDQ} monitor; courtesy [108]	174
4.7	One cell of the Philips 64 Kb SRAM; courtesy [120]	177
4.8	Defect models for I_{DD} testing; courtesy [157]	180
4.9	Circuit design for an I_{DD} testable SRAM; courtesy [157]	182
4.10	Structure of HEMT; courtesy [113]	184
4.11	A Basic GaAs SRAM cell; courtesy [113]	185

4.12	Equivalent circuit for HEMT; courtesy [113]	186
4.13	Simplified equivalent circuit for HEMT; courtesy [113]	187
4.14	Devices involved in the read operation; courtesy [113]	187
4.15	Variation of V_1 with threshold voltage; courtesy [113]	188
4.16	Read error as a function of pull-up(PU) and access transistor(TG) βs; courtesy [113]	188
4.17	Simplified circuit for analyzing the write operation; courtesy [113]	189
4.18	Write error as a function of threshold voltages; courtesy [113]	189
4.19	Effect of parameter variations: simulation results; courtesy [113]	190
4.20	Canonical set of resistive paths; courtesy [113]	190
4.21	Simulation of cell with missing load: V12 and V13 are the cell storage nodes. The figure shows a write followed by 3 read operations the last one causing the cell to change state; courtesy [113]	191
4.22	Parametric pattern-sensitive fault due to leakage current; courtesy [113]	192
4.23	Bit-line structure for (a) an SRAM cell and, (b) a DRAM cell	208
4.24	Single-ended write; courtesy [97]	210
4.25	Algorithm 1: Checkerboard test to detect single-ended write and transmission line faults	212
4.26	Algorithm 2: Walking test to verify whether the weak-inversion current causes a bit-line voltage imbalance (serial version)	213
4.27	Algorithm 3: Bit line and word-line decoder test (serial version)	214
4.28	Algorithm 4: Power supply voltage transition test – serial version	215

Chapter 5

5.1	A RAM test configuration	224
5.2	Multiple bit architecture	225
5.3	Multiple array architecture	226
5.4	A self-testable LFSR	227
5.5	Block diagram of a DRAM with an integrated PSA; courtesy [155]	230

List of Figures

5.6	Functional diagram of a parallel signature analyzer; courtesy [155]	230
5.7	A tree RAM organization; courtesy [66]	235
5.8	DFT features of a TRAM chip; courtesy [66]	235
5.9	Pseudocode for the SMarch algorithm	239
5.10	Pseudocode for the SMarchdec algorithm	240
5.11	Pseudocode for the SGalpat algorithm	241
5.12	Pseudocode for the SWalk algorithm	242
5.13	Translation from logical space to physical space; courtesy [65]	245
5.14	BIST without and with delay fault detection; courtesy [144]	248
5.15	A testable DRAM system organization; courtesy [97]	252
5.16	Modified bit line decoder; courtesy [97]	253
5.17	Operation of modified decoder; courtesy [97]	253
5.18	Parallel comparator and error detector; courtesy [97]	254
5.19	3D trench-type memory cell; courtesy [97]	256
5.20	Bit-line voltage imbalance; courtesy [97]	256
5.21	Precharge voltage level degradation due to leakage current; courtesy [97]	257
5.22	Bit-line to word-line crosstalk; courtesy [97]	258
5.23	Single-ended write; courtesy [97]	259
5.24	Pseudocode for the parametric checkerboard test (Algorithm 1)	260
5.25	Pseudocode for the parallel parametric walking test	262
5.26	Bit-line and word-line decoder test	263
5.27	Power-supply voltage transition test	264
5.28	Parallel NPSF test in RAM	266
5.29	Effective neighborhood size; courtesy [101]	268
5.30	Cell type assignment; courtesy [101]	271
5.31	State space graph for neighborhood with $k = 4$; courtesy [101]	273
5.32	Hamiltonian cycles on subgraphs of symmetric 4-cubes; courtesy [101]	274
5.33	Parallel test of PSFs over type-2 neighborhood	276
5.34	Pseudocode for parallel testing for faults in the modified bit-line and the word-line decoders	277
5.35	Pseudocode for parallel testing for faults in the parallel comparator	278

5.36 The column address maskable parallel test architecture (CMT); courtesy [114]   279
5.37 The error detection circuit used in CMT; courtesy [114]   280
5.38 Basic organization of a dynamic CAM; courtesy [99]   283
5.39 Dynamic CAM cell; courtesy [99]   284
5.40 Testable CAM circuit; courtesy [99]   286
5.41 Algorithm 1: test of NPSFs in embedded CAMs   288
5.42 Parallel testing algorithm for stuck-at faults   290
5.43 BIST implementation of the circuit; courtesy [99]   292
5.44 Transparent test algorithm derived from Marinescu's algorithm; courtesy [126]   297
5.45 Deterministic vs. pseudorandom testing   301
5.46 An embedded random-access memory   302
5.47 Markov model for a stuck-at-zero fault; courtesy [103]   304
5.48 Quality of detection & number of samples vs. test length coefficient; courtesy [103]   306
5.49 State probability diagram for stuck-at faults; courtesy [103]   306
5.50 Signal probability vs. test length coefficient; courtesy [103]   307
5.51 Markov model for a CFid $\langle \uparrow, 0 \rangle$; courtesy [103]   308
5.52 Markov model for various CFids; courtesy [103]   310
5.53 Markov model for various CFins; courtesy [103]   311
5.54 Markov model for an SNPSF $\langle \uparrow, S \rangle$; courtesy [103]   312
5.55 Quality of detection vs. test length coefficient; courtesy [103]   313
5.56 Markov model for a data-line fault or a pre-logic fault; courtesy [9]   315
5.57 Markov model for the non-common and common addresses; courtesy [9]   316
5.58 Modified circuit with write enable latch; courtesy [103]   318
5.59 Design of write-sensing latch; courtesy [103]   318
5.60 Random-test algorithm for parallel DFT architecture   319
5.61 Scan path + comparator for output response verification; courtesy [127]   320
5.62 Single scan path for test data application and output response verification; courtesy [127]   321
5.63 Modified double scan path; courtesy [127]   321
5.64 Bus-interface devices used in combined boundary-scan/DFT; courtesy [27]   326

5.65	Modified bit-line decoder	333
5.66	Four types of tessellations	334
5.67	Test generator circuit	336
5.68	Waveforms at the RAM control input	337

LIST OF TABLES

Chapter 2

2.1	The TMS4016 - an SRAM chip; courtesy [115]	51
2.2	Values of DC parameters of the TMS4016; courtesy [115]	51
2.3	AC & DC operating characteristics for the TMS4016; courtesy [115]	53
2.4	Standby current (I_{sb}) conditions for a dual-port SRAM; courtesy [137]	66

Chapter 3

3.1	A fault-detection experiment for a 2-cell memory modeled by a Mealy automaton	81
3.2	A comparison of testing times for coupling fault tests; courtesy [24]	100
3.3	ANPs and PNPs; courtesy [165]	115
3.4	Optimal write sequence to generate DANPs for TLAPNPSF1G; courtesy [165]	125
3.5	Hamiltonian sequences used by advanced TLSNPSF1G; courtesy [165]	129
3.6	Test sequence to locate SNPSFs for advanced TLSNPSF1G; courtesy [140]	130
3.7	A comparison of testing times for NPSFs; courtesy [24]	131
3.8	Defect classes in DRAMs; courtesy [131]	148
3.9	Relationship between functional faults and reduced functional faults	154

Chapter 4

4.1	Simulation parameters for three generations of process technology; courtesy [70]	171

4.2	An actual I_{DDQ} test; courtesy [108]	175
4.3	Read and write errors caused by different failure modes; courtesy [113]	200

Chapter 5

5.1	Serialized march elements	237
5.2	Comparison of three schemes for parallel memory testing	251
5.3	Algorithms and their coverage of parametric faults	261
5.4	All possible SSPSFs and SDPSFs	271
5.5	Truth Table for CAM; courtesy [99]	285
5.6	Content of Memory Cells; courtesy [99]	289
5.7	Test Size Optimality versus BIST Hardware	293
5.8	Test length coefficient for stuck-at faults; courtesy [103]	307
5.9	Test length coefficient for coupling faults; courtesy [103]	309
5.10	Test length coefficient for coupling faults; courtesy [103]	309
5.11	Test length coefficient for coupling faults; courtesy [103]	310
5.12	Test length coefficients for static NPSF: $\langle \uparrow, S \rangle$; courtesy [103]	314
5.13	Test length coefficients for active (dynamic) NPSF: $\langle \uparrow, S \rangle$; courtesy [103]	314
5.14	Boundary-Scan versus Boundary-Scan + DFT; courtesy [27]	324
5.15	Comparison of Different PSF test algorithms	337

Appendix B

B.1	**DRAMs - Manufacturer: Texas Instruments**	355
B.2	**DRAMs - Manufacturer: Toshiba**	356
B.3	**DRAMs - Manufacturer: NEC**	356
B.4	**DRAMs - Manufacturer: IBM**	357
B.5	**DRAMs - Manufacturer: Mitsubishi**	357
B.6	**DRAMs - Manufacturer: Hitachi**	358
B.7	**DRAMs - Manufacturer: Matsushita**	358
B.8	**DRAMs - Manufacturer: Fujitsu**	359
B.9	**DRAM - Manufacturer: Siemens**	359
B.10	**DRAMs - Manufacturer: NTT**	360
B.11	**DRAM - Manufacturer: AT&T**	360
B.12	**DRAM - Manufacturer: Intel**	360

List of Tables

B.13	**DRAM - Manufacturer: Samsung**	361
B.14	**DRAM - Manufacturer: Motorola**	361
B.15	**SRAMs - Manufacturer: Sony**	362
B.16	**SRAMs - Manufacturer: Toshiba**	362
B.17	**SRAM - Manufacturer: Philips**	362
B.18	**SRAMs - Manufacturer: Mitsubishi**	363
B.19	**SRAMs - Manufacturer: Hitachi**	363
B.20	**SRAMs - Manufacturer: Fujitsu**	363
B.21	**SRAMs - Manufacturer: NEC**	364

ABOUT THE AUTHORS

Pinaki Mazumder received a B.S.E.E. degree from the Indian Institute of Science in 1976, an M.Sc. degree in Computer Science from the University of Alberta, Canada in 1985, and a Ph.D. degree in Electrical and Computer Engineering from the University of Illinois at Urbana-Champaign in 1987. Presently, he is an Associate Professor in the Department of Electrical Engineering and Computer Science, The University of Michigan, Ann Arbor. Prior to this, he worked for two years as a research assistant at the Coordinated Science Laboratory, University of Illinois, and for six years as a Senior Design Engineer at BEL, India's premiere and largest electronics organization, where he developed analog and digital integrated circuits for consumer electronics products. During the summers of 1985 and 1986, he worked as a Member of Technical Staff in the Naperville branch of AT&T Bell Laboratories, where he started the CONES synthesis project. He is a recipient of Digital's Incentive for Excellence Award, National Science Foundation Research Initiation Award, Bell Northern Research Laboratory Faculty Award and BFGoodrich Collegiate Inventors Award. His research interests include VLSI memory testing, physical design automation, and ultrafast circuit design using quantum well devices. He has written over 100 archival journal and rigorously reviewed conference papers, and numerous industrial technical reports and memoranda while he worked at BEL and Bell Labs. He was a Guest Editor of the special issues on **multimegabit memory design and testing** of the following two journals: *IEEE Design and Test of Computers* (June 1993) and *Journal of Electronic Testing - Theory and Applications* (August 1994). Dr. Mazumder is a senior member of IEEE, Sigma Xi, Phi Kappa Phi and ACM SIGDA.

Kanad Chakraborty is a Ph.D. candidate in Computer Science and Engineering at the University of Michigan, Ann Arbor. He received a B.Tech. degree in Computer Science and Engineering from the Indian Institute of Technology, Kharagpur, in 1989, and an M.S. degree in Computer and Information Science from Syracuse University, New York, in 1992. Mr. Chakraborty's interests are in the areas of VLSI design and testing, design of CMOS layout generation tools for fault-tolerant VLSI circuits, hardware neural network design and applications, and board-level design automation. He has published three papers in archival journals and two in well-known international conferences. His journal papers are on layout-related testing of SRAMs, and neural net applications in time-series analysis, feature recognition and object classification in a manufacturing paradigm. His conference papers are on efficient testing of busses for complex board-level circuits, and application of neural nets in algebraic equation-solving. He is also the author of a CMOS layout generation tool for built-in self-repairable static RAMs (called **BISRAMGEN**); this work forms the major part of his Ph.D. thesis. Mr. Chakraborty worked as a Design Automation Engineer from May through August 1994 in the area of board-level circuit synthesis at Omniview Inc., Pittsburgh, PA.

FOREWORD

"It is not in the interest of business leaders to turn public schools into vocational schools. We can teach [students] how to be marketing people. We can teach them how to manage balance sheets," stated Louis V. Gerstner Jr. of IBM at the recent Education Summit meeting in New York. He continued, "What is killing us is having to teach them to read and to compute and to communicate and to think." (*TIME*, April 8, 1996, page 40).

The last sentence is most significant because it sets requirements for education and hence gives the specification for a textbook. The textbook should contain all the necessary scientific information that the reader will need to practice the art in the technological world. In addition to the scientific detail, illustrative examples are necessary. The book should teach science without restricting creativity, and it should prepare the student for solving problems never encountered before.

In pursuing our goal of advancing the frontiers of test technology, we must cover applications, education, and research. This is the first textbook in the "Frontiers" series. Semiconductor memories represent the frontier of VLSI in more ways than one. First, memories have always used more aggressive physical design rules and higher densities than other VLSI chips, thus advancing the semiconductor technology. Second, the availability of low-cost memory chips makes numerous software applications possible by fueling the demand for all types of chips.

Testing of semiconductor memories presents challenges that are quite different from testing of logic chips. The state of the art includes specialized fault models, tests, design for testability methods, and test equipment. The knowledge of memory test techniques, therefore, cannot be neglected. In this fast-moving technology we find that the last textbook, *Testing Semiconductor Memories: Theory and Practice* by A.J. van de Goor (Wiley, 1991), is already five years old. Now Mazumder and Chakraborty provide an up-to-date coverage in Chapters 1 through 4.

Very large memory chips have been mass produced for several years. In some ways semiconductor memory technology is a mature art. Yet, memory test presents a new challenge. Modern systems have memory blocks embedded among digital or mixed-signal circuitry. This makes memory design for testability methods, similar to built-in self-test, more useful than before. This topic is covered in Chapter 5.

This book is a basic need for all those who want to learn and practice memory technology. It is also essential for those who design and manufacture modern electronic systems. The latter should, however, supplement the material with literature on general concepts on logic test and design for testability.

I am confident the present volume will serve a very useful purpose irrespective of whether it is used as a textbook in an engineering course on computer memories, as additional reading material in a course on testing and fault tolerance, or as an information source by a practicing engineer.

Vishwani D. Agrawal
Consulting Editor
Frontiers in Electronic Testing Series

PREFACE

This book deals with the study of fault modeling, testing and testable design of semiconductor random-access memories. It is written primarily for the practicing design engineer and the manufacturer of random-access memories (RAMs) of the modern age. It provides useful exposure to state-of-the-art testing schemes and testable design approaches for RAMs. It is also useful as a supplementary text for undergraduate and graduate courses on testing and testability of RAMs.

Organization

The book is organized as follows.

Chapter 1 introduces random-access memories, describing the technological advances made in the past two decades of semiconductor memory design.

Chapter 2 describes electrical testing performed at the interface (i.e., input and output terminals) of memory chips during post-manufacture inspection and field use.

Chapter 3 describes functional fault modeling and testing of memories. It deals with conventional design-independent functional fault models and also more recent design and layout-dependent ones.

Chapter 4 deals with technology and layout-related electrical (or parametric) testing of memories. It describes fault models and test algorithms for electrical faults that require a thorough analysis of cell design, processing technology and layout.

Chapter 5 introduces the concepts of built-in self-test (BIST) and design for testability (DFT) and describes traditional and modern approaches for BIST and DFT.

Chapter 6 concludes the book with an overview of the previous five chapters and describes the importance of testing modern-day SRAMs and DRAMs.

The book includes three appendices:

Appendix A includes a glossary of terminology on RAM testing and testable design.

Appendix B gives tables of commercial DRAM and SRAM data against the names of the manufacturing companies; these tables contain useful information on technological aspects of commercial RAMs.

Appendix C describes commercial RAM markets using pie-charts.

Focus

The book presents an integrated approach to state-of-the-art testing and testable design techniques for RAMs. These new techniques are being used for increasing the memory testability and for lowering the cost of test equipment. Semiconductor memories are an essential component of a digital computer — they are used as primary storage devices. They are used in almost all home electronic equipment, in hospitals, and for avionics and space applications. From hand-held electronic calculators to supercomputers, we have seen generations of memories that have progressively become smaller, smarter, and cheaper. For the past two decades, there has been vigorous research in semiconductor memory design and testing. Such research has resulted in bringing the dynamic RAM (DRAM) to the forefront of the microelectronics industry in terms of achievable integration levels, high performance, high reliability, low power, and low cost. The DRAM is regarded as the technological driver of the commercial microelectronics industry.

A number of books have been written on testing of semiconductor RAMs. The focus of most of these books is on fault models and test algorithms that address a wide range of memory devices. For example, many books describe a generic pattern-sensitive fault model without specific details of the accompanying cell design or RAM layout. In reality, however, many faults are design-dependent. DRAMs, for instance, have a cell design that causes them to have some faults (for example, bit-line voltage imbalance faults) not commonly observed in SRAMs. There are other faults that are technology- and layout-dependent. We have written our book with an eye on the requirements of the design engineer and the manufacturer, who, we feel, may need a more detailed perspective than the researcher. In our view, a manufacturer or a design engineer is concerned with how a fault model maps into a physical defect model at the lowest level of cell design and layout. The fault model is thereby a function of the circuit design and processing technology.

Exhaustive testing is an intractable problem. The total number of possible faults in an N-cell memory is exponential in N and the total number of possible memory cell defects that may or may not produce these faults is even higher. To simplify the problem of testing, a *fault model* is commonly used. A fault model represents a set of commonly occurring faults, such that testing algorithms can be designed with the objective of detecting faults belonging to a specific fault model. This book provides an extensive description of various fault modeling approaches, spanning different levels of abstraction, and covering a wide spectrum of testing characteristics and requirements. A more ambitious approach to testing is via *defect models*. A defect model represents a set of physical defects that may or may not produce observable faults. We have examined various defect modeling approaches and have studied the relationship between the cell design, the technology and memory array layout, and the various testing approaches.

Design of a fault model (or defect model) is itself a difficult task. Most fault models designed in the past have been process and layout independent. Some of these algorithms have a poor fault coverage because they do not use any layout information. For example, neighborhood pattern-sensitive faults (NPSFs) are observable between *physically neighboring* cells, and test algorithms which try to detect these faults without a proper knowledge of the *physical neighborhood* of cells in a memory array are not of much use. Layout-independent models are often found to give a poor *defect* coverage even if they achieve a high fault coverage, because not all defects produce faults. This is especially true for deep submicron processes of the present age. Some recently developed fault and defect modeling techniques for SRAMs require a thorough understanding of the cell type, cell technology and associated processing and layout. For example, the technology and cell design for a GaAs SRAM are quite different from those of a silicon MOS SRAM, and consequently, there are fault types peculiar to GaAs SRAM that are not present in silicon SRAMs. We have dealt with such models in addition to the more conventional ones, and have characterized their testing requirements.

Usefulness of this book

This book is intended for design engineers and manufacturers of random-access memories. We believe that it is at the appropriate level for our intended readership. It deals with real-world examples that should be useful to readers. This book also provides college and university students a systematic exposure to a wide spectrum of issues related to RAM testing and testable design. Last but not the least, each chapter in the book comes with a comprehensive array of problems designed to stimulate readers to read material from outside sources on their own. Our solution to one sample problem is provided in each chapter. These problems are intended to reinforce and further develop readers' understanding of basic concepts in testing and testability of RAMs. Some of these problems are experiments that encourage readers to derive the personal satisfaction of actually *performing* memory testing in an industry-like environment, instead of merely reading descriptive accounts of such experiments. A hint has been occasionally provided for a problem by pointing readers to a journal or conference article that deals with the problem directly or indirectly.

In order to keep the book within a reasonable page limit, we have chosen a concise style of presentation. Important practical issues on memory testing are described in a balanced fashion. These issues are relevant to the manufacturer, the practicing design engineer, and the researcher. The breadth and depth of issues presented justifies using this book as a state-of-the-art reference in semiconductor memory design and testing.

The book has a number of line drawings, most of which have been borrowed from recent papers, but redrawn by us, with suitable modifications. Also, we have presented some algorithms using a cryptic pseudocode format and others in a more verbose style, to make the overall presentation as lucid as possible.

To readers

This book presents a compendium of state-of-the-art literature on diverse aspects of testing and testable design of random-access memories, spanning over 180 research papers. Though we invested a lot of effort in writing an error-free book, a research-oriented book like ours is likely to contain technical and typographical errors. Please contact us at the following address if you notice any errors in the book.

Pinaki Mazumder & Kanad Chakraborty
2215 EECS Building
Department of Electrical Engineering & Computer Science
1301 Beal Avenue
The University of Michigan
Ann Arbor, MI 48109-2122
E-mail: *mazum/kanad@eecs.umich.edu*,
Phone: 313-763-2107, Fax: 313-763-1503.

Solution manual and memory test software

A solution manual for the problems in our book will be made available to instructors who wish to adopt this book as a supplementary text in a graduate-level course on digital and analog testing. Also, we are developing test software for useful electrical, functional and layout-related tests for RAMs. Diskettes (in both IBM and Macintosh formats) containing this software written in C, will be distributed to interested readers who are requested to contact us at the above address.

SYMBOLS AND NOTATION

ABF	And bridging fault
AF	Address decoder fault
ANPSF	Active neighborhood pattern sensitive fault
APNPSF	Active and passive NPSF
ATD	Address transition detection
ATS	Algorithmic test sequence
B_i^L	Left half of a DRAM bit-line
B_i^R	Right half of a DRAM bit-line
BCR	Boundary control register
BF	Bridging fault
BIST	Built-in self-test
BRAM	Battery RAM
BSI	Boundary-scan interconnect (tester)
BSR	Boundary-scan register
CAM	Content addressable memory
CAS	Column address select
CF	Coupling fault
CCF	Capacitive coupling fault in DBM
CFid	Idempotent coupling fault
CFin	Inversion coupling fault
CMOS	Complementary metal oxide semiconductor
DBM	Double-buffered memory
DCF	Dynamic coupling fault
DFT	Design for testability
DR	Data register
DRAM	Dynamic random-access memory
FIFO	First-in first-out
GaAs	Gallium arsenide
GALCOL	Galloping pattern with column read
GALPAT	Galloping pattern of 0s and 1s
GALROW	Galloping pattern with row read

HEMT	High electron mobility transistor
I_{DD}	Normal power supply current
I_{DDQ}	Quiescent power supply current
IFA	Inductive fault analysis
IIF	Input-to-input fault
IOF	Input-to-output fault
IR	Instruction register
LCid	Linked idempotent coupling fault
LCin	Linked inversion coupling fault
LFSR	Linear feedback shift register
LTCid	Linked transition and idempotent coupling faults
LTCin	Linked transition and inversion coupling faults
M	Transition matrix of a Markov chain
MAMB	Multiple-array multiple-bit
MASB	Multiple-array single-bit
MISR	Multiple-input shift register
MOS	Metal oxide semiconductor
MSCAN	Memory scan test
NPSF	Neighborhood pattern sensitive fault
OBF	Or bridging fault
OOF	Output-to-output fault
PAL	Programmable array logic
PAM	Pointer-addressed memory
PRPG	Pseudo-random pattern generator
PSA	Parallel signature analyzer
Q_D	Quality of detection
$p_L(S)$	Probability of reaching state S of a Markov chain after L transitions
r	Read operation
RAM	Random-access memory
RAS	Row address select
RBF	Random bridging fault
RDE	Response data evaluation
ROM	Read-only memory
RUT	RAM under test
SAF	Stuck-at fault
SAMB	Single-array multiple-bit
SASB	Single-array single-bit
SCF	State coupling fault

Symbols and Notation

SDD	Soft defect detection
SDI	Serial data in
SDO	Serial data out
SNPSF	Static neighborhood pattern-sensitive fault
SOI	Silicon-on-insulator
SRAM	Static random-access memory
STRAM	Self-timed RAM
SSPSF	Symmetric static PSF
SDPSF	Symmetric dynamic (active) PSF
SEM	Scanning electron microscope
TAP	Test access port
TDG	Test data generation
TF	Transition fault
UCid	Unlinked idempotent coupling fault
UCin	Unlinked inversion coupling fault
UPSF	Unrestricted pattern-sensitive fault
V_{DD}	Power supply voltage
VDRAM	Video RAM
v_{th}	Threshold voltage
ΔV	Change in voltage
W^L	Left-half word line
W^R	Right-half word line
w	Write operation
XOR	Exclusive-OR
$\Uparrow$	Up addressing order
$\Downarrow$	Down addressing order
$\Updownarrow$	Up/down addressing order
α-particle	Alpha particle (= Helium nucleus)
$\Psi(x)$	Transition write operation on cell x
σ_y	State of cell y
1 Mb	1 megabit = 2^{20} = 1048576 bits
1 Kb	1 kilobit = 1024 bits
1 KB	1 kilobyte = 1024 bytes
1 Gb	1 gigabit = 1024 megabits

ACKNOWLEDGEMENTS

This book is the product of nine years (1986-1994) of research done by Dr. Pinaki Mazumder in diverse aspects of testing, diagnosis, repair, reconfiguration and error-correcting circuitry for semiconductor random-access memories. Mr. Kanad Chakraborty who is now a Ph.D. candidate, joined Dr. Mazumder's research group at the University of Michigan in 1992, and started doing research in testing and fault-tolerance of embedded RAMs. Dr. Mazumder would like to express his heartfelt thanks to his Ph.D. thesis advisor, Prof. Janak H. Patel, of the Coordinated Science Laboratory, University of Illinois, Urbana-Champaign. Prof. Patel's inspiration and encouragement immensely contributed to this author's interest in this area and were instrumental in writing this book. In addition, he would like to thank Prof. Jacob A. Abraham (University of Texas, Austin), Prof. W. Kent Fuchs (University of Illinois, Urbana-Champaign), and Prof. Sudhakar M. Reddy (University of Iowa, Iowa City), who have written seminal papers on testing and fault-tolerance of RAMs, and also encouraged him immensely ever since he started working on his doctoral thesis in 1986, and Prof. John P. Hayes (University of Michigan, Ann Arbor), who is a pioneer in the field of pattern-sensitive fault testing in RAMs. Dr. Mazumder would also like to express his sincere gratitude to Prof. George I. Haddad, who as director of the Center for High-Frequency Microelectronics, took an active interest in his research on memories and provided partial support through the URI (Army) program when he started working as an Assistant Professor at the University of Michigan. Several people who took keen interest in Dr. Mazumder's research and provided him with funding for the various research work described in this book include, Dr. Ralph Kevin of the Semiconductor Research Corporation, Dr. Bernard Chern, Dr. Robert Grafton, Dr. Mike Foster and Dr. Robert Jump of the National Science Foundation, and Dr. Mike Stroscio of the Army Research Office; their support is gratefully appreciated.

The authors acknowledge the significant contributions made by the reviewers who read the manuscript meticulously and promptly provided their valuable feedback: Dr. Vishwani Agrawal of AT&T Bell Laboratories, Dr. Bruce Cockburn of the University of Alberta, Canada, Dr. Kewal K. Saluja of the University of Wisconsin, Madison, and Dr. Yervant Zorian of AT&T Bell Laboratories. Their suggestions and recommendations have improved the quality of the book to a large extent. The authors also thank each member of their research group — Alejandro, Anurag, Mahesh, Mayukh, Mohan, Sanjay and Shriram for the fine job they did in revising the book chapters, and improving the appearance, readability, accuracy, and thoroughness of the book.

The task of writing this advanced book reflecting the authors' efforts (especially, the first author's nine-year research in this area) has required an enormous amount of sacrifice by family members. Dr. Mazumder would like to thank his parents and Swami Vikashanandaji for inspiring him, from early childhood, to strive for excellence; his wife, Deepika, for providing the much-needed emotional and moral support during the years that he burnt the midnight oil at the workplace, somewhat unfairly placing the brunt of family responsibilities on her; and his doting children, Monika (9) and Bhaskar (13), for the joy and pleasure they bring to him. Mr. Chakraborty would like to thank his parents for the love, support and inspiration they have always given him and his wife, Romita, a Ph.D. student in Economics at the University of Maryland, College Park, for her love, care, understanding and patience, despite the 600 miles of geographic separation.

Pinaki Mazumder
Kanad Chakraborty
Ann Arbor, Michigan

1
INTRODUCTION

Chapter At a Glance

- Static and dynamic RAM cell implementations with peripheral circuitry

- Memory cell technologies

- Application-specific DRAMs and SRAMs
  -- VDRAM, BRAM, Multi-port RAM, FIFO, CAM

- Circuit Techniques for Improved Performance
  -- Speed, Low Power, Sense-amplifier design, ATD, Synchronous RAM, Flash write

- Organization of the Book

- Primer on memory testing terminology

1.1 INTRODUCTION

Over the last two decades, semiconductor memories have been the fastest growing market in the semiconductor industry. Their compound annual growth rate in terms of dollars in world sales has been about 7% on average. During this time, they have found their way from mainframes to telecommunication systems. They are also used by households worldwide on a daily basis — those in automobiles record mileage and time even after the ignition is turned off, those in cable TV systems remember the number of channels watched, and those in pocket calculators are used to keep track of intermediate results of calculations. The semiconductor memory market, expanding rapidly as more people become aware of the diversity of use of memory chips, is currently about 35% of the total semiconductor market.

Semiconductor memories came into existence in the early 70s, and rapidly replaced the slow, expensive and bulky magnetic core memory that was used till then. The price per bit of these memories has been decreasing monotonically since the mid-70s. N-channel MOS devices have always had the lowest price per bit in the MOS family. Price per bit of CMOS devices has shown a sharp downward trend through the 70s and 80s, by more than an order of magnitude every five years, starting at about eight cents per bit in 1974.

The various semiconductor memories have been conventionally divided into five basic types: bipolar, charge-coupled devices (CCD), magnetic bubbles, complementary MOS (CMOS), and N-channel MOS (NMOS). Nowadays, NMOS and CMOS are most commonly used. The others are not as widely used because of a variety of reasons.

Even though bipolar devices represent a limited section of the market, they are worth mentioning for their high speed. Bipolar technologies that are currently in use are ECL (emitter coupled logic), which has very high speed but also high power consumption, TTL (transistor/transistor logic) or I^2L (integrated injection logic) for a more moderate speed/power performance. Besides, hybrid bipolar-MOS technologies (BiCMOS) have been recently implemented. Bipolar memories are inferior to MOS memories with respect to size, density and organization. Charge-coupled devices (CCDs) work by pulsed charges stored under depletion-biased electrodes along the chip surface. They are much slower than MOS memories, with the added disadvantage of being serial access devices. They are currently used in a very limited range of applications that demand neither the performance nor the flexibility of random-access memories, for example, simple solid-state imaging systems such as video cameras. Because of

Introduction 3

their analog storage capability, they are also useful in some telecommunication transmission systems. Magnetic bubbles (of which 4 Mb devices have been built) are non-volatile, serial-access and slow compared with random-access memories, and were mostly used for diskette replacements in computer systems. Most of the NMOS manufacturers like Texas Instruments, Rockwell, National Semiconductor, Motorola, Siemens, Fujitsu and Intel, started into the magnetic bubble market but stopped production by 1982 because of high cost and low yield. NMOS and then later, CMOS, dynamic random-access memories (DRAMs) started pervading the market because of their low cost, high performance, high density, low power dissipation, random access, non-volatility, ease of testing and high reliability. The devices manufactured nowadays using MOS technologies are RAMs (random-access memories), ROMs (read-only memories), EPROMs (erasable programmable read only memories), EEPROMs (electrically erasable programmable read only memories), and EAROMs (electrically alterable read only memories). This book will focus solely on the testing and design for testability of RAM circuits.

1.2 CELL DESIGNS

A random-access memory (RAM) is a memory which supports both read and write operations of data (hence often called a read/write memory), with a read access time that is independent of the physical location of the data. Serial access memories, for example, magnetic tapes used for secondary storage, and bubble memories, have some location-dependent latency associated with access.

Structurally, RAMs are of two types — static and dynamic. A static RAM, also called an SRAM, stores a bit of data in a latch formed by a pair of cross-coupled inverters. A dynamic RAM, also called a DRAM, stores a bit of data as a tiny quantity of charge on a capacitor (a few fF). Since this charge decays away to a very small value within a few milliseconds, a periodic refresh is needed to restore the charge.

The earliest RAM cells consisted of six to eight transistor flip-flops or latches. Continued technological advances introduced the first dynamic RAM (DRAM) cell composed of three transistors and a storage capacitor. This arrangement causes a logic-1 or logic-0 to be represented in the form of a tiny quantity of charge stored or not stored in a capacitor. The development of the DRAM has decreased the size of each storage cell and has caused a higher storage density. However, an additional design complexity was introduced by the necessity of refreshing. The early developments introduced the two different RAM types — static and dynamic, depending on whether the charge is stored by a cross-coupled inverter or a capacitor. These cell designs are shown in Figures 1.1 and 1.2. For static RAMs (SRAMs), the six transistor cell is the most widely used design, whereas single-transistor cells have been designed for DRAMs, as shown.

1.3 CIRCUIT IMPLEMENTATIONS

A RAM cell array consists of rows and columns of either SRAM or DRAM cells. The internal organization is shown in Figures 1.3 and 1.4.

There are row and column decoders to choose a row and column. These decoders, under normal conditions, choose only a single row or a single column for a read or write access. However, there exist some test techniques that modify the design of the decoder to perform multiple access; they will be described later.

The main components of a RAM are described below.

1. **Decoders:** CMOS row and column decoders can be implemented by using NAND gates that stack N-channel devices. Design of such decoders is shown in [175].

2. **Cells:** The cell design depends on the layout ground rules. There are usually five wires associated with each cell: the power and ground lines, one word line and two bit lines. Most word lines are polysilicon lines which form the gates of the access transistors whereas most bit lines are either diffusion lines which can easily branch into cells to form pass devices, or metal lines. The power and ground rails are solid, to avoid the coupling noise. They are usually formed with metal. The choice of wires is also dependent on minimum geometries, cell reliability, and feasibility of sharing common structures among adjacent cells. For example, metal contacts are best placed on flat surfaces, and many cells share contacts to the power and ground buses. Stability and power dissipation dictate a large W/L ratio for the inverters.

3. **Read/write control:** This is a simple control input pulse that selects either read or write operation. This pulse is not latched, and hence should be held constant for the entire cycle of the operation.

Introduction

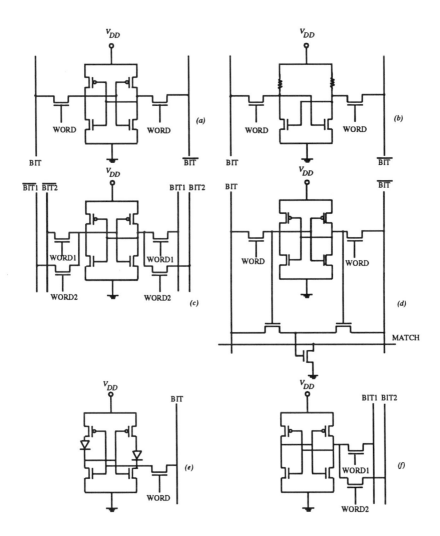

Figure 1.1 Static RAM cell designs: (a) six-transistor full CMOS, (b) four transistor with resistor load NMOS, (c) dual-port SRAM with double-ended access, (d) static content-addressable memory cell (static CAM or SCAM), (e) five-transistor CMOS with a single access transistor, (f) dual-port SRAM with single-ended access

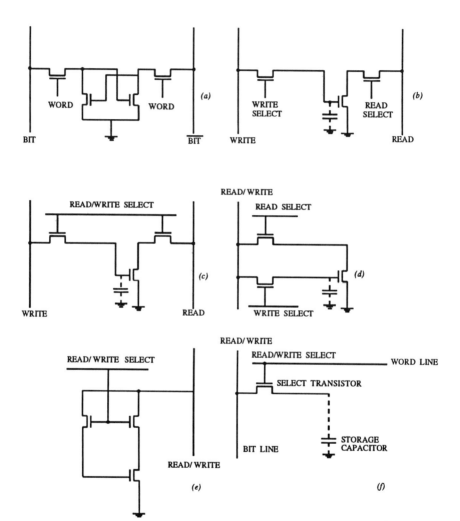

Figure 1.2 Dynamic RAM cell designs: (a) four-transistor cell, (b) three-transistor cell with two control and two I/O lines, (c) three-transistor cell with one control and two I/O lines, (d) three-transistor cell with two control and one I/O lines, (e) three-transistor cell with one control line and one I/O line, (f) one-transistor cell with one control line and one I/O line

Introduction

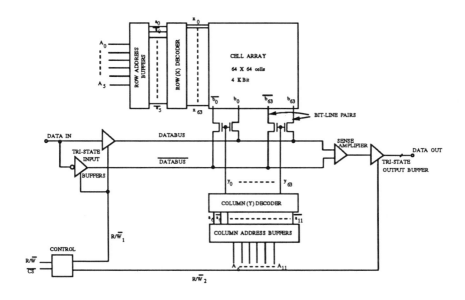

Figure 1.3 Block diagram of basic internal organization of a 4 Kb SRAM; courtesy [136]

4. **Sense amplifiers:** Signals on bit lines are latched up by sense amplifiers during the read operation. During a read operation, the column (bit) address is first decoded to select a $bit\text{-}\overline{bit}$ pair which are both initially precharged to a high voltage. Then the row (word) address is decoded to select a particular storage cell. When a cell is selected, the data stored in it will cause either the bit or the $\overline{bit}$ line to be pulled to a low voltage. The differential voltage signal is detected by the sense amplifier, amplified, and then latched into a cell of the output buffer.

DRAMs have, in addition, the following features.

5. **Specialized sense amplifiers:** Figure 1.5 gives a high-level view of the sense amplifier in a DRAM. The sense amplifier in a DRAM reads the contents of the cell and also replenishes the cell with fresh charge. Most DRAMs refresh their cells with forced read, which is done either with built-in circuitry or off-chip by the user. On one side of the sense amplifier is a memory cell that is either fully charged or completely discharged, depending on its stored voltage level. On the other side is a dummy cell which stores half the maximum charge. Each of these cells is connected

8 CHAPTER 1

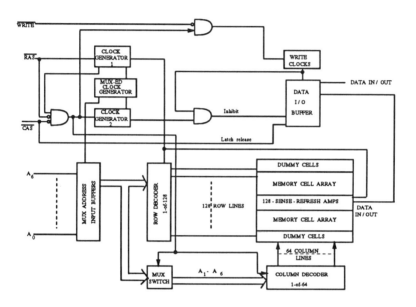

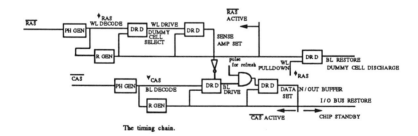

Figure 1.4 Block diagram of basic internal organization of a 16 Kb DRAM; courtesy [136]

Introduction

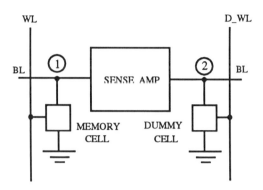

Figure 1.5 A DRAM sense amplifier

to the sense amplifier through a *half-bit* line. When the DRAM cell is not being accessed, both half-bit lines are precharged to the same potential, say $V_{DD} - V_{th}$, where V_{DD} is the power-supply voltage and V_{th} is the gate threshold voltage. During a read operation, the word lines WL and D_WL go high, causing charge-sharing between the memory cells and dummy cells. After this redistribution of charge, the point labeled **1** in Figure 1.5 has either a higher voltage or a lower voltage than the point labeled **2**, depending on whether the memory cell stores a 1 or a 0. This voltage difference is amplified by the sense amplifier, which will store value 1 or 0 in the output latch accordingly.

7. **Multiplexed addressing:** The simple structure of a 1-transistor DRAM cell allows an extremely high packing density. Manufacturers use most compact layout rules to maximize the number of cells in the DRAM array. This causes a large number of address inputs (for example, a 1 Mb DRAM needs 20 address bits). Hence, almost all practical designs use time-multiplexed row and column addresses on the same pins.

8. **Timing chains:** Most circuits in DRAMs are dynamic, and hence no DC power is dissipated even when a cell is being accessed. Since dynamic circuits must use clocks, the design of the internal timing plays an important role in DRAM performance. In practice, the number of on-chip clocks or 'phases' varies from 6 to 30. These phases are generated sequentially in a driver chain with appropriate delays. Since the loads have very high capacitances, powerful dynamic drivers are used for the clock phases. These driver chains, called timing chains, control the operation of the whole chip. The driver chains are interlocked at various critical points to avoid race

among signals generated from different chains. Some of these timing generators and their role in read/refresh and write are described below.

9. **Phase and restore generators:** The memory cycle of a DRAM begins when a clock signal, $\overline{RAS}$ goes low. A *phase generator* PH_GEN (Figure 1.4) inverts and adjusts the control signal; and the output of this generator drives the word address decoders, and begins word decoding. These decoders are fed with address input values produced by *true/complement* generators. Once the RAS timing chain has been initiated, a high-power delayed driver (DR_D, described a little later) is also driven by the $\overline{RAS}$ pulse. This driver sets a pulse that activates the sense amplifiers.

   Once the sense amplifiers are activated, the $\overline{CAS}$ goes low, and the CAS timing cycle starts. The first clock pulse of the CAS drives the true-complement generators of the column (bit) decoders, and also drives the corresponding DR_D. This, in turn, generates a pulse that connects the sense amplifier to the I/O buses by turning on the selected bit-line pair. The read/write ($W/\overline{R}$) control comes into effect as soon as the sense amplifier is connected to the I/O buses. During a read operation, the data latched in the sense amplifier will be output, with the help of a clock pulse, the output being valid after the I/O latch has settled down; during a write operation, the $WRITE$ latch contents are fed to the bit lines. After the output data is valid, the $\overline{CAS}$ pulse remains low for a period of time, and then goes high, causing the valid data period to end with some delay and the output to return to a high impedance state.

   After all sense amplifiers have settled down, RAS timing is terminated. All drivers in this timing chain remain set, however, until $\overline{RAS}$ goes high. As soon as this signal goes high, a strong *restore* pulse is generated by the *restore generator* R_GEN. This pulse resets all DR_Ds in the chain and pulls all word lines low. This is followed by another pulse that precharges the bit lines and discharges the dummy reference cells in preparation for the next memory cycle.

10. **Delayed Driver (DR_D):** The delayed driver DR_D is a set/reset driver with dominant reset control. Its function is to generate pulses to drive the word line and dummy cells during the RAS cycle and to connect the sense amplifier to the I/O buses during the CAS cycle.

Introduction

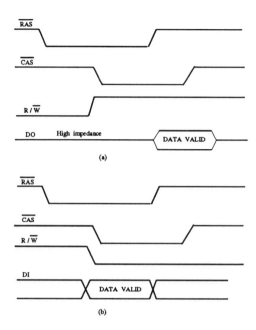

Figure 1.6 Chip I/O signal timing: (a) read, (b) write; courtesy [174]

1.4 BASIC READ/WRITE MECHANISMS IN RAMS

Figure 1.6 shows a series of waveforms that describe the chip cycle time for an SRAM during read and write operations.

For writing into a six-transistor static RAM cell as shown in Figure 1.1(a), first the cell has to be selected by making the word line active (high). Then a high voltage is applied to the *bit* line and a low voltage to the $\overline{bit}$ line, or vice versa, depending on whether a 1 or a 0 is to be written. The cell is a flip-flop, or a bistable multivibrator, which means that it has two stable states, defined as logic 1 and logic 0. Writing into a 1-transistor DRAM cell involves selecting the cell with the word line and then applying the data to be written (a high or a low voltage) on the bit lines. This causes the drain capacitance of the MOS transistor to either store a charge if a 1 is applied, or no charge if a 0 is applied. Since the data is stored as charge on a capacitor, the problem of charge leakage with time is encountered. A dynamic RAM, therefore, suffers from the disadvantage of being *volatile* even when the power supply is on, unlike a static RAM. A static RAM has two stable states and gets charged from the power supply through the load devices. Refresh circuitry is needed to restore the charge periodically in DRAMs. Refreshing involves reading the contents of a DRAM cell and then writing back the read value. DRAMs also need external address counters to keep track of cells that are being refreshed. DRAM design requires engineering sophistication and DRAMs have, therefore, been developed only by manufacturers of large systems who can afford the cost of the engineering before exploiting cost/performance advantages.

For reading out the contents of an SRAM or DRAM cell, the system selects the cell by turning on various control signals, such as chip enable, correct word line address, and output enable, and waits for a certain minimum time to receive the data. The column decoder output selects a pair of bit lines. These bit-line contents are then transferred to sense amplifiers. A sense amplifier senses the voltage difference between the bit lines and computes whether the cell stores a 1 or a 0. The appropriate value is latched into an appropriate storage cell of the data register.

The sense amplifier used during reading a memory cell compares the charge stored in a memory cell with a known charge or reference level in a dummy cell. A signal of 200-300 mV is typical on the bit lines. This combined with a high minimum sense amplifier voltage sensitivity of 50 mV gives acceptable noise immunity. Two approaches that have been used in the design of sense amplifiers by Motorola are open and folded bit sense lines. Open bit sense lines align the sense amplifiers in the middle with cell arrays on both sides. This has the disadvantage of introducing unbalanced noise in the layout, which might cause wrong sensing. In the folded bit-line scheme, the column decoders are at one end of the array and sense amplifiers are at the other end. The bit lines are parallel, folded and have a separation of about 9 μ. Thus, any fluctuations in the column decoders are not transmitted to the bit lines directly. Substrate or junction noise are not likely to affect the sense operation because they will be common-mode noise.

1.5 A BRIEF HISTORY OF RAM PROCESSING AND CIRCUIT TECHNOLOGY

Random-access memories (RAMs) are at the forefront of commercial microelectronic design. The most spectacular achievements with regard to high integration in a cost-effective manner have been witnessed in the design of RAMs, particularly that of DRAMs. The DRAM has been the technological driver for the semiconductor industry.

The first of the modern DRAMs was made in the older P-channel MOS technology, and had a maximum storage capacity of 4 Kb. Philips came up with a design in 1973 that operated with a regulated timing arrangement and an on-chip shift register. This design was rather bulky by today's standards (minimum feature width was 10 μ, and the cell size was 864 μ^2) and the chip-area utilization was rather small (array to chip-area ratio was about 26%). Mostek, now a part of SGS-Thomson, introduced the first address-multiplexed DRAM chip as a 16 pin package, which had a capacity of 4 Kb. With time-multiplexed addressing, each of these devices used six address pins to select one of 64 rows followed by one of 64 columns, instead of using twelve address pins. The six multiplexed address lines were clocked into the row and column decoders using two strobes, the row address strobe $\overline{RAS}$ and the column address strobe $\overline{CAS}$. Also, latches were used for storing input and output data, and this simplified design caused a significant reduction in the pin count of the total package.

The era of innovative technology dawned with the design of the 16 Kb DRAM that had the basic structure of the next few generations of DRAM design up to and including the 1 Mb DRAM. This new technology started out as a scaled version of the existing production technology to achieve higher densities and smaller chip sizes. In 1976, Intel introduced the 16 Kb DRAM and this device rapidly became the industry standard at this density level, following the standard of address multiplexing established by Mostek for 4 Kb DRAMs. Since 16-pin packages were still in use, the systems could easily be upgraded without changing the board size. The first 16 Kb DRAM had a typical access time of 150 ns and a cycle time of 350 ns, drawing a current of 40 mA for the 500 ns cycle and a standby current of 1 mA. Moreover, it required three different power supplies: +12 V for V_{DD}, +5 V for the output driver, and -5 V for the substrate bias.

The Intel 16 Kb DRAM used a 5 μ N-channel technology. It also used the double-poly single-transistor cell which became the industry standard for DRAMs of up to 1 Mb storage capacity. This cell became known as a **planar** cell because it utilizes a two-dimensional storage capacitor which is placed horizontally beside the single transistor, occupying about 30% of the area of the cell. The next three generations of DRAM production tried to scale this basic cell down while retaining sufficient charge necessary for correct operation. Details of a double-poly cell are given in [136]. In the process of scaling down the basic cell, several technological advances took place. Efficient array organization to achieve reduced refresh time, such as the folded bit-line architecture [136], and more sensitive and faster sense amplifiers using the 1/2 V_{CC} sensing scheme, were developed. Also, the requirement of three different power supply voltages was eliminated in favor of a single 5 V power supply. This resulted in more efficient pin usage than before. Moreover, ways to reduce noise-related problems caused by alpha particles from cosmic radiation and/or nearby package materials were developed. Companies like Fairchild, Mitsubishi, Inmos, NTT, NEC and Toshiba started the production of the next three generations of DRAMs — the 64 Kb, the 256 Kb and the 1 Mb DRAM.

A new age of innovation was ushered in by the multi-megabit ($>$ 1 Mb) DRAMs in 1984. The basic approach was the replacement of the planar technology by more efficient charge storage technologies such as the buried **trench** capacitor and the **stacked** capacitor storage plates. The major departure from conventional planar storage was the introduction of new forms of vertical storage to reduce the size of the horizontal capacitor. A trench capacitor is a capacitor cell folded vertically into the surface of the silicon in the form of a trench, to obtain greater area. Isolation of the vertical trenches between adjacent devices allows the transistors to be moved closer together and minimizes the problem of channel width control in two distinct ways. First, it eliminates the field implant; second, it eliminates the formation of a bird's beak produced by lateral growth of oxide that causes current leakage. There were some initial difficulties in implementing the trench technology, hence industries came up with an alternative method, called the stacked capacitor. This method stacked the capacitor over the bit line, above the surface of the silicon, instead of burying it inside the silicon like a trench capacitor. The trench and stacked capacitor technologies are illustrated in Figure 1.7.

The first 4 Mb DRAMs were reported in 1984. Three 4 Mb DRAMs were discussed at the 1986 International Solid-State Circuits Conference (ISSCC) by Texas Instruments, NEC and Toshiba. Five more 4 Mb DRAMs were reported in 1987 by Mitsubishi, IBM, Hitachi, Matsushita and Fujitsu, and one in 1988 by Siemens. Eight of these devices used trench capacitor cells. These new

Introduction

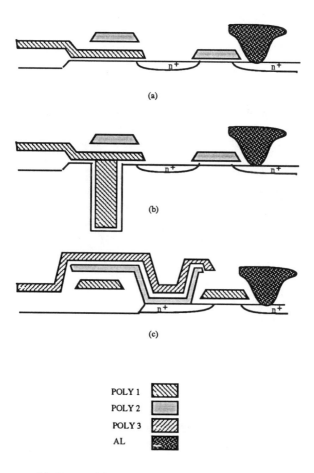

Figure 1.7 (a) Planar, (b) trench and (c) stacked capacitor cells; courtesy [136]

DRAMs began to ramp into volume production in 1989. By early 1990, at least eight manufacturers — NTT, Toshiba, Matsushita, Hitachi, Mitsubishi, Samsung, Texas Instruments, and IBM — had described the design of 16 Mb DRAMs at various technical conferences. In 1990, early samples of the 16 Mb DRAM, manufactured by Samsung, Texas Instruments, and others, began to appear in the market. Their volume production began only in 1991. In 1993, an IBM-Siemens-Toshiba combined manufacturing team presented the design of a 256 Mb trench-cell DRAM at the International Electron Devices Meeting in Washington, D.C..

There were problems, however, with these new storage techniques. Two problems discovered with the early trench capacitors were trench-to-trench current leakage and reduction of alpha-particle induced soft-error immunity. Toshiba proposed a triple-well structure to suppress trench to trench punch-through current. This structure consists of an N-well placed between two P-wells in its own P-well, such that each well and substrate can be independently biased. This improves the body effect so that high performance transistors can be obtained. For 16 Mb DRAMs, Hitachi proposed a different technique for solving these problems — they isolated the storage node from the substrate, and used an isolated storage plate similar to the buried trench capacitor used for 4 Mb DRAMs.

In 1988, non-volatile ferroelectric RAMs were first reported [38]. A ferroelectric RAM consists of a standard SRAM or DRAM cell over which a small 'battery' consisting of a ferroelectric material sandwiched between two metal electrodes is deposited.

The ferroelectric material is a dielectric with three useful characteristics — its dielectric constant is about 100 times larger than that of silicon dioxide or silicon nitride, it has two stable states between which the material can switch when a voltage is applied, and it retains some charge even when the voltage is brought down to zero after charging the cell to saturation (this phenomenon is called *hysteresis*). Hence a ferroelectric device can be used as a non-volatile memory cell with high storage capacity. Figure 1.8 gives the cell design and hysteresis curve of ferroelectric memory cells.

In 1993, DRAM manufacturers started experimenting with new high dielectric-constant ferroelectric materials such as **barium strontium titanate (BST)** [146]. The primary reason why such new materials are useful for memory cell capacitors is that existing dielectrics cannot be shrunk much further without sacrificing the capacitance level for proper device operation. High dielectric materials allow greater storage capacitance within the same physical space.

Introduction

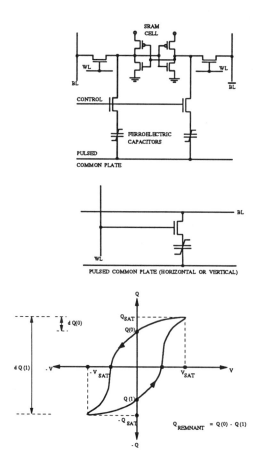

Figure 1.8 Memory cells using ferroelectric capacitors and their hysteresis curves; courtesy [38]

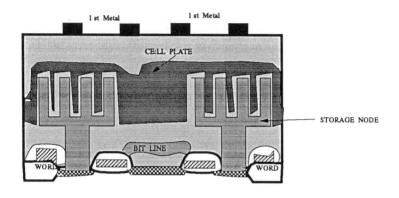
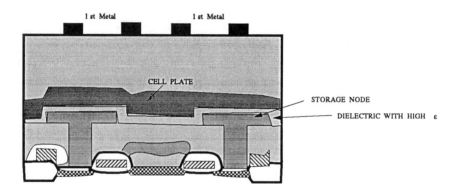

Figure 1.9 A DRAM cell structure gets simpler if a high dielectric constant material such as BST is used instead of standard oxinitride dielectric; courtesy [146]

The capacitor material most commonly used nowadays in the industry is ONO (oxynitride oxide) which has an effective dielectric constant of about 5.0. By comparison, the dielectric constants of Ta_2O_5 and BST are 25 and greater than 500, respectively. Figure 1.9 illustrates how the DRAM cell becomes smaller if a high dielectric material is used.

As observed by Prince [136], a single 4 Mb DRAM chip about the size of a thumbnail contains more storage capacity nowadays than a room-size computer did thirty five years ago. Besides, the cost of producing DRAMs has gone down quite impressively: from $8000 per megabyte in 1975 to $500 per megabyte in

Introduction

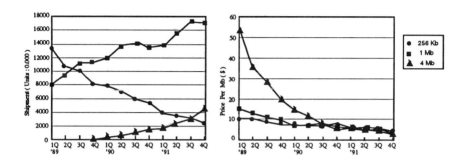

Figure 1.10 Variations in DRAM demand and price/bit over a 3-year period; courtesy [123]

1985. Nowadays, the production cost per megabyte is about $10. The cost of a DRAM manufacturing facility doubles every three years. A factory for 4 Mb DRAMs can cost as much as $350 million and a factory for 16 Mb DRAMs can cost about $700 million [136]. Prince [136] estimates that a company whose DRAM sales grow at the average rate of the DRAM industry will have an almost constant revenue to capital investment ratio. The demand and price per bit for DRAMs over a 3-year period is illustrated in Figure 1.10.

The bit density of DRAMs (number of bits per unit wafer area), however, has quadrupled about every three years. As a result, the price per bit has fallen continuously with time. The motivation, therefore, is to increase the number of bits on a single wafer and this, in practice, translates into increasing the number of bits per chip. Over the years, the rate of increase in the chip size has been about an order of magnitude less than the rate of decrease in the cell size. As we have just seen, DRAMs expanded from 4 Kb storage capacity to 16 Kb from 1973 to 1980 when PMOS technology was used. The first CMOS DRAM, which had a capacity of 64 Kb, was introduced by Intel in 1983.

SRAMs have a smaller chip density but have the advantages of faster operation (access times for high-performance silicon SRAMs and DRAMs are about 2 ns and 20 ns, respectively) and a wider range of application than DRAMs. For example, RAMs embedded within application specific ICs (ASICs) and microprocessors are mostly static. Historically, SRAMs have lagged behind DRAMs by more than a generation. For example, the first 16 Mb DRAMs were presented in 1987, but the first 4 Mb SRAMs were not reported until 1989.

Early SRAMs used bipolar, NMOS and CMOS technologies, and these products had widely different characteristics and targeted markets. Bipolar SRAMs that form less than 1% of the total SRAM market, are still used for very fast applications. Commodity SRAMs use the 'mixed-MOS' technology which combines CMOS for low power and NMOS for high density. Some ultra-fast SRAM technologies are being used these days, such as BiCMOS, CMOS ECL-compatible, and Gallium Arsenide (GaAs). In this book, the relationship between various processing technologies and fault and defect types will be thoroughly examined in Chapter 4.

1.6 APPLICATION-SPECIFIC DRAMS AND SRAMS

A large number of practical board-level circuits use a few memory chips and a wide-bus microprocessor. Therefore, wide-word memories are needed. Since the conventional design techniques have been biased toward SRAMs, system designers of modest board-level circuits found it easier to work with SRAMs than with DRAMs. DRAM designers, in turn, tried to manufacture 'SRAM-like' DRAMs to satisfy the system designer. They came up with concepts like wider-word DRAMs with non-multiplexed addressing (for example, the 128 Kb × 8 DRAM from NEC, 1984), DRAMs with external control as simple as that of SRAMs, built-in refresh and address functions, DRAMs with lower standby power characteristics, and so on.

1. **Pseudostatic and Virtually Static DRAMs (PSRAMs and VS-RAMs):** These DRAMs were provided with built-in refresh control and could be used in SRAM sockets. Depending on the function implemented, they were called either pseudostatic DRAMs (PSRAMs) or virtually static DRAMs (VSRAMs).

    A pseudostatic DRAM (PSRAM) is a single-transistor cell DRAM with non-multiplexed addressing, an on-chip refresh address counter, and in most cases, an external 'refresh' pin. The 8-bit wide input and output ports make the device resemble a slow static RAM. The device is synchronous and operates from a system clock applied to the chip enable pin. The address access time is normally 1/3 to 1/2 of the cycle time. Refresh occurs during the part of the cycle time when the memory is not being accessed. Since refresh is externally controlled, the processor is usually not stalled by a slow cycle. Some examples of PSRAMs are a 256 Kb

Introduction   21

PSRAM (1987) and a 1 Mb PSRAM (1989) from Hitachi [57], and a 1 Mb PSRAM from Toshiba (1989) [143].

A virtually static DRAM (VSRAM) has the refresh totally transparent to the user. These are built out of pseudostatic cells with embedded refresh timing circuitry that generates a refresh signal intermittently. A VSRAM manufactured by Hitachi in 1989 uses a background refresh with an access time overhead of about 29%.

2. **DRAM macrocells embedded in integrated circuits:** Historically, SRAMs have been preferred for embedded memories, because of refresh and timing requirements of DRAMs. The development of built-in refresh and timing control has opened up a new range of potential applications for devices embedding the smaller-sized DRAM cells. A 32 Kb RAM chip from LSI Logic Corporation (1989) is a good example of a DRAM that was designed to be embedded. This was designed for use as an ASIC (application-specific IC) macrocell and can be reconfigured into various array organizations. It found immediate use with a 100 Kb gate logic array. The use of DRAM instead of SRAM reduced the chip size by about 50%.

3. **Battery back-up DRAMs:** In the 90s, we observe a trend towards battery operated systems, and a consequent demand for low cost and low power memories. In 1990, Mitsubishi described a 4 Mb DRAM with a battery backup facility. This enabled automatic retention of the data with reduced power consumption. The design is such that if $\overline{CAS}$ is held low for over 16 ms without an $\overline{RAS}$ refresh cycle, the auto-refresh mode will trigger, continuing as long as the $\overline{CAS}$ is held low, and switching back to normal mode when it goes high. This part is, however, not compatible with the conventional $\overline{CAS}$-before-$\overline{RAS}$ (cbr) refresh operation, but can be refreshed by any other conventional refresh operations, for example, extended cbr, nibble-mode, and $\overline{RAS}$. Hitachi (1989) designed a low voltage 16 Mb DRAM suitable for battery back-up applications. This 1.5 V DRAM has a reduced bit-line voltage swing and a pulsed cell plate, and an average operating current of about 10 mA.

4. **DRAMs as silicon files in battery-backup mode:** Unlike the usual DRAMs, disks need non-volatility and serial access. An example of a DRAM targeted at the disk market was the silicon file memory produced by NEC in 1989. It was a 1 Mb × 1 asynchronous DRAM with on-chip self-refresh. This device can operate in page mode, as follows: the first word is accessed in the usual manner, with both $\overline{RAS}$ and $\overline{CAS}$ clocks needed, subsequent column addresses are generated only by the $\overline{CAS}$ clock. It uses a standard $\overline{CAS}$-before-$\overline{RAS}$ (cbr) refresh. Though much slower

than ordinary DRAMs (random-access time being 600 ns), it was designed primarily with the aim of reducing leakage of the DRAM capacitor at high temperatures, thereby reducing the refresh frequency required. Since the power consumption is much lower than ordinary DRAMs, there is much lower leakage from thermal generation of carriers. Hence, the retention time of this DRAM is long enough for it to be considered static.

5. **Multilevel high-density storage techniques with DRAMs:** In 1988, Hitachi [59] proposed a multilevel high-density storage technique to design a DRAM file memory with on-chip refresh control. This was a 1 Mb DRAM that could provide sixteen levels (4 bits) of storage to produce a 4 Mb serial file memory. The DRAM cells were divided into 4 Kb serial-access blocks. The DRAM circuitry was equipped with some built-in peripheral circuits such as a staircase-pulse generator for multilevel storage operation, a voltage regulator to safeguard against a power supply voltage surge, an error-correction circuit to protect the data from alpha-particle-induced soft errors, and timing generators to simplify testing.

6. **DRAMs with speeds approaching those of SRAMs:** High-speed DRAMs with speeds approaching those of SRAMs have been popular since portable microprocessor-based systems started being manufactured. In some small and medium-sized circuits, the switch to fast DRAMs has started taking place, so that the CPU can operate without wait states or use of additional cache memory. Another major application has been the serial graphics applications in these small systems where the focus is on fast serial access. High-speed operation has been accomplished in two ways: use of technology such as BiCMOS, and special high-speed operating modes and design techniques. An example of a high-speed operating mode is the nibble mode, used, for example, for the design of a two-chip DRAM (by IBM, 1984) that gave a serial access time of 20 ns. Another such DRAM was the IBM 512 Kb DRAM chip (1988) that had a 20 ns random-access time and a high speed 12 ns page-mode access time. This chip used various smart design techniques such as segmented lines with large RC constants, wide buses for power, and separate sets of row and column address buffers.

7. **Graphics Applications for multi-port DRAMs — the video DRAM (VDRAM):** High resolution color graphics systems of today require multiple memory planes for the frame buffer to achieve an acceptable color capability. High resolution and multiple color planes necessitate a high rate of screen refresh. The time to update the large frame buffer memory is a major determinant of speed of such applications. Pixel data require more manipulating than display characters, causing a greater amount of

memory access in less time. A high resolution (for example, 1024× 1024) graphics display, as used in many CAD systems, needs a refresh rate of between 75 and 125 MHz in most cases. This is not dependent upon the graphics controller's need to access the refresh buffer periodically in order to update the stored image.

The first such memory was a 64 Kb DRAM manufactured by Texas Instruments. This chip provides a large video bandwidth for a medium to high-resolution video system and its wide-word architecture allowed more bits to be accessed per device, simplifying the display hardware.

A serious problem in a graphics system is bus contention between the graphics controller and the display refresh. One method of improvement of performance is to use a double display frame buffer to avoid contention. In this system, one buffer provides information for display and the other is available for updates to the graphics controller. When a new image has been completed, the system switches the two buffers. Two major disadvantages of this system are: (a) more interaction time with memory is achieved at the expense of doubling the memory, and (b) when the buffers are switched, the graphics controller does not have a copy of the data for the most recent image.

For providing fast access to bitmapped graphics systems, Inmos came up with the 'nibble mode' of accessing in 1981. This mode performs a burst transfer to reduce the access time when a block of four contiguous memory locations is addressed from the DRAM array. This technique still suffers from the problem of bus contention. In 1983, the first DRAM with static column architecture was introduced to eliminate the $\overline{CAS}$ precharge time and thereby reduce the memory access time over that of page mode. This mode is particularly useful since a graphics processor tends to make localized, but not necessarily sequential accesses. However, this mode also has the problem of bus contention.

To deal with the bus contention problem, multi-port video RAMs (VDRAMs) were introduced. For a dual-port VDRAM, the basic configuration includes a random-access port and a serial access port. The supporting dual memory array includes a random-access memory and a set of serial data registers. The VDRAM allows the system processor and the display refresh to work independently and access memory simultaneously. The need for doubling memory is thereby removed. The earliest dual-port VDRAM was a 64 Kb×1 DRAM provided with a 256-bit shift register, and was introduced by Texas Instruments [135] in 1983. It is provided with a set of four 64-bit shift registers designed to allow shift lengths of 64, 128, 192 or 256 bits. In this device, each row in the DRAM array consists of 256 bits and can be loaded into the shift register in a single cycle. After loading the data, the shift register can be internally decoupled from the DRAM array and accessed serially while the DRAM is accessed simultaneously and asynchronously via the random-access port, by the graphics processor.

Many other manufacturers, such as Mitsubishi, Hitachi, NEC, Texas Instruments and Fujitsu have marketed commercial multi-port VDRAMs. Texas Instruments introduced 1 Mb multi-port VDRAMs, implemented as 256 Kb× 4 arrays (the so-called JEDEC format).

SRAMs have also found a wide range of applications. They are described below.

1. **Battery RAMs (BRAMs):** Battery-operated memory systems require: (*a*) very low operating power and battery-back-up data retention current levels, and (*b*) circuitry to protect data if the battery voltage becomes low or if the power supply fails. In the early 1980s, Mostek, now called SGS-Thomson Microelectronics, incorporated the circuitry to protect against accidental writes during power down. They designed a family of byte-wide SRAMs called 'zero power RAM'. These parts contain a lithium battery and a temperature-compensated power failure detection circuit which monitors V_{CC} and switches from the line voltage to the back-up battery when the voltage fell below a specified level and switches back to the line voltage when the voltage was restored. In 1983, NEC manufactured a 64 Kb SRAM chip with similar characteristics.

2. **Dual-port RAMs:** Dual-port SRAMs are implemented in systems to reconcile speed mismatch between a processor and a peripheral device, or between a main memory and a set of distributed processors. They are also used in systems with multiple servers and processors communicating with each other. To avoid bus contention, two ports are provided to allow two independent and simultaneous memory accesses. Older designs use discrete components to implement dual-port architecture, with bus contention being handled by arbitration logic configured to decide among different access requests. The discrete implementation, therefore, increases the hardware complexity because it requires additional address buffers, latches, data latches, arbitration logic, and an SRAM cell array. For systems requiring large amounts of memory, dual-port architecture is still implemented with discrete components in the traditional way. This is because the cost of adding the extra circuits is still found to be less than that of replacing the memory with a dual-port version [136]. For smaller memory buffers, however, dual-port devices are preferred.

   Earlier dual-port RAMs used exclusive access states, to prevent bus contention. This meant that when one of the two ports was provided access, the other had to wait. In 1983, the first SRAM with built-in contention arbitration was described by Bell Laboratories. This 512 × 10 NMOS RAM reportedly replaced fifteen to twenty logic and single-port memory chips in a system.

Various 1 Kb × 8 dual-port SRAMs have been introduced by Synertek, Cypress Semiconductor and Integrated Device Technology Inc. Most of them ranged in size between 8 Kb and 16 Kb.

3. **FIFOs and line buffers:** Small first-in-first-out (FIFO) memories are widely used in communication buffering type of applications. They are generally made from static RAM cells if the data has to be stored in the FIFO, and dynamic RAM cells if the data simply passes through. Larger FIFOs are mostly application-specified DRAMs. They are used in television application for frame storage, for example, a 910 × 8 bit FIFO made by NEC, for line interleaving. This device reduces flicker in television sets.

4. **Content-addressable memories (CAMs) and cache tag RAMs:** Content-addressable or associative memories have been known since the 1960s. Most of them are used as embedded memory in larger VLSI chips and as stand-alone memory for specific applications. The CAM associates an address with data. When data is provided at the inputs of a CAM, it searches for a match for the data without regard to address; when a match is found, the address is output. CAMs have a wide range of applications from database management to pattern recognition. An example of a content-addressable memory is a 256 × 48 organized CAM designed by AMD for filtering addresses in an Ethernet network and performing network address look-up functions. Another example is a self-testable and reconfigurable 256 × 64 CAM from Bellcore, designed for high-speed table look-up applications.

Cache tag RAMs store a tag that represents a few bits of the stored data. It functions like a CAM, matching the input data tag with the addressed RAM location tag. Detection of a match and reporting of the matched address is done as discussed above. This address may be used for retrieving stored data.

1.7 CIRCUIT TECHNIQUES FOR IMPROVED PERFORMANCE OF RAMS

1.7.1 Techniques for improved speed and low power

In the mid-80s, various approaches were taken to make DRAMs faster. Besides the nibble and page modes, new, faster access modes were implemented. In the 1 Mb DRAM, two commonly used access modes were the fast page mode and static column mode. Sense-amplifier designs were improved to achieve higher speeds, and new sensing schemes, such as the half-V_{CC} bit-line sensing scheme, were devised. Several innovative design techniques led to reduced power dissipation and increased device stability. These techniques and their results are described below.

1. **New timing modes:** The *fast-page* mode was first used in the Intel 64 Kb CMOS DRAM. In this method, after the row address has been latched (at an edge of the $\overline{RAS}$ signal) into the row decoder, external signals can be fed into the input port of the address buffer. As in a standard DRAM, the column address is latched at the falling edge of the $\overline{CAS}$ signal. Since the address buffer port remains open during the read operation, this scheme allows rapid access and random readout of the 256 individual column bits on a selected row. The IBM 1 Mb DRAM further increased the data rate of the fast page mode by reducing the precharge time for columns. This is achieved by an internally generated early restore phase. Another innovative fast access mode is the *static column mode*. Using this mode, the long precharge time normally associated with conventional DRAMs during a page mode operation is eliminated, and the access time becomes comparable to that of a static RAM. This mode is also generally adopted for multi-megabit (> 1 Mb) DRAMs. In conventional DRAMs, the CAS signal has three functions — it activates the column circuitry, it acts as a strobe for column address inputs, and it also acts as a strobe for the output enable signal; in contrast, the static column method allows elimination of the column circuitry-triggering function of the $\overline{CAS}$. This is done with an internal read operation of the column circuitry triggered by address transition detection (ATD, described later) rather than by the edge of the $\overline{CAS}$ signal. AT&T suggested a fast column mode clocked by chip enable for their 1 Mb DRAM in 1985.

2. **Improved sense amplifier design:** High-density DRAMs suffer from the problem of increased RC delays due to increasing bit-line lengths. This necessitates the design of new sense amplifiers.

   The evolution of multiplexed DRAM sense amplifiers is shown in Figure 1.11. Conventional *shared* sense amplifiers have a mechanism for selectively disconnecting the right (left) bit-line pair when the left (right) bit-line pair is selected. The voltage in each segment is sensed (amplified) and restored by an NMOS latch at the center and a PMOS latch on the same side as the segment. The switching devices on the two segments act both as cutoff and reconnect transistors in each cycle, causing increased access time.

   To reduce the access time, Mitsubishi (1987) came up with a 1 Mb DRAM that uses an alternative to the shared sense amplifier, called distributed sense and restore amplifier. The basic difference from the shared sensing scheme is that the sense circuitry halves the bit-line length, thereby reducing the capacitive coupling effects, and the latches on each side of the amplifier independently sense and restore the bit-line pair on their side. This allows early reconnection of the left and right segments and a corresponding smaller access time.

   To solve the RC delay and capacitive coupling problems, Mostek (1985) introduced the *divided* bit-line architecture, a combination of folded-bit-line and the shared sense amplifier concepts. In this approach, the long columns are divided into sixteen polysilicon bit-line segments with 64 cells per segment. Eight segments are arranged end to end in a line on either side of a central column decoder. Eight memory blocks of 128 Kb each are formed by adjacent segments grouped into pairs of open bit-lines. A block comprises bit lines running parallel to the segments as folded bit lines, thus disconnecting about 75% of the matrix capacitance in any one cycle so that low power consumption and high signal levels are obtained.

3. **Half-V_{CC} bit-line sensing schemes:** In conventional sensing schemes, bit-lines are precharged to $V_{CC} - V_T$. After the sense amplifier has been set, there is degradation of the 1-level signal that is restored into the cell. To make up for the charge loss, active PMOS pullup transistors are used in CMOS technology. To improve access times, however, bit-lines are precharged approximately half-way between V_{CC} and ground. This reduces the peak currents at both sensing and bit-line precharge by almost a factor of two, thereby reducing the electromigration problem and the voltage drop across resistors. The overall effect is an increase in chip reliability and better speed. It also causes a lower transient power dissipation and decreases voltage bounce noise due to the inductance of the wiring. A drawback of this approach is that a longer period during which bit lines

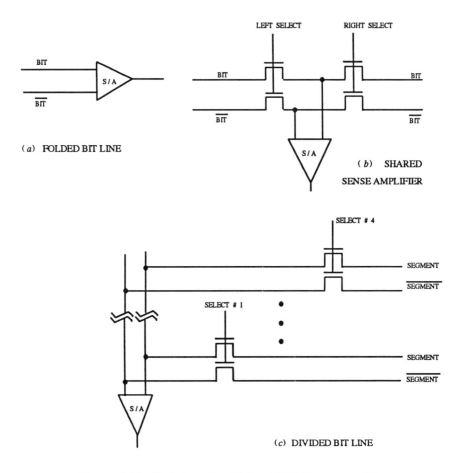

Figure 1.11 Evolution of multiplexed DRAM sense amplifiers

Introduction

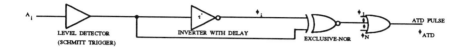

Figure 1.12 An ATD circuit; courtesy [174]

float makes them more susceptible to radiation-induced soft errors. This provides a reason for use of dummy cells to generate reference potential so that the absolute bit-line precharge level and the word-line to bit-line coupling noise are low.

IBM [81] discussed a half-V_{CC} bit-line sensing scheme in N-well CMOS DRAMs in 1984. Lu and Chao [81] made the observation that half-V_{CC} sensing provides higher performance for high-density CMOS DRAMs with N-well arrays than conventional techniques.

4. **Half-V_{CC} biased cell plates:** Often the cell plate is provided with a half-V_{CC} bias, that reduces the electric field across the thinner oxide of the storage capacitor, and thereby reduces the influence of fluctuations in the supply voltage (V_{CC} bumps). From 1985 onwards, 1 Mb DRAMs manufactured by Mitsubishi and Toshiba have used this technique.

5. **Address transition detection (ATD) for power reduction:** The 1 Mb DRAM manufactured by Toshiba in 1985 uses clocked circuitry with address transition detection, shown in Figure 1.12, for high-speed operation. The philosophy of address transition detection is described later. The Hitachi CMOS 1 Mb DRAM also uses ATD to reduce power dissipation during long cycles, by limiting the time for which the column driver clock, the column select signal and the main amplifier were active.

SRAMs with sub-10 ns access times are extremely sensitive to factors such as the turnaround time for a common I/O line, the delay in multiplexed addressing, and inductance contributions from packages to bonding wires. A combination of various techniques is used to achieve a satisfactory power-delay product. We next describe some of these techniques.

1. **Separate I/O lines:** On a common system I/O bus, there is a turnaround time for data, between a read and a write access. This problem can be solved by having separate input and output lines. This improves speed but causes a higher pin count for the package. An example of such a chip is a 16 Kb × 4 SRAM from Cypress Semiconductor, described in [29].

2. **Synchronous or self-timed SRAMs (STRAMs):** High performance systems have very fast system clocks. Specially designed memories which can operate without clock skews are needed for such systems. This necessitates the addition of simple latches to the inputs of SRAMs, thereby making them *synchronous*. Asynchronous SRAMs, which are the most common, require no external clocks and have a very simple system interface. Synchronous SRAMs are faster than asynchronous SRAMs because all inputs are clocked into the memory on the edge of the system clock, whereas in asynchronous SRAMs, various inputs are set up in the memory after some internal timing delays. However, synchronous SRAMs require a larger package to accommodate the clock input and latches.

   The input data in a synchronous SRAM is controlled by the level of the clock signal (input). When the clock level is low, the input data is not sampled, and the data already present in the latch is maintained. When the clock turns high, the input data is latched in. Registers, which are edge-sensitive latch structures, consist of two D-latches in series, one being controlled by the edge of the clock (CLK) input and the other being controlled by the edge of the $\overline{CLK}$ input. The output of the register, therefore, remains stable during a full clock cycle even though the data at the input of the latch may change. When a new cycle starts, the data from the previous cycle is available very early, allowing a fast system read of data from the previous cycle. Registers may be used to store addresses, input and output data, and control signals, for example, the write-enable signal.

   At the system level, a synchronous SRAM is much faster than an asynchronous SRAM since the system address drive can go on to the next address while the previous address is being latched into the memory. Fujitsu made a whole family of STRAMs — latch-latch, latch-register, and register-register combinations at the input and output. A Motorola 16 Kb × 4 synchronous static RAM [116] used input registers for address and control, but latches for outputs. A pipelined SRAM from AT&T Bell Laboratories, reported in 1989 [132], has fully synchronous design with cycle times between 4 ns and 7 ns and synchronous data access times under 2 ns. Its address access time during asynchronous operation is 5 ns.

3. **Reducing supply line noise inductance:** The current drawn by a chip increases suddenly and momentarily when one or more outputs of the chip switch following a clock transition. This changing current is primarily due to charging and/or discharging of parasitic capacitances which cause a voltage drop, referred to as *ground bounce*, across the parasitic inductances and resistances in the power supply system. If the voltage drop seen by

the chip or other chips sharing the same power supply becomes too large, logic values in the chip may be erroneously modified.

The noise inductance in the power supply line and ground bounce caused by transient output changes become more pronounced with larger number of outputs switching simultaneously. This, in turn, depends on the word size; in other words, these problems are more significant for byte-wide memories than single bit or 4-bit wide memories, for instance. Since the chip has no valid output until the power line voltages have become stable, noise inductance has the effect of slowing down the effective access time. Also, the current drawn from the supply increases with the number of outputs that simultaneously go high. For chips that use address transition detection, parasitic inductances might cause a large voltage drop if a sudden change in driver current takes place. This would cause a false detection of address transition by the ATD circuitry. One approach taken by Philips in their 1987 256 Kb SRAM uses an output buffer to drive the load such that the rate of change of current with time is constant. Another approach adopted by Motorola in 1987 for 256 Kb SRAMs involved separate power buses for output drivers.

4. **Pinout characteristics:** Multiple power and ground pins and centrally located power pins reduce the ground bounce problem for very fast SRAMs and thereby reduce the access time. The adjacent power and ground lines have a mutual inductance that reduces the self-inductance causing ground bounce when the outputs swing. Laying out power lines at the center of the package reduces the length of the bond wires to the power supply pins, and thereby reduces the self-inductance. As mentioned earlier, ground bounce effect increases with increase in the number of outputs that change states simultaneously. Hence, wide bus memories are more vulnerable to this effect and need the features mentioned.

5. **Flash write features:** In 1984, Motorola [10] described a bulk write feature to enable a write operation of the entire memory in one extended write cycle. With this operation, the entire array can be written with the data supplied to the data-in pin during an extended cycle.

6. **Address transition detection (ATD) for synchronous internal operation and power reduction for SRAMs:** Address transition detection (ATD) circuits are used to provide the initial pulse by which asynchronous SRAMs can be operated as if they were synchronous, and this synchronous operation remains transparent to the user. ATD pulses can be used to generate two types of synchronous pulses, *equalization* and *activation* pulses. Equalization pulses reduce delay by restoring differential

nodes before they are selected. Activation pulses selectively turn on particular parts of the circuitry. The ATD generates a monoshot pulse when at least one of the inputs such as addresses or chip selects have changed. The pulse is used, in turn, to activate subsequent internal clock signals that control the timing of various internal operations. In 1979, the RCA 16 Kb SRAM used internal precharge circuitry in CMOS SRAMs. This device was designed to have internally generated precharge signals transparent to the user. ATD is also used for automatic power-down in CMOS circuits. This concept originated in the late 1980s. Since CMOS circuits dissipate power in proportion to the speed of operation, there is only a low standby power when the chip is not selected. On most conventional SRAMs, the low standby power is controlled with the chip enable. Clocks derived from ATD pulses can turn internal circuits on and off by internal timing signals so that power down could be automatic without the need to turn off the chip enable signal. ATD has also been shown to effect power reduction during read and write.

Several techniques are used to reduce power dissipation during read. They are: *bit-line equalization, latched column*, and *pulsed word line*. Techniques used to reduce power dissipation during write include *variable impedance bit-line loads, tri-state word lines* and *data transition detection*.

The *bit-line equalization* technique during read uses an ATD pulse to reset the bit line before new data can appear, offering increased speed. However, it also causes high current through memory cells. In this technique, the bit line, which has a low impedance, always carries some small signal, and thereby draws current. The *latched column* technique achieves lower power dissipation during read by latching the signal to the column, pulling one of the bit lines to a completely low state in the process, and resulting in long delay due to long bit-line recovery time. The cell current can go to zero before bit-line equalization, which builds up peak currents that are larger than in the bit-line equalization scheme. The *pulsed word line* technique operates cells the same way as the conventional techniques, with a small impedance on the bit line that keeps the voltage swing low. After the data is read out, the word line is switched off and the cell current becomes zero. The read data is stored in a latch. This technique causes the read operation to be asynchronous. Internal clocks activate memory cells for a very short period and then turn them off. This scheme has a high address noise immunity.

Power reduction during write is implemented within the ATD framework using one of several possible schemes. *Variable impedance bit-line loads* consist of an NMOS transistor bit-line load controlled by one clock and a PMOS transistor bit-line load controlled by another clock. The clocks are

Introduction

used to set the load to either a high or a low impedance. This technique suppresses the peak current during write recovery and optimizes the read current over a wide power supply range. Hitachi [177] used this scheme in 1985 for the design of a 256 Kb SRAM. Another scheme, used to reduce power consumption during write, requires a normal two-level word line during the read cycle, but a *tri-level word line* during the write cycle. During the write cycle, the word line is taken to its normal power supply voltage and then held at a middle level of 3.5 V. This level is high enough for a static write operation but saves both D.C. column current and transient current after write. This scheme was implemented by Mitsubishi in 1985 [150] for a 256 Kb CMOS SRAM. Philips (1987) proposed another approach to power reduction during write operation, called *data transition detection*, in their 256 Kb SRAM design [51]. They use an automatic power-up generator to generate an activation pulse whenever a transition occurs on either address or data lines. Data transition detection (DTD) pulses are triggered only when the write enable signal is active.

Sony proposed the use of ATD to generate equalizing pulses for the sense amplifier output level and the data bus level, thereby reducing the access time, in their 128×8 SRAM (1987) [77] by about 7 ns. Other manufacturers like Toshiba, Hitachi, Fujitsu, and Philips also used the ATD technique for bit-line equalization, precharging, data path activation, and so on. Mitsubishi used ATD to precharge the output buffer circuitry to reduce ground bounce effects, in their 1 Mb SRAM [171], increasing access time by 10% and reducing peak current on the power supply lines by 30%.

1.7.2 Multimegabit cell technology for high-density DRAMs and SRAMs

For the first four generations of SRAM design, the load devices were made of polysilicon, but gradually they were replaced by P-type transistors. This allowed operation at lower power and higher speed at the 4 Mb generation. The low power and large noise margin of the six-transistor cell was necessary, but at the same time, small chip size demanded that the memory cell be made smaller than the scaled poly-load resistor cell; the six-transistor cell was too large. There are two mutually opposing factors that play a very important role in scaling the poly-load cell. If the polysilicon resistor has high resistivity, then the standby power, and hence, the standby current, will increase. The low standby current requirement, therefore, necessitates the resistance to be greater than or equal to a certain minimum value. On the other hand, the resistance must be small enough to provide enough current for restoring sub-threshold leakage charge in the storage transistors. The sub-threshold leakage of the storage transistors increases with decreasing threshold voltage caused by shrinking feature widths and the resultant increase in densities. The difference between the minimum and the maximum values of the resistance, therefore, decreases as the density of the chips increases, making the process difficult to control. Moreover, the power supply of the memory cell array especially under the modern sub-0.6 μ technologies must be less than 5 V for reliability reasons, even though the external chip power supply is 5 V. At lower voltage values, the poly-load cells have reduced stability and lower noise margins. The industry has been confronted with the problem of finding a cell as small as a poly-load cell and as stable, reliable, and low-power as a six-transistor cell. The following process innovations and cell-level design achievements were found to be necessary to this end.

1. **Deep submicron transistors:** Advanced processing of deep submicron transistors in the sub-0.5 μ range has been carried on by many major manufacturers. For example, Motorola (1989) [52] investigated a 0.4 μ process with reduced size inter-well isolation. Their technique allowed the N^+ and P^+ space to be scaled down to less than 2 μ, and improve the transistor design to reduce the bird's beak effect. IBM (1989) [173] examined a CVD gate oxide for 0.25 μ channel length NMOS and PMOS transistors which could be deposited at low temperatures. IBM (1989) [21] also investigated the scaling properties of trench isolation used with ECL transistors. Matsushita (1989) [58] designed 0.25 μ transistors with drain structures implanted at an angle to minimize gate overlap.

2. **Refinement and scaling of planar CMOS cells:** Scaling down the planar poly-load SRAM cell was investigated by Sony, NEC, Mitsubishi, Motorola, and other companies. Mitsubishi (1989) [184] designed a 0.5 μ geometry cell for a 4 Mb SRAM, in which they use a four-level polysilicon and two-level aluminum CMOS technology. The load resistor was formed in the thin fourth level of polysilicon. This resistor is narrow and has high resistivity, and 0.6 μ trench isolation is used. The cell is quite complex because of the measures taken to increase the noise margin under low supply voltages. Memory cell stability is enhanced by increasing the threshold voltage of the driver transistor using a salicide ground line to reduce changes in V_{SS} below 0.04 V, and by decreasing the access transistor conductance. In addition, an extra P^+ buried barrier layer is used in the P-well because of the soft-error sensitivity of the cell. NEC and Sony also proposed such poly-load cells for their 4 Mb SRAMs.

3. **Stacked transistor CMOS cells:** The poly-load cell has certain drawbacks such as large standby current and hence, greater power dissipation, lower stability as manifested by smaller noise margins and lower immunity to soft errors. The need was felt for a cell having the size of a four-transistor poly-load cell but retaining the advantages of a full CMOS cell. One solution was the development of a 4 Mb SRAM with six-transistor cells of four NMOS transistors in the silicon substrate and two PMOS load transistors formed in the thin film polysilicon layers above the cell in much the same way as the loads were stacked in the poly-load cell. In 1988, NEC described a 1 Mb SRAM, built using 0.8 μ technology, that stacks the transistor in a polysilicon cell [7]. Also in 1988, Hitachi [178] designed a 25.4 μ^2 cell in 0.5 μ minimum geometry for use in 4 Mb SRAMs that uses PMOS load transistors in the third-level poly with the gate in the second-level poly. This cell was too large to be manufactured commercially with reasonable yield. In subsequent reports ([64] 1989, [142] 1990), Hitachi described very fast multi-megabit SRAMs that use 0.5 μ and double-metal technology but triple and quadruple-poly, and achieve smaller cell size. In 1991, Mitsubishi [118] proposed a design for a 4 Mb cell using a thin-film polysilicon PMOS transistor to attain very low (0.4 μA) standby current. The cell size is 19.5 μ^2 and is intended for use in 3 V battery-operated systems. Six-transistor cells are necessary for high noise margin and low current leakage.

4. **Silicon-on-Insulator (SOI) and buried layers:** Further advances beyond the controlled poly-load cell technology were required for the 64 Mb generation because of the requirements of high speed and reduced chip size. Manufacturers like Philips, NEC and Mitsubishi looked into silicon-on-insulator (SOI) technology. SOI transistors are attractive for deep submicron technologies because of improved subthreshold slope, smaller short-channel effects, reduced electric fields, improved soft error immunity, latch-up elimination, and total dielectric isolation.

5. **Vertical transistors:** Memory designers have considered new types of vertical cell structures and transistors for the 16 and 64 Mb SRAM generations. Apart from stacking additional interconnect layers with transistors over the surface as in poly-load PMOS or burying layers under it (SIMOX), another approach is to dig vertical trenches or etch vertical pillars. Toshiba proposed the *surrounding gate transistor* (SGT) in 1988 [161]. This cell has a gate electrode surrounding a pillar silicon island and the occupied area is reduced by about 50%. Conventional DRAM trench techniques are used for these SRAMs. In SGT, the source, gate and drain are arranged vertically. The side-walls of the pillar silicon island form the channel region. The side-walls are surrounded by the gate electrode, whose length

depends on the pillar height. SGT potentially reduces short and narrow channel effects and thereby prevents the threshold voltage from becoming too low. Toshiba continued their work on SGT in a DRAM memory cell. In 1988, Toshiba [160] described a trench-type capacitor called the *double LDD concave* transistor for a 0.5 μ technology. It uses source and drain junctions above the channel, and thereby improves the short channel effects. It can also be operated at 5 V.

6. **Combination of vertical and stacked transistors:** Vertical and stacked transistors have been combined for the 64 Mb generation. In 1989, Texas Instruments [40] proposed a six-transistor cell with a thin film polysilicon PMOS transistor and a vertical sidewell trench NMOS transistor. Since the thin film transistor can be stacked over the vertical NMOS, the potential for scaling this cell to enable higher packing densities is high. Other such combination transistors were explored by Hitachi in 1989 (the *delta* transistor [56]).

These trends will probably last till the 1 Gb SRAM generation, after which newer techniques will be required.

1.8 TESTING OF RAMS

During the last two decades, the increasing complexity and speed of DRAMs and SRAMs has necessitated the rapid advance of testing techniques. The complexity of testing and cost of test equipment have also gone up considerably. In the early days of memory testing, only electrical measurements were performed at the interface. As memory devices became faster, it became necessary to implement rigorous speed tests such as access time tests to verify correct functionality. These tests came to be classified under AC parametric tests. With increased complexity, it also became necessary to perform algorithmic functional testing as well as electrical testing. These two techniques were often employed together to analyze faults associated with electrical and functional behavior, which were often correlated. Both these techniques use very simple fault models. With the advent of deep submicron technology, the need was felt to perform testing based upon a thorough analysis of the circuit design, layout and processing technology. It has been found that, depending on the type of memory, some faults are more common than others — DRAM cells, for example, are more susceptible to pattern-sensitive faults caused by bit-line voltage imbalance (described later) than SRAM cells, because they store charge in bit-line capacitors instead of latches. Some peculiar fault types are caused by the detailed layout characteristics and can best be understood and tested for, only if a technology and layout-based fault model is used. A GaAs SRAM, for example, has a different set of canonical fault types than a silicon SRAM, because of some peculiar processing and device defects that are not observable in silicon SRAMs.

Memory testing is costly, especially for high-density memories. A high capital investment is required to maintain a sophisticated test area comprising equipment, software and a skilled engineering work-force. A popular cost-effective approach these days is to design chips that are *self-testable*, that is, have special built-in hardware for applying test patterns and reading responses. *Built-in self-test (BIST)* is not only cost-effective, but is also very important for some special applications, such as memories embedded within integrated circuits or circuit boards. Since the input/output pins of embedded memories are not directly accessible for external testing, BIST is often the only feasible test strategy.

Memory testing is performed for a variety of reasons — it is performed by a device manufacturer to verify that the device operates within predefined specifications, by a bulk purchaser to monitor quality and performance before using it in his/her own product assembly, or by an end equipment manufacturer to screen assembled boards or systems according to his/her needs. Also, if a new processing technology has been introduced, detailed functional and layout-related testing must be performed to verify the correct functioning of the device. Moreover, before a new device type is introduced in the market, or when some established device or process type has been upgraded, a *characterization program* is usually run on a representative sample. This program not only performs go/no-go tests, but also measures AC parameters and the worst case operating voltage margins (DC parameters, described in Chapter 2). Before assembly of the new device, a *probe test* is performed. This involves testing each die on a wafer before assembly to weed out the gross rejects. A probe test examines DC and AC parameters and also provides information relating to device speed. The 'early warning' it provides is important for a reliable design of the finished product and gives valuable feedback to the wafer processing plant. Assembly is followed by a final test which comprises the following: a *contact* test to verify that contacts are made properly and electrical connections between pins and bonding pads are proper — this test is done without applying power to the device, and products that show discontinuities are screened off; a *DC parametric* test to verify the performance of voltages and currents at the device interface, a *functional and AC parametric* test to monitor the functionality and verify that

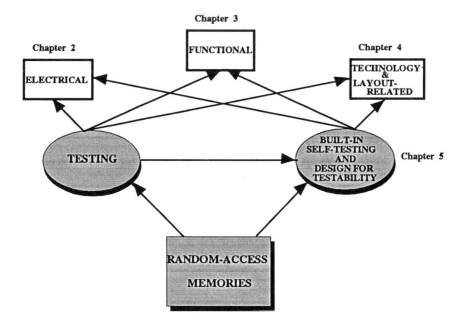

Figure 1.13 A flow chart of the organization of the book

the delays in the device functioning for example, access times, refresh time for DRAMs and cycle times, are within acceptable limits.

1.9 SUMMARY OF THE FOCUS OF THE BOOK

The book focuses on the testing and testable design techniques that are suitable for RAMs. Chapter 2 deals with a simple view of electrical testing and presents a case study to bring out the important aspects of such testing. Chapters 3 and 4 deal with more detailed techniques for rigorous testing of RAMs — namely, functional testing and technology and layout-related testing approaches. Chapter 5 deals with techniques for built-in self-testing (BIST) and design for testability (DFT) of RAMs.

1.10 A PRIMER ON MEMORY TESTING TERMINOLOGY

We briefly mention some terms used in the context of testing. More such terms are given in the Glossary (see Appendix 1).

- **Fault modeling:** This deals with describing a failure mode of a memory device with the help of a simplified model of the corresponding faulty behavior; this model is then used to come up with testing strategies. For example, a commonly occurring failure mode prevents the writing of a 1 (or a 0) in memory cell, this causes a *stuck* fault. Fault models are usually classified as: functional, electrical, and layout-dependent. Functional fault models describe faulty and fault-free behavior in terms of signal *levels* (i.e., logic 1 or 0). Electrical fault models describe this behavior in terms of signal values (i.e., current and voltage) or various timing parameters (for example, access times) measured at the input/output *interface* of the device. Layout-dependent fault models describe the memory behavior in terms of defects at the cell layout caused by variations in the processing technology; for example, a short between the bit and $\overline{bit}$ lines may be physically modeled as a spot of extra metal in the cell layout, a layout fault in a Gallium Arsenide (GaAs) SRAM cell may be described by the parametric characteristics of the process.

- **Defect:** A physical failure in the cell layout that may cause a visible faulty behavior is called a defect. An example is a broken P-load in an SRAM cell that might cause a stuck value on the cell.

- **Functional, electrical, and technology and layout-related testing:** Testing algorithms based on functional, electrical, and technology and layout related fault models are known by their respective names.

- **External testing:** This refers to the mechanism of testing memory chips and boards using an external test fixture and a host computer for applying test patterns and examining test responses.

- **Built-in Self-Test (BIST):** This is a set of algorithms and hardware that are built into the memory system and make it self-testable. These systems eliminate the need to perform external testing.

- **Design for Testability (DFT):** This is a set of algorithms and memory testing hardware that may be partly built into the memory system and partly external.

- **Diagnosis algorithms:** These algorithms not only detect faults in the memory array, but also find the location of faulty cells. This information is in turn useful for two reasons — (*a*) to repair the memory array, and (*b*) to provide feedback to the fabrication station for improvement of the manufacturing process. For example, a mask misalignment, or an over- or under-etching might consistently cause a certain specific region of the array layout to have faults.

- **Built-in Self-Diagnosis (BISD):** This is a collection of algorithms and hardware built into the memory array to make it self-diagnosable.

Introduction

SUMMARY

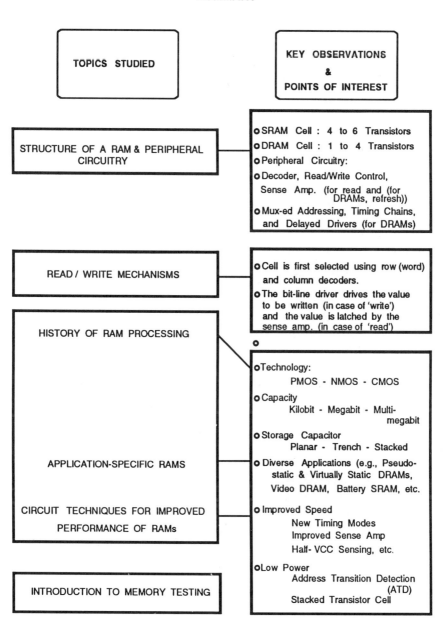

1.11 PROBLEMS

1.11.1 Worked-out Example

1. Give examples of some non-volatile semiconductor memories, and briefly highlight their differences.

   (**Sample Answer Sketch:**) In this book, we have focused primarily on volatile memories; that is, memories that lose their data if the power is turned off. In many systems, it is important for the memory to retain its data permanently. For example, for some computers, a part of the operating system may reside permanently in a non-volatile region of the main memory.

   For such applications, a class of memory devices called the ROM, or read-only memory, has been developed. ROMs come in a number of versions. The mask-programmable silicon ROM is preprogrammed by the memory manufacturer. It consists of a set of locations in the memory where single transistors are either present or have been blown out with a laser-activated fuse. Once written, the contents of a ROM cannot be altered by the user.

   An EPROM (Erasable Programmable ROM) is a device consisting of one-transistor cells that can be taken out of the system and then reprogrammed by the user using ultraviolet light. An EEPROM (Electrically Erasable Programmable ROM) can be reprogrammed electronically by the user a certain number of times, by writing into each word. The difference with EPROM is that ultraviolet light causes the entire information within the EPROM to be erased, whereas for EEPROMs, information is erased on a word-by-word basis by writing new data.

   A flash memory uses a technology similar to that of the EPROM and has a one-transistor cell. However, unlike EPROM, it is electrically reprogrammable *in situ* a limited number of times. This makes it suitable for those applications in which the memory data is expected to remain largely unchanged over a long period. Electrical erasure in the system is for either the entire memory array or for blocks of the array — hence the name 'flash.'

   Non-volatile SRAMs with ferroelectric capacitors have been described in the text. These capacitors have 100 times the dielectric constant of silicon nitride or silicon dioxide and exhibit hysteresis (that is, retention of charge even when the voltage drops to zero).

1.11.2 Exercises

1. Derive the access time of a 6-cell CMOS SRAM cell in terms of the bit-line capacitance C_B, input capacitance C_i of the pulldown transistor, and transconductances of the pull-down and pass transistors (g_p). Plot the voltages at each node at the output of the cross-coupled inverter pair for different values of transconductances, as a function of time.

2. Show that the phase trajectories for the memory read process for a static RAM cell can be given by
$$x_0^2 = x_1^2 - c,$$
where c is a constant. Draw the phasor diagram as a function of V_1 and V_0, the voltages at the two output nodes of the inverters.

3. Explain, with the help of timing diagrams, the detailed mechanism by which (a) page mode, (b) fast page mode, (c) nibble mode, and (d) static column mode of operation cause faster access times for DRAMs. Use suitable examples of real DRAM chips.

4. Refer to Figure 1.1 (d) which illustrates a static CAM cell. Explain how the associative read operation is performed, causing MATCH to go high or low.

   Assume a 4-bit CAM containing the data 1001. Explain what happens when the associative search operation is performed for each of the following data applied to the cells:

   (a) 1001;
   (b) 1011.

5. Derive the equations for the change in voltage of the precharged nodes of an SRAM cell storing a 0, as a function of the ratio of the access transistor drain capacitance to the output node capacitance.

6. Describe the detailed mechanism of the $\overline{CAS}$-before-$\overline{RAS}$ refresh cycle of DRAMs, and draw and explain the corresponding timing waveforms.

7. Give a detailed implementation of an ATD (address transition detection) circuit and explain the mechanism of its operation.

8. Display in a tabular fashion, how the cell geometry, access time, process, capacitor type, cell area and chip area of DRAMs have increased from the 64 Kb to the 256 Mb age.

9. Draw the block diagram of a synchronous SRAM and explain its read cycle operation with appropriate timing diagrams.

10. What advantages are obtained by having

    (a) a half-V_{CC} sensing scheme;
    (b) a half-V_{CC}-biased cell plate;
    (c) a trench-type cell instead of a planar cell

    for DRAMs ?

2
ELECTRICAL TESTING OF FAULTS

Chapter At a Glance

- Electrical Fault Modeling -- Characterization & Production Tests

- A Commercial Chip (TMS4016) Test Specifications

- DC Parametric Testing
    -- Contact, Power Consumption, Leakage, Threshold,
       Output Drive Current & Voltage Bump Tests

- AC Parametric Testing
    -- Rise and Fall Time Tests, Set-up, Hold and Release Time Tests,
       Propagation Delay, Running Time, Recovery Fault Tests

- Electrical Testing of Dual-Port SRAMs

2.1 INTRODUCTION

The order of tests in a typical external (host-driven) test program is as follows: contact test, functional and layout-related test, DC parametric test and AC parametric test. Contact, DC parametric and AC parametric tests have been classified as electrical tests in this book. These tests are used merely to verify that the device meets its specifications with regard to electrical characteristics (voltage and current levels, and timing parameters), and measure the operating limits. In this chapter, we shall describe some electrical tests that are performed at the input-output interface of the RAM chip by an external tester to verify whether the chip has satisfactory electrical behavior.

2.2 ELECTRICAL FAULT MODELING

With regard to the memory block diagram given in Figure 3.1 in the next chapter, an electrical fault can be described as a fault that modifies the external electrical behavior of a memory device, for example, observed voltage/current levels and delays noticed externally on input and output pins, and affects the input/output interface of the memory IC with other chips. Hence, the **fault model** in this type of testing consists of two kinds of faults: (a) a significant deviation of a circuit parameter (such as a voltage, current or timing parameter) from its prescribed (data-book) value; (b) unsatisfactory limits of operation of a device. Testing such faults can usually be done *without* a thorough analysis of the underlying process technology or layout; it is usually sufficient to measure various voltage, current, and/or various timing parameters at the chip input/output interface with an external tester. Technology and layout-related parametric faults such as sleeping sickness and voltage imbalance in the divided bit-line architecture for DRAMs, cause abnormal electrical behavior during operation or at the quiescent state. Testing for such faults necessitates complex fault and defect models and a thorough analysis of the circuit design and layout. Such tests are *technology and layout-related*, comprising both electrical and functional testing approaches and detailed layout analysis. These approaches will be described in Chapter 4.

Electrical and layout-related fault models consider *time* as a parameter; functional models do not. Faults observed are assumed to be time-dependent, and their testing often involves taking various timing measurements. In the next two sections, we shall discuss these electrical fault testing mechanisms.

Electrical Testing of Faults

As mentioned before, an electrical test may be performed for one of two purposes — to determine exact values of the operating limits of a device, that is, exact values of AC and DC parameters at which a device fails; or, to determine whether or not a device meets its specifications. The former is known as a characterization test and the latter is known as a production test.

1. **Characterization tests**

    When several environmental factors vary independently, the worst case is of greatest interest in testing because (a) it is easier to evaluate than the average case, and (b) it is obvious that a circuit that works satisfactorily under worst case conditions will work satisfactorily for any other conditions. Unfortunately, determining the worst case may still be quite difficult at times, because different fault modes may become dominant under different combinations of the independent variables, leading to several "worst cases." In fact, failure rates may actually be *higher* for a combination of variables which does not include any one variable at its limit.

    A good example of the multivariate worst case problem is multiple power supply voltages used in some memories. Other variables of interest may include access times, pulsewidths, temperature, and so on. Statistical analysis can be used after a significant number of samples have been taken, but the task of organizing the data to produce a recognizable result is a problem. Characterization tests deal with the above problem by using the following test philosophy:

    (1) Select a test sequence that results in a pass/fail decision.

    (2) Take a statistically significant sample containing devices to be tested, and repeat the test for each combination of two or more environmental variables and plot the test results. (Such tests are easily visualized for two variables, but are certainly possible for more than two).

    Thus a characterization test typically applies functional tests repetitively and takes measurements of various DC or AC parameters resulting in a pass/fail decision at each stage; the test is repeated for different values of variables taken pairwise such as the level of V_{CC} and some other parameter. The pass/fail decision for each pair of parameters chosen is presented in the form of a two-dimensional plot called a *Schmoo plot* [165] (named after the comic strip characters created by Al Capp). This plot shows the regions corresponding to correct and incorrect operation of the chip as determined by the values of the pair of parameters chosen, using '*' for correct operation and '•' for failure. A generic Schmoo plot is shown in Figure 2.1.

48   CHAPTER 2

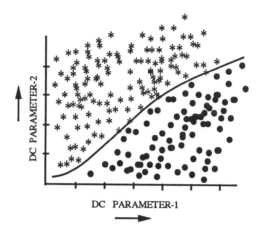

Figure 2.1 A Schmoo plot

These tests consist, therefore, of a repetitive sequence of measurements to estimate the operating limits of the device. These tests are usually done during production to control the process. Since these tests are repetitive, they usually take a long time to perform.

2. **Production Test**

A production test is typically performed by the manufacturer of a memory chip as the 'outgoing inspection test' and also by the user as an 'incoming inspection test.' These tests are also referred to as *Go/No-Go* tests. The main difference with characterization tests is that here the exact values of the operating limits are not very important provided they satisfy the (broad) specifications; therefore, the measurements involved are not repetitive in nature.

A production test is quite simple and fast; it consists of a test that verifies whether some chosen parameters are in agreement with the device specifications, while operating the device under nominal conditions. Devices are tested either at the speed used in the application, or the speed guaranteed by the supplier. This may cause problems if these tests are performed near the margins of proper operation of the device, because under conditions that slightly differ from the test conditions, for example, ambient temperature changes, power supply voltage spike, input currents and output loads, the device could fail the electrical test and operate slower than expected.

The reason why production tests should be performed at 'the outgoing inspection stage' of fabrication can be explained via a curve called the *bathtub*

Electrical Testing of Faults

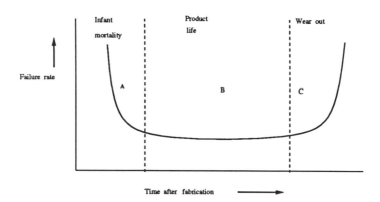

Figure 2.2 The bathtub curve

curve. The bathtub curve (Figure 2.2) shows the failure rate of a memory device as a function of time after fabrication. In an initial period, labeled A, the failure rate is relatively high, because a minor or undetectable failure can blow up into a detectable fault. The middle period, labeled B, is characterized by a low failure rate and consequently a high reliability; this is the period when the device is released in the market (the *outgoing stage*). Each device has an expected operating lifetime under normal operating conditions. At the end of this lifetime, the failure rate starts to increase again because of slowly developing fault mechanisms, for example, corrosion, mechanical stress, electromigration of metallization, and so on.

Manufacturers compensate for the relatively high initial failure rate by performing production tests at the outgoing stage under worst case conditions, called '*burn-in*.' During burn-in, the memory chips are subjected to continuous operation under high-temperature and over-voltage conditions, an environment designed to accelerate premature failure mechanisms. Measurements to verify the operating specifications of the device are done during this phase, so that faulty devices can be promptly identified and eliminated. Also, in applications requiring a high reliability for a sufficiently long period of time, a large safety margin should be allowed, for example, we may consider using an 80 ns DRAM where an 120 ns DRAM would suffice.

The nature of tests performed can be divided into two basic categories — DC parametric testing and AC parametric testing.

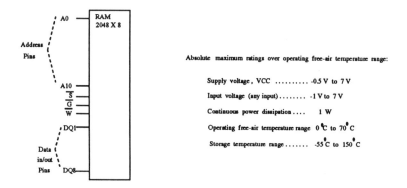

Figure 2.3 The TMS4016 - an SRAM chip; courtesy [115]

We shall first describe an SRAM chip called TMS4016 and then explain how various electrical tests are performed. We chose this memory device because its AC and DC parameters can be easily understood and span a wide range of parameters studied by chip designers.

2.3 DESCRIPTION OF THE TMS4016

The TMS4016 (Figure 2.3 [115], and Tables 2.1 and 2.2 [115]) is a fully static RAM (no clocks and no refresh) organized as 2048 words of 8 bits each, fabricated using an N-channel silicon-gate MOS technology, and comes packaged in a 24-pin 15.2 mm dual-in-line package (DIP). It needs a single 5 V power supply. It is guaranteed for operation from 0°C to 70°C. It operates at high speeds and requires less power per bit than 4Kb static RAMs. It dissipates under 385 mW maximum power, and has a guaranteed DC noise immunity of 400 mV with standard TTL loads. All inputs and outputs are fully TTL compatible. Access times for the TMS4016 series vary from 120 ns to 250 ns. The function of the different pins is described in the next section. All the pins are fully TTL-compatible.

Electrical Testing of Faults

Table 2.1 The TMS4016 - an SRAM chip; courtesy [115]

$\overline{W}\ \overline{S}\ \overline{G}$	DQ1-DQ8	Mode
L L X	Valid data	Write
H L L	Data output	Read
X H X	Hi-Z	Device disabled
H L H	Hi-Z	Output disabled

Table 2.2 Values of DC parameters of the TMS4016; courtesy [115]

Parameter	Min	Nom	Max	Unit
Supply voltage, V_{CC}	4.5	5.0	5.5	V
Supply voltage, V_{SS}		0.0		V
High-level input voltage, V_{IH}	2.0		5.5	V
Low-level input voltage, V_{IL}	-1.0		0.8	V
Operating free-air temperature, T_A	0		70	°C

2.4 OPERATION OF THE PINS IN TMS4016

1. **Address Pins**(A_0 through A_{10}): The eleven address pins select one of the 2048 ($=2^{11}$) 8-bit words in the RAM. When a write operation is performed, the address inputs must be stable for the entire duration of the write cycle. These inputs can be driven directly from standard series 54/74 TTL without using pull-up resistors.

2. **Output Enable**($\overline{G}$): The output enable terminal configures the bidirectional data-in/data-out port for reading (i.e., as data-out) when required. When output enable is high, the data-out terminals are tristated; when output enable and chip select are both low, the data-out terminals are active. Output enable provides greater flexibility in controlling the output, thereby also simplifying the data bus design.

3. **Chip Select**($\overline{S}$): The chip select terminal can inhibit the input data during a write operation by being high, thereby causing the data-in/data-out port to be tristated — during normal read and write, the chip select should be low.

4. **Write Enable($\overline{W}$)**: The write enable signal controls the read or write mode; a high logic level selecting the read mode and a low logic level selecting the write mode. Address changes should be accompanied by a high level at the write enable terminal to prevent erroneously writing data into an unintended memory location.

5. **Data-in/data-out**($DQ_1 - DQ_8$): These data terminals have a fan-out of one Series 74 TTL gate, one Series 74S TTL gate, or five Series 74LS TTL gates, when data is output. Data can be written into a selected device when $\overline{W}$ is low; during the write operation (and also when chip select and output enable lines are disabled), these terminals are in the high impedance state.

2.5 DC PARAMETRIC TESTING

DC parametric tests measure the *steady-state* electrical characteristics of a device using Ohm's law. The DC parameters are tested by forcing a terminal voltage and measuring terminal current, or vice versa. This is done using a *parametric measurement unit*, also called PMU. The forced voltage/current is applied suitably so as to ensure a certain expected response of the circuit. For example, in order to measure output low current (I_{OL}), the terminal voltage is adjusted so as to cause the device to output a logic-0 voltage. The actual DC parameter values will vary from chip to chip; hence, in the data sheet for the chip, the manufacturer publishes the margins within which the actual values of these parameters should lie.

There are various types of DC parametric tests — *contact test, power consumption test, leakage test, threshold test, output drive and short current tests* [165], and *voltage bump test* [136]. We shall characterize the manner of performing the first six tests for the TMS4016, described earlier. Standby current tests will be described with regard to the design of a dual port SRAM, and voltage bump tests which are very easy to perform will be briefly described at the end of this section.

Electrical Testing of Faults

Table 2.3 AC & DC operating characteristics for the TMS4016; courtesy [115]

DC Parameters:

PARAMETER		TEST CONDITIONS	MIN	TYP	MAX	UNIT
V_{OH}	High level voltage	$I_{OH} = -1$ mA, $V_{CC} = 4.5$ V	2.4			V
V_{OL}	Low level voltage	$I_{OL} = 2.1$ mA, $V_{CC} = 4.5$ V			0.4	V
I_I	Input current	$V_I = 0$ V to 5.5 V			10	μA
I_{OZ}	Off-state output current	$\overline{S}$ or $\overline{G}$ at 2 V or $\overline{W}$ at 0.8 V, V_O 0 to 5.5 V			10	μA
I_{CC}	Supply current from V_{CC}	$I_O = 0$ mA, $T_A = 0°C$, $V_{CC} = 5.5$ V		40	70	mA
C_i	Input capacitance	$V_I = 0$ V, $f = 1$ MHz			8	pF
C_o	Output capacitance	$V_O = 0$ V, $f = 1$ MHz			12	pF

AC Parameters:

	TMS 4016-12		TMS 4016-15		TMS 4016-20		TMS 4016-25		
PARAMETER	MIN	MAX	MIN	MAX	MIN	MAX	MIN	MAX	UNIT
Read cycle time	120		150		200		250		ns
Write cycle time	120		150		200		250		ns
Write pulse width	60		80		100		120		ns
Address setup time	20		20		20		20		ns
Chip select setup time	60		80		100		120		ns
Data setup time	50		60		80		100		ns
Address hold time	0		0		0		0		ns
Data hold time	5		10		10		10		ns

PARAMETER	MIN	MAX	MIN	MAX	MIN	MAX	MIN	MAX	UNIT
Access time from address		120		150		200		250	ns
Access time from chip-select low		60		75		100		120	ns
Access time from output-enable low		50		60		80		100	ns
Output data valid after address change	10		15		15		15		ns
Output disable time after chip-select high		40		50		60		80	ns
Output disable time after output enable high		40		50		60		80	ns
Output disable time after write enable low		50		60		60		80	ns
Output enable time after chip select low	5		5		10		10		ns
Output enable time after output enable low	5		5		10		10		ns
Output enable time after write enable high	5		5		10		10		ns

1. **Contact Test:** This test is done as the first test in the test sequence to verify that the tester is in contact with the internal circuitry and that there are no open circuits and shorts at the device pins. It is performed by drawing current out of the device and measuring the voltage at the input pin. The current drawn out at a pin is expected to result in forward

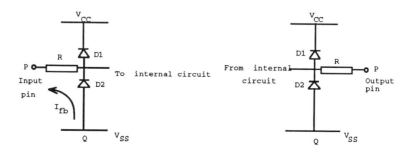

Figure 2.4 Protection diodes for input and output pins

biasing a protection diode connected to the pin (causing a potential drop of 0.7 V across it) as shown in Figure 2.4. In order to protect a device against electrostatic discharge caused by α-particles and other such factors, all input and output pins are connected to the V_{CC} (power) line and the V_{SS} (ground) line via protection diodes that are reverse biased under normal operating conditions. By getting forward-biased during abnormal conditions, these diodes guarantee that inputs and outputs cannot assume a voltage level of 0.7 V (one diode drop) below V_{SS} or 0.7 V above V_{CC}.

The tester is assumed to make contact with the device through a resistance R. There are three possibilities — (a) there is a contact *short*, which would cause R to be close to 0Ω; (b) no problem exists at the contact, and R has a certain nominal value; and (c) there is an *open circuit* at the input terminal of the internal circuit, causing R to have a very large value (infinity in the ideal case).

Let us now examine the manner in which a contact test will be performed on each input pin of the TMS4016.

(1) Set all input pins to 0 V.

(2) Force a predetermined I_{fb} from point Q (held at V_{SS}) to the input pin (see Figure 2.4). This current would cause the diode D_2 to be forward-biased. Typical values for I_{fb} are in the range of 100 μA to 250 μA.

(3) Measure the voltage V_{pin} at the input pin resulting from the forcing current I_{fb}. The value of R will be proportional to the potential drop across R, and, under normal contact, would consequently affect V_{pin} as given by the equation

$$V_{pin} = V_{SS} - 0.7 - I_{fb}R$$

Suppose, $R = 0.5k\Omega$ and $I_{fb} = 150\mu A$. Then, from the above equation, $V_{pin} = -0.775$ V under normal conditions. If a contact short occurs, the diode drop would be very small, and consequently, V_{pin} would be about -0.075 V. If there is an open circuit at the contact, the value of I_{fb} would be large (an open circuit is noticed when too much current is forced through a diode with a smaller current rating) hence V_{pin} will also be large (about -1.5 V).

2. **Power Consumption Test:** This test is done to determine the worst case power consumption under static and dynamic conditions (i.e., with the inputs at a steady logic value, and the inputs changing dynamically during operation, respectively). One way of performing this test is to measure the maximum current drawn by the device from the power supply (called I_{CC}) at the specified value of the power supply voltage V_{IN}. If this maximum current does not lie within expected limits, the chip fails this test.

   For the TMS4016 SRAM chip, the following sequence of operations performs a static power consumption test, as deduced from the chip description in Table 2.3 [115]:

   (1) Set the power supply voltage V_{CC} (also known as V_{IN}) to 5.5 V, the ambient temperature to 0°C (worst case conditions), and the output current I_O to 0 mA (i.e., the outputs should be open).

   (2) Measure I_{CC}. If I_{CC} is around 40 to 70 mA, the chip passes the test; if it is greater than 70 mA, the chip fails the test.

3. **Leakage Test:** Conceptually, CMOS circuits and tristated outputs are to be regarded as open circuits; in reality, they are high impedance circuits. The maximum current drawn by the device is called the *leakage* current, for inputs it will be *input leakage* and for tristated outputs it will be *tristated leakage*. To detect unacceptable input leakage, a certain voltage is forced across the device by the PMU and the current is measured. The chip is diagnosed as faulty if the current is greater than a certain maximum value. This measurement is repeated for each input pin.

   For tristate output leakage, the PMU forces a device output to become tristated and then measures current drain from the output. To pass this test, the leakage current from the device should be less than a certain maximum value.

   For the TMS4016 SRAM chip, the following steps would comprise a leakage test:

   (1) Apply a high voltage (> 2 V) to the input $\overline{S}$ (chip select). This causes the chip not to be selected. The value of I_I (the input leakage current) is

measured for the chip select pin. If it is less than 10 μA, the chip passes this step of the test.

(2) Keeping the chip select ($\overline{S}$) high, apply a high voltage (> 2 V) to the output enable ($\overline{G}$) and a low voltage (0.8 V) to the write enable ($\overline{W}$); this causes the data-out lines ($DQ_1 - DQ_8$) to be tristated. Two sets of tests are performed:

(a) force a high voltage (5 V) on each of the data-out lines and measure the value of I_{OZ} (the output leakage current); the chip passes the test if this current is less than 10 μA.

(b) force a low voltage (0.4 V) on each of the data-out lines and measure the value of I_{OZ} (the output leakage current); the chip passes the test if this current is less than 10 μA.

(3) Now select the chip (by making $\overline{S}$ low). Two sets of tests are again performed:

(a) set the chip in read mode (by setting $\overline{W}$ high), force a high voltage (5 V) on each of the address lines (A_0 through A_{10}) and measure the value of I_I (the input leakage current) for each address line; the chip passes the test if this current is less than 10 μA.

(b) set the chip in write mode (by setting $\overline{W}$ low), force a high voltage (5 V) on each of the data-in lines (DQ_1 through DQ_8) and measure the value of I_I (the input leakage current) for each data line; the chip passes the test if this current is less than 10 μA. Repeat with a low voltage on each data input line. A low voltage can be presented at the address input lines and the value of I_I measured for these lines.

4. **Threshold Test:** This test determines the maximum and minimum input voltages (V_{IL} and V_{IH}) required to cause the output voltage of the device to switch from high to low, and from low to high, respectively. A characterization test for each of these measurements can be done by performing a functional test repeatedly while raising or lowering the level of the input voltage; similarly, a production test for each of these can be done by setting input voltages to V_{IL} or V_{IH} while performing a functional test. Typically, $0 < V_{OL} < V_{IL} < V_{IH} < V_{OH} < V_{CC}$. The procedure to perform this test on the data in/out pins is as follows:

(1) For each of the pins DQ_1 through DQ_8, write a 0 by applying a low voltage (starting at say 0.3 V) and then read it out. Increase the voltage in steps of 0.1 V and stop when an erroneous read occurs. A chip is regarded as good if an erroneous read occurs above 0.8 V.

(2) For each of the above pins, write a 1 by applying a high voltage (say 3 V) and then read it out. Decrease the voltage in steps of 0.1 V and

stop when an erroneous read occurs. The chip is regarded as good if an erroneous read occurs around 2 V.

5. **Output drive current Test:** This test verifies that the output voltage level is maintained for a specified output driving current. For CMOS devices, a simple way of doing this would be to force the output voltage to low or high and measure the resulting current in each case. If the currents are greater than some minimum specified value, the chip passes this test.

    For the TMS4016, the following steps can be performed:

    (1) The chip is selected by setting $\overline{S}$ low and put in write mode, by setting $\overline{W}$ low. The cells are filled with a pattern of 0s.

    (2) The chip is put in read mode, by setting $\overline{W}$ high.

    (3) While reading out 0s, the output voltage is forced to 0.4 V (a low logic level) and the output current under low level output voltage, I_{OL}, is measured. This current should be greater than 2.1 mA.

    (4) Steps (1) through (3) are repeated with the all-1 pattern. The output voltage is forced to 2.4 V this time and the output current under high level output voltage, I_{OH}, is measured. This current should be greater than -1 mA.

6. **Output short current Test:** The aim of this test is to verify that the output current drive capability is sustained at high and low output voltages.

    The following steps will perform an output short current test for the TMS4016:

    (1) The chip is selected for writing ($\overline{S}$ and $\overline{W} = 0$) and filled with the all-1 pattern.

    (2) The chip is set up for a read operation (by setting $\overline{W}$ to 1) and the output of the PMU is shorted to 0 V.

    (3) The short current is measured, it should typically be 40 mA or above.

    The output should not be shorted for a long time because it might cause the output drivers to burn out.

7. **Voltage Bump Test:** *Voltage bumping* [136] is the phenomenon of power supply voltage fluctuations that cause erroneous data to be read out of the RAM. The voltage bump problem demands some special testing and the voltage bump test is a pretty rigorous one for DRAMs.

    A positive voltage bump is said to have occurred when V_{CC} during the write operation exceeds V_{CC} during the read operation. This causes the readout voltage to be lowered and this may cause a read error. In high density DRAMs, charge is often stored in ion implanted capacitor cells with capacitances in the 40 to 50 fF to prevent leakage caused by alpha-particle related soft errors and other kinds of noise injection from the V_{CC} power supply. If V_{CC} is directly connected to these capacitor cells, any fluctuations of the power supply voltage V_{CC} might cause erroneous data to be read out.

A positive voltage bump raises the stored level in the cell — thereby often causing a 0 to be erroneously read as a 1. A stored 1, however, is unaffected. This reasoning led to the development in the early 1980s of the half-V_{CC} biased cell plate which minimizes the effects of a voltage bump.

A voltage bump test would consist of the following steps:

(a) Fill the memory with 0s.

(b) Slowly increase the supply voltage above V_{CC} in steps of, say, 0.01 V, and for each voltage setting, read the memory. Stop as soon as a 1 is read from any location. Record this voltage as V_{high}.

(c) Fill the memory with 1s.

(d) Slowly decrease the supply voltage below V_{CC} in steps of, say, 0.01 V, and for each voltage setting, read the memory. Stop as soon as a 0 is read from any location. Record this voltage as V_{low}.

(e) Compare V_{high} and V_{low} with the prescribed (data-book) values.

There are advantages and disadvantages of DC parametric tests. They are simple, inexpensive and fast, and measure output levels with worst case DC loading; however, they give inadequate or no data on functional tests, switching characteristics, and pattern (sequence) information. Complex circuits, made from thousands of transistors, diodes and resistors, must be tested as thoroughly as possible. The DC parametric test performs a DC test on a few devices connected to the terminals, leaving thousands of internal transistors untested and providing practically no information concerning switching characteristics or pattern sensitivity.

2.6 AC PARAMETRIC TESTING

AC parameters for SRAMs and DRAMs are tested by applying an alternating voltage at some frequency or frequencies and measuring the terminal impedance

Electrical Testing of Faults

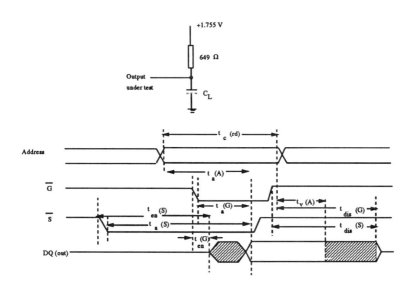

Figure 2.5 Read cycle timing waveforms for the TMS4016 SRAM chip; courtesy [115]

(dynamic resistance or reactance). The tests are performed at some DC bias level or possibly at several levels. Such tests are done to obtain engineering information pertaining to various kinds of delays caused by input and output capacitances. These tests are relatively simple and provide information on input loading characteristics. However, they provide no functional data, switching information, information on pattern sensitivity, or DC data; moreover, as in DC testing, data are gathered for only a few devices near the terminals and the test provides no information about the thousands of internal transistors. Both static and dynamic memories are subjected to various sensitivity tests using AC inputs. Some of these tests are: access time sensitivity, address set-up time sensitivity, write recovery time sensitivity, and data set-up and release-time sensitivity. Some of these tests are performed together with functional tests like Thatte and Abraham's test for coupling faults [163] (1977).

We shall describe the AC parametric tests for the TMS4016 SRAM chip with regard to Table 2.3 that shows the timing requirements during operation. This chip has simple AC parameters that are easy to understand. The various timing waveforms associated with the read cycle for this chip are shown in Figure 2.5.

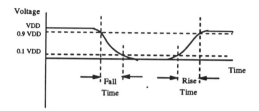

Figure 2.6 Determination of rise and fall times

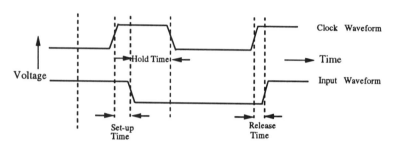

Figure 2.7 Determination of setup, hold and release times

1. **Rise and fall time tests:** These times are determined by measuring the time interval between two voltage levels of either a rising edge or a falling edge of the output voltage. This is shown in Figure 2.6. These timing parameters have not been specified for the TMS4016.

2. **Tests for set-up, hold and release times:** Tests for set-up and hold times measure the delay involved in propagating input signals to internal latches with respect to the arrival of the control signal (clock) at these latches. The *set-up time* is the time for which the input data must be present *before* the edge of the clock arrives; the *hold time* is the time for which the input data has to remain unchanged *after* the edge of the clock arrives in order to ensure proper sampling of the input. These time intervals are measured from the 50% point of the input signal to the 50% point of the clock signal, as shown in Figure 2.7. The *release time* is the maximum time for which the input data is allowed to be present during the clock period without being sampled.

Set-up times at address pins are an important consideration. A test for evaluating address set-up time sensitivity for the TMS4016 is as follows:

(1) Write a 1 on the first cell of the memory.

Electrical Testing of Faults 61

(2) Flip each address bit and write a 0 at the resulting address (this will cause access to the last cell if the number of cells is a power of 2).

(3) Flip each bit back again and read the first cell.

These steps are done for each cell, and the entire procedure is repeated with complementary data. If at any stage, a discrepancy occurs between the data read from a cell and the data stored by the most current write operation, an excessive address set-up time will be witnessed. The reason for flipping each address bit is to force the device to undergo the maximum possible delay in setting up the new address. From the data given in Table 2.3, the minimum address set-up time for a good chip should be 20 ns.

Release times for addresses is also an important criterion. This time should be quite high for satisfactory performance.

3. **Access time tests:** These tests are done to find out if the memory can perform its read/write operations within a specified time, given that it is functionally correct at reduced speeds. Three main approaches are employed for measuring access times of memory devices:

    (a) **Characterization test using an external tester**: An example of this test would be to measure the access time from address (i.e., the time delay between setting up a new address and getting the data during a read operation from that address). The following steps can be followed for the TMS4016.

    (1) Split the memory into two halves.

    (2) Write 0s in one half and 1s in the other.

    (3) Read the entire memory — this is done by accessing a cell c_1 belonging to one group, reading it and comparing it with the expected value, then going to a cell c_2 of the other group and reading it next, comparing it with the expected value, and eventually, reading c_1 again, to compare the time for a change in the data outputs with respect to the change in the address inputs.

    For the TMS4016, the access time from address is seen to have a maximum value of 120 ns for the TMS4016-12 chip, and this value is greater for the other devices in this series, the maximum being 250 ns for the TMS4016-25 chip.

    (b) **Characterization test using a well-known functional test**: A functional test such as MATS++ can be repeated, each time with a shorter access time, until the memory fails. The access time thus lies between those corresponding to the last and the next to last tests.

(c) **Production test**: The memory is operated at the specified access time and the read/write operations are verified for correctness, any discrepancy being regarded as an indication for a faulty chip.

An example of an AC parametric test using this approach is MOVI (Moving Inversions) [32] (1976). This test determines the access time of the chip by performing two successive read operations at two different addresses with different data. The access time measured is a function of (a) the address decoder delay, and (b) the read logic delay, for which the worst case corresponds to reading opposite data values on two successive read operations.

The procedure for access time testing using MOVI is outlined below:

(a) Initialize the memory with 0s.

(b) For $i = 0$ to N-1, where N is the total number of address bits, (the address bits being denoted as $a_{N-1} \ldots a_1 a_0$, do the following:

   i. Set address bit a_i to 0.
   ii. Starting from the lowest address (with each address bit equal to 0), read a memory cell, write back its complementary value, and then read this value out. Increment the address by 2^i, adding any carry out that results from this incrementing with the least significant bit (LSB), (this is known as an end-around carry) and repeat this step until the last address location has been accessed. It should be noted that incrementing the address by 2^i with provision for end-around carry would flip the address bit a_i.
   iii. At this point, the memory (under fault-free conditions) should be filled with the all-1 pattern. Repeat step ii.
   iv. At this point, the memory (under fault-free conditions) should be filled with the all-0 pattern. Repeat step ii. but start from the highest address and keep decrementing it by 2^i this time till all the address bits become 0.
   v. Likewise, repeat step ii. again following the reverse sequence of addressing, to verify read and write operations for complementary data.

A small example can be used to illustrate the addressing sequence of MOVI. Suppose we have a 2-bit address $a_1 a_0$. Then the first pass through the *for*-loop above will address the memory locations back and forth according to the addressing sequence $00 \to 01 \to 10 \to 11$, followed by the second pass that will access the locations back and forth according to the sequence $00 \to 10 \to 01$ (an end-around carry will occur when the address transition from 10 to 01 takes place) → 11. For the first addressing sequence, the address is incremented by $2^0 = 1$ each time and for the second sequence, it is incremented by $2^1 = 2$ each time.

The basic idea behind the electrical test part of MOVI is simple. It may be noted that the ith address bit ($0 \leq i \leq N-1$) is forced to switch back and forth during the sequential memory addressing at the ith iteration. Hence, any delay in the address decoder that involves the ith bit or any delay in the read logic that causes a sluggish response to a change in the ith bit would result in data being retrieved from a previously encountered memory location; since all the previously encountered locations have the opposite value as the current one (brought about by the write operation in step (b)ii.), this access delay error will be detected.

The complexity of this algorithm is in $O(Nn)$, where N is the number of address bits and n is the number of memory locations — since $N = O(\lg n)$, the complexity of this algorithm is in $O(n \lg n)$. This algorithm has a very impressive fault coverage — it detects all address decoder faults (AFs), stuck-at faults (SAFs), transition faults (TFs), unlinked inversion coupling faults (CFins) (because the march notation of MOVI is a superset of March X), and most unlinked idempotent coupling faults (CFids) (because MOVI is almost a superset of March C-). It also has a smaller time complexity than GALPAT. These algorithms are discussed in Chapter 3.

4. **Propagation delay tests:** Propagation delay time is the time that elapses between a change in state (edge) on an input and a resulting change in state on an output. The time is measured from a specified voltage level of the input edge (usually a 50 % mark) to a specified voltage level on the resulting output edge (50 %). Propagation delay times are specified in terms of a minimum or maximum time, a minimum time indicating that the response should *not* be observable before the specified time has elapsed, and a maximum time indicating that the response must be observable within the specified time.

   A manner of measuring propagation delays is with the use of a technique called *pulse testing* — pulse testing consists of applying an input pulse with a controlled rise and fall time, measuring the propagation delay from input to output and the transition time of the output pulse. The test is conducted with a standard output load (RC or RL). The test is simple and provides data on switching times of internal devices, but doesn't give any data on input loading (DC or AC), pattern sensitivity or output DC measurements. Also, hardware required for such tests may not always produce satisfactory results with regard to accuracy, controlled rise and fall times at the input, and flexibility in changing timing parameter measurements.

For the TMS4016, the parameters $t_{en}S$ and $t_{dis}S$ (output enable time after chip select is low and output disable time after chip select goes high, respectively) determine the acceptable propagation delays. These values are in the neighborhood of 5 ns (minimum) and 40 ns (minimum) respectively, as seen in Table 2.3.

5. **Running time tests:** These tests determine the fastest running speeds of devices with regard to read/write operations. One algorithm, called the 'read cycle time test' verifies the read cycle time (which is in the ballpark of 150 ns for the TMS4016, see Table 2.3) performs read operations reading 0s and 1s from alternating addresses very rapidly (at a specified speed) and makes a pass/fail decision for the device under test to yield a production test. Another algorithm runs a characterization test on the device by alternating read operations at progressively increasing speeds until at least one such operation fails.

6. **Tests for sense amplifier recovery fault:** This fault is caused when the sense amplifiers get 'saturated' after having to read a long string of bits of the same value (0/1) and then become too sluggish to read the opposite value. This fault may also be caused when a long string of bits of a particular value is stored in the memory and then a read operation is performed on a cell with the opposite value. This fault is caused because the input lines of the sense amplifiers are also the lines on which the data bits to be written are placed.

Such a fault can be tested in two passes — in one pass, the algorithm should read a long string of 0s (or 1s), followed by reading a single 1 (or 0); in the other pass, the algorithm should write a long string of bits of the same kind and read the complementary bit. In this test, read and write operations should not be mixed, as that may mask the sense amplifier recovery fault. A simplified version of such an algorithm is given below:

(1) Write a certain background pattern $dddddd\overline{d}d$ (First time through, d is 0 and then d is made 1) in memory. The address k at which $\overline{d}$ is written is to be noted.

(2) Read d from all the memory locations where d was written in the previous step — any discrepancy indicates a read-error.

(3) Read cell k — if a d is read, then a sense amplifier recovery fault is detected.

(4) Write a pattern of all ds in the memory, excluding cell k. Read cell k for a $\overline{d}$, and report an error if a discrepancy is noticed.

7. **Test for write recovery fault:** This fault may occur when a write operation is followed by another read/write operation at a different address;

thus there are two kinds of such faults: (a)*read-after-write* fault, in which a read operation on cell c_2 after a write on cell c_1 causes cell c_1 to be erroneously read; or (b) a *write-after-write* fault, in which a write operation on cell c_2 immediately following a write operation on cell c_1 causes an erroneous write into c_1. In any case, a recovery fault is reported.

(1) Write a 0 at an address c_1.

(2) Complement all address bits to access c_2 and write a 1 in it.

(3) Complement all address bits to access c_1 and read its contents. A 1 indicates that a write recovery fault has occurred.

(4) Repeat steps 1 through 3 for each cell in one half of the memory (such that complementing its address bits would cause a cell to be accessed in the other half).

(5) Repeat steps 1 through 4 with complementary data.

To test all possible pairwise address combinations for write recovery faults, a functional test algorithm such as GALPAT (described in the next chapter) can be used, at the cost of an increased computation time ($O(n^2)$).

2.7 ELECTRICAL TESTING OF DUAL-PORT SRAMS

A dual-port SRAM [137] is a single memory array with two totally independent sets of interfaces for address inputs, control lines, and I/Os (see Figure 2.8). The most common task performed by dual-port SRAMs is message/data passing, due to the ability of these devices to be operated by two non-compatible systems; for example, one system could possibly have a CMOS voltage level while the other one could have a TTL level.

2.7.1 Standby current test

Thus there are 4 possibilities (CMOS vs. TTL) for the voltage level combinations at the two ports of a dual-port SRAM. Relative to these, there are 4 standby current test conditions, as shown in Table 2.4 [137].

A standby current test must thereby verify that the current specifications are met for each of the above cases.

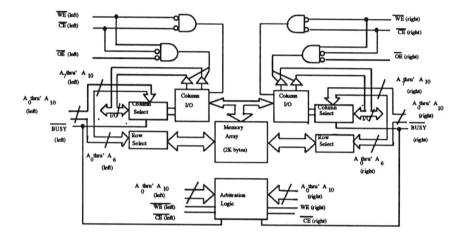

Figure 2.8 Block diagram of a dual-port SRAM; courtesy [137]

Table 2.4 Standby current (I_{sb}) conditions for a dual-port SRAM; courtesy [137]

Symbol	Parameter	Test Condition
I_{sb1}	Standby Current: both ports have TTL level inputs	$\overline{CE}_{left}$ and $\overline{CE}_{right} \geq Vi_{high}$
I_{sb2}	Standby Current: one port has TTL level inputs	$\overline{CE}_{left}$ and $\overline{CE}_{right} \geq Vi_{high}-$ active port outputs open $F = F_{max}$
I_{sb3}	Full Standby Current: both ports are all CMOS level inputs	$\overline{CE}_{left}$ and $\overline{CE}_{right} \geq V_{CC} - 0.2V$ $V_{in} \geq V_{CC} - 0.2V$ or $V_{in} \leq 0.2V$
I_{sb4}	Full Standby Current: one port all CMOS level inputs	One port is $\overline{CE} \geq V_{CC} - 0.2V$ $V_{in} \geq V_{CC} - 0.2V$ or $V_{in} \leq 0.2V$ Active port outputs open $F = F_{max}$

Circuit-dependent tests

As described earlier, the memory array of a dual-port SRAM can be addressed by two independent sets of addresses and control lines [137]. The fundamental problem in testing these devices is the exact manner of accessing the entire array individually from both sides. In the absence of some sort of *dual pattern generation* and *dual timing arrangement* (which are *not* provided by most existing memory testers), three possible methods can be applied, as described below.

1. Tie address lines from both ports together, so that both ports receive the same address.

The disadvantage of this method is that although no additional hardware is needed and the ports can operate independently with selective enabling, they *cannot* be tested simultaneously, because enabling both ports will force the device into address contention.

2. Complement address.

This method is similar to 1, except that the address fed into the left port is *inverted* and fed into the right port. This removes the problem of the two ports being in contention, because their addresses will necessarily be different. Also, this scheme is quite simple and requires little additional hardware (only inverters are needed). However, the disadvantages of this scheme include: impossibility to test the arbitration circuits (because they will never be activated because there is no address contention), unacceptable timing delays introduced by the inverters used for complementing the address bits, and thereby reduced accuracy.

3. X/Y Separation of Ports.

In this method, the X address drivers are used to manipulate the left port, and the Y address drivers to manipulate the right port. Thus the tester has twice as many I/Os as the device, and it is also required to have at least six clock drivers. The block diagram for this scheme is shown in Figure 2.9. This ensures totally independent addressing and hence simultaneous testing of both ports vis-a-vis ability to test the arbitration circuit (contention testing). There are two disadvantages: it may be difficult to write topologically correct patterns in the memory array with this scheme, and to meet the requirements of address format control, to adequately perform contention testing. Nevertheless, this approach

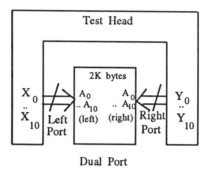

Figure 2.9 X/Y separation of the two ports in a dual-port SRAM system; courtesy [137]

is apparently the best of the three and with some additional components forms the most widely used testing approach for dual-port SRAMs. The additional components are: (a) software to ensure that topologically correct patterns are written, and (b) special tester hardware to handle the format control necessary for adequate contention testing. The purpose of the software is to translate logically correct patterns into topologically correct ones. This is necessitated because in the X/Y separation scheme, the tester's X drivers (or Y drivers) are assigned the combined row and column addresses. Thus in Figure 2.9, the row addresses A_{0L} through A_{6L}, and the column addresses A_{7L} through A_{10L} have all been assigned to the X driver pins X_0 through X_{10}. A similar observation holds for the Y driver pins Y_0 through Y_{10} also.

The different kinds of circuit-dependent tests to be performed under this scheme on the different hardware components of a dual-port RAM are described below.

Interrupt test

This is performed by simultaneously writing data into the interrupt location and monitoring the $\overline{INT}$ output of the opposite port. Hence, the test system should either be provided with sufficient comparators so that one may be dedicated to the $\overline{INT}$ output, or else the comparator may have to be switched from a data output to the $\overline{INT}$ output. This can be achieved with relays or FET switches.

2.7.2 Arbitration test

Testing of arbitration circuitry is a two-step process — first, testing the contention circuitry which continuously monitors the addresses applied to the two ports, activating a busy flag ($\overline{BUSY}$) when they match; and second, testing a set of *semaphore* flags that control the concurrent address bus accesses in an orderly fashion.

1. Contention Testing: This is the most difficult element of dual-port testing. A state of contention between the two ports arises when both of them try to address the same cell simultaneously. The contention circuitry will then arbitrate between the two ports, granting write access to one and read-only access to the other. This arbitration is dependent solely upon the signal timing relationships, with the win/lose status being indicated by the busy flags ($\overline{BUSY}$).

Raposa [137] used the IDT7132 chip as a benchmark. A timing parameter known as *arbitration priority set up time* (T_{aps}) is associated with this chip. This is defined as the minimum time for which the address applied to one port should remain stable before the address arrives at the opposite port, for the first port to receive write priority. For the IDT7132 dual-port RAM chip, this time is 5 ns. So for contention testing of this chip, we must ensure that the address is applied to one port at least 5 ns before the address arrives at the opposite port.

If the tester is provided with separate X and Y timing generators, then we can simply specify separate timing for each; however, if there is a single timing generator for all address drivers (as with most existing memory testers), we must apply the same signal at the same time to both ports, and yet have one port receiving its address before the other. This is achieved using *address formatting*, a trick that causes one address to appear to arrive before another. This is demonstrated in Figure 2.10, which shows single timing combined with address formatting. The whole idea is to disable the Y-complementer at the moment when the address has stopped arriving, so as to cause the Y-address to *lead* the X address by T_{aps}, as shown in Figure 2.10.

2. Semaphore Testing: Semaphores are a set of latches accessible to either port. Once a semaphore has been requested and granted, it can only be released by the port it has been granted to. These latches are not physically tied to the memory array but may use the same address and data lines. The simplest method of testing them is to request, verify, and release each latch.

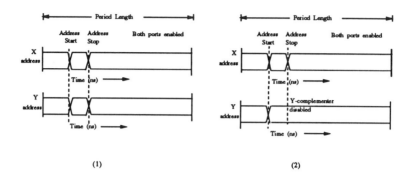

Figure 2.10 Single timing combined with address formatting to achieve a time difference between X and Y addresses (the addresses applied to the two ports in an X/Y separation scheme); courtesy [137]

2.8 CONCLUSION

We have looked at AC and DC parametric testing paradigms for simple RAMs and more complex electrical testing paradigms for dual-port RAMs in this chapter. It may be recalled that all these paradigms (fault models and test techniques) deal with a superficial abstraction of the chip I/O interface behavior without reference to the underlying layout, technology and detailed cell design. These paradigms therefore achieve a good fault coverage in most cases but not necessarily a good *defect* coverage. Often, defects in the memory array do *not* manifest themselves as faults, but nevertheless, may warrant some corrective action, such as cell array repair or chip replacement. In Chapter 3, we shall describe functional testing in which conventional fault modeling, for example, stuck-at, coupling and transition faults, and layout-related fault modeling (for example, scrambled pattern-sensitive faults) are employed. In Chapter 4, we examine test algorithms that use much finer process-related defect and fault modeling approaches and perform the RAM test comprehensively by employing a combination of functional and electrical test algorithms. As multi-megabit RAMs increase in density, new defect types which were not known before are being observed and studied. These new defect types are the direct result of new processing techniques and changes in the structure of the basic storage cell. We shall examine these issues in the next two chapters.

Electrical Testing of Faults

SUMMARY

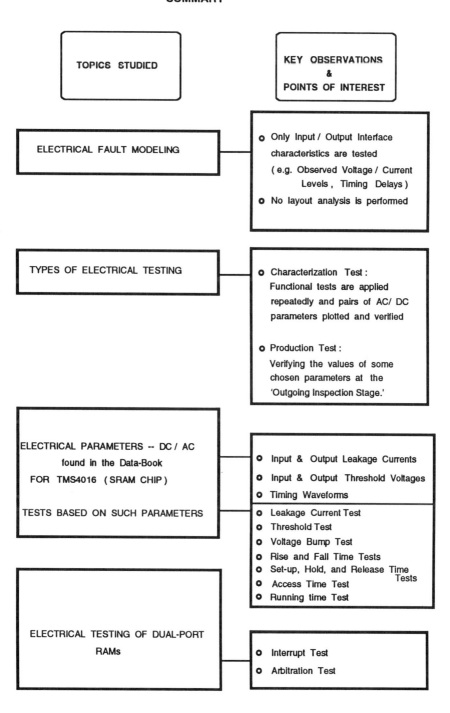

2.9 PROBLEMS

2.9.1 Worked-out Example

1. Give a detailed description of an electrical test for the **refresh** mechanism for a dynamic RAM chip. Illustrate it with a suitable example.

   (**Sample Answer Sketch:**) For a DRAM chip, the refresh operation must be performed periodically to retain data. The prescribed period can be obtained from a memory data book. For the TMS4164 64 Kb DRAM chip manufactured by Texas Instruments, this operation must be performed every four milliseconds. To perform a refresh test, we need to verify that the refresh operation is fault-free and causes the original data to be written back in the memory. For the TMS4164, the $\overline{RAS}$-only refresh sequence may be used, keeping the $\overline{CAS}$ high throughout the refresh sequence to conserve power. This will cause the output buffer to be in the high-impedance state and no data will be output. Strobing each of the 256 row addresses (A0 through A7) with $\overline{RAS}$ causes all bits in each row to be refreshed. The 256 row addresses may be sequentially applied using an up-counter, so that the output bits of the counter are strobed with the $\overline{RAS}$ signal. Afterwards, for each applied address, a read operation is performed (with $\overline{CAS}$ turned low) to verify that the correct data was written back during refresh. This procedure is repeated with various data backgrounds: all-0, all-1, checkerboard patterns, etc., written into the memory.

2.9.2 Exercises

1. Examine the literature on EPROM testing and list the electrical failure mechanisms of EPROMs. Suggest electrical tests for EPROM failures.

2. A RAM is required only to store and retrieve data comprising 0s and 1s. Why is it not sufficient to test merely the correctness of the data stored? Give practical applications where the RAM data integrity is not the only issue.

3. Describe the design of state-of-the-art (a) dedicated memory tester, (b) LSI tester, and (c) general purpose tester for ICs.

4. Use a high-speed digital tester to test a commercial SRAM or DRAM chip. Measure the values of the various AC and DC parameters described in this chapter.

Electrical Testing of Faults

5. Describe the various forms in which memory test equipment present test results. What is the difference between **wafer mapping** and **bit mapping**? Give a diagram of a typical wafer map and a typical bit map for a 16 Kb RAM and describe their uses.

3
FUNCTIONAL FAULT MODELING AND TESTING

Chapter At a Glance

- **Layout-independent Functional Testing**
  -- Mealy Machine modeling; Functional Fault Models: Stuck-at, Coupling, Pattern-Sensitive; Test Algorithms for functional faults : GALPAT, GALROW, GALCOL, Butterfly, Sliding Diagonal, March, Test for NPSFs

- **Layout and Circuit-dependent Functional Testing**
  -- IFA, test for DBMs & PAMs, Scrambled neighborhood

- A brief introduction to Transparent Testing

- Fault Simulation based on Memory Cell Layout
  -- Oberle & Muhmenthaler's Simulator, Defective Hashing

3.1 INTRODUCTION

The purpose of functional testing is to verify the logical behavior of a memory device. A functional fault model describes the logical behavior in terms of logic-0 and logic-1 voltage levels as opposed to voltage and current values. Such voltage levels are observed at the I/O interface of the memory device consisting of the address inputs, the data input/output lines, and the control signals.

Most of the functional testing approaches in the late seventies and early eighties, with a few notable exceptions like the work done by Hayes [53, 54], dealt with layout-independent functional fault models. In other words, these models were mathematical abstractions of some generic *fault types* (such as stuck-at-zero, coupling, and others) and did not describe the mapping between the physical layout defects in a RAM chip and the logical faults. These models were found to be useful because they achieved a good fault coverage, even though the precise relationship between fault types and physical defect types was not completely understood.

However, as new processing technology is developing at a steady pace, it has become necessary to address testing problems from a layout-oriented perspective, for two reasons. First, a wide variety of layouts for memory chips are being created by various manufacturers. These layouts are often associated with peculiar processing defects, for which the exact faulty behavior is difficult to predict. Second, not all defects caused by circuit design errors and minor process variations are associated with faults, and conventional algorithms that achieve a high fault coverage do not necessarily achieve a high *defect* coverage. In view of this, the objective of memory testing has been expanded to include fault coverage *and* design and process-related defect coverage.

At present, therefore, there are two functional testing approaches that are commonly used — those that consider fault models based on layout defects due to process variations and design-related errors, and those that do not consider the circuit design or layout at all. However, regardless of the fault modeling approach used, the functional test algorithms are designed to verify *only* voltage *levels* of various signals (like data, address, control, and others) and not the exact values of voltages, currents, or timing delays between signals. It may be emphasized here that other kinds of test algorithms exist that are based on thorough analysis of the layout and circuit techniques, but these algorithms are designed to test not only voltage levels, but also current and voltage values, and sometimes, timing delays. Such fault models and test algorithms are henceforth

Functional Fault Modeling and Testing

described as **technology and layout-related**, and are discussed in Chapter 4. These layout-related tests are comprehensive in nature and typically consist of both functional and electrical tests. Hence most of the layout-related functional tests described in this chapter and the electrical tests described in the previous chapter are included in the testing approaches described in Chapter 4.

In this chapter, we shall first give an account of functional testing of faults from a logical perspective, without reference to the detailed layout of the memory chip and associated circuitry. Even though these fault models are layout-independent, the tests derived from them are found to have a reasonably good fault coverage. Later, we shall present fault models and testing paradigms based upon the detailed physical layout of the chip. Such tests have a good defect coverage also, in addition to a good fault coverage, but take greater time to be developed.

A block diagram of a memory system is shown in Figure 3.1. A memory system consists of six subsystems — row and column address buffers, for driving the address input lines before they are decoded; the row and column decoders, which select, for a given address at its input, a unique memory location for reading or writing into; sense amplifiers and write drivers, which read data out of, and write data into, a memory cell, respectively; an optional memory data register (not shown in the figure) which temporarily stores data after a read operation or before a write operation; the array of memory cells; and some control logic for driving the signals (such as $R/\overline{W}$ and $\overline{CS}$) that cause memory access. Faults can occur in any part of this system, but we shall concern ourselves mainly with faults in the decoder, the memory cell array, and the memory data register. Stuck-at faults in the memory address register are indistinguishable from those at the decoder inputs. Also, stuck-at faults occurring in other physical parts of the system can be regarded as faults in the above three subsystems, using a suitable fault modeling. Stuck-open faults can give rise to spurious sequential behavior, especially in the decoder, which may be hard to test. We shall describe various fault models and their testing approaches in this chapter.

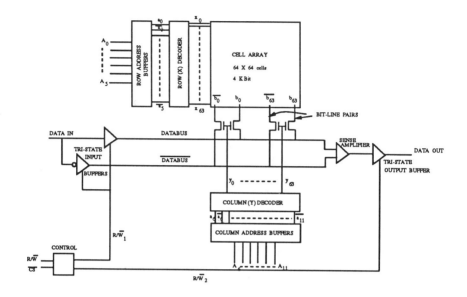

Figure 3.1 A RAM system organization

3.2 LAYOUT-INDEPENDENT FUNCTIONAL TESTING OF RAMS

Functional testing of RAMs is performed both during their production and during field use. It consists of several steps. First, the design and layout of the RAM circuitry are examined, and a set of likely layout and design defects predicted. This produces a so-called *defect model* for the RAM under test. In a layout-independent testing approach, a simplified *fault model* is then computed from the defect model, and the actual defect model is subsequently ignored. The fault model is then used to construct a test involving read and write operations of logic-0 and logic-1 voltages. One common approach is to model both the faulty and fault-free memory devices as finite state machines, such as Mealy automata, with states denoting memory contents and transitions denoting changes in the memory contents (which can only be due to transition writes in the fault-free case). The problem of constructing a functional test, therefore, reduces to computing a checking sequence for such a machine. An alternative approach is to study the manner in which the faults can be triggered, and formulate tests appropriately, without using a Mealy machine model of the memory. Often, this problem is simplified by collapsing the set of faults in the fault model into a reduced set of non-equivalent faults. Tests are then

Functional Fault Modeling and Testing

designed to achieve a high fault coverage. This approach simplifies the task of fault detection by employing knowledge regarding dominance of one class of faults by another.

3.2.1 Deterministic Mealy machine modeling

The Mealy machine modeling approach is described in [53, 54] (1975, 1980) for diagnosing *pattern-sensitive faults*, and has subsequently been extended formally to include the detection of other kinds of faults (especially, coupling faults, described later) by Cockburn [24] (1990). This approach is based upon the general framework developed and formalized by Brzozowski, Jürgensen, and Cockburn for the testing of finite state automata [15, 16, 23].

In this scheme, the memory (consisting of n 1-bit cells) is modeled as a deterministic Mealy automaton $M_0 = (Q, X, Y, \delta_0, \lambda)$, where $Q = \{0,1\}^n$, $X = \{r_i, w_0^i, w_1^i | 0 \leq i \leq n-1\}$, and $Y = \{0, 1, \#\}$. A memory state q is denoted by a vector $\langle c^0, c^1, \ldots, c^{n-1}\rangle$ which specifies the contents, denoted by the variables c^i, $0 \leq i \leq n$ of each of the n memory cells. We shall use the symbol $\langle q \rangle_i$ to denote the i-th component of q. The output function λ is defined as follows: $\lambda(q,x)$ is equal to $\langle q \rangle_i$ if $x = r_i$, and is equal to $\#$ otherwise. The state-transition and output functions can be easily deduced from the state-transition/output graphs in Figure 3.2.

3.2.2 Fault modeling for the Mealy automaton

The effect of single faults on the structure and behavior of this Mealy machine is also modeled in Figure 3.2. In Figure 3.2 (i) through (iii), a simple 1-cell memory is considered. This simple cell has two states — 0 and 1. A read operation at any state should keep the system in the same state, whereas a write operation that writes a 1 at state 0 should cause a transition to state 1, and a write operation of 0 at state 1 is expected to bring about a transition to state 0. In Figure 3.2(i) and (ii), the effect of a stuck-at fault has been modeled. A stuck-at-zero fault will cause the machine to be modified into one that has a single state 0 instead of two states, and a stuck-at-one fault will convert the machine into one that has a single state 1. Likewise, a transition fault in this simple machine would cause the machine to have both states but fewer edges (transitions) between these states. Thus any fault will alter the

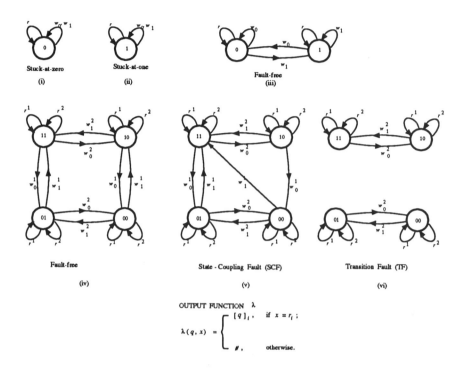

Figure 3.2 Transition graph of Mealy machines modeling (i)&(ii) stuck-at-zero and stuck-at-one faults; (iii) a fault-free 1-cell memory, with 2 states 0 and 1; (iv) a fault-free 2-cell memory; (v) a state-coupling fault in which an 'up' ('down') i.e., $0 \to 1$ ($1 \to 0$) transition in cell i causes an up (down) transition in cell k; (vi) a transition fault in cell i that forces this cell to remain in its initial state (state upon switching on the power supply) always; courtesy [15, 16, 23]

state transition graph of the Mealy machine. In general, therefore, a stuck-at-one fault in memory cell i of an n-cell memory (having 2^n states) will replace all transitions to states q with $\langle q \rangle_i = 0$, with spurious ones to states r with $\langle r \rangle_i = 1$, thereby modifying the state transition graph. Modifications of the state transition graph resulting from two other kinds of faults have been shown in Figure 3.2(v) and (vi).

Functional Fault Modeling and Testing

Table 3.1 A fault-detection experiment for a 2-cell memory modeled by a Mealy automaton

No.	Instruction	State of cells 1,2
1	Write 0 in cell 1	
2	Write 0 in cell 2	0 0
3	Read 0 from cell 1	0 0
4	Write 1 in cell 1	1 0
5	Read 0 from cell 2	1 0
6	Write 1 in cell 2	1 1
7	Read 1 from cell 1	1 1
8	Write 0 in cell 1	0 1
9	Read 1 from cell 2	0 1
10	Write 0 in cell 2	0 0
11	Read 0 from cell 2	0 0
12	Write 1 in cell 2	0 1
13	Read 0 from cell 1	0 1
14	Write 1 in cell 1	1 1
15	Read 1 from cell 2	1 1
16	Write 0 in cell 2	1 0
17	Read 1 from cell 1	1 0
18	Write 0 in cell 1	0 0

3.2.3 Design of fault detection experiments with a Mealy automaton

Consider the Mealy automaton given in Figure 3.2(iv), which depicts a fault-free 2-cell memory. The following treatment can be extended to an n-cell memory in a fairly straightforward manner.

Let us first describe this machine and then characterize the different kinds of interesting input sequences which will be important in its fault detection experiment. For a definition of these sequences, see [76].

This machine is reduced and strongly connected. It is reduced as no two states are equivalent. This is easily seen by applying the input sequence $r_1 r_2$ (read both cells). Also, the state encoded as (x,y) is reachable from any other state by applying the input sequence $w_1^x w_2^y$. Hence the machine is strongly-connected.

The fact that the machine is reduced is important because we want to *distinguish* between every pair of states in the fault-detection experiment. As shown in the literature [76], being reduced is a *necessary* condition for the existence of a *distinguishing sequence*. Being strongly connected makes the machine capable of undergoing a transition from any state to any other state, and hence we can eventually check that the machine does indeed possess the required number of states (namely, four), and that it exhibits correct behavior in each state.

The minimum length **homing sequence** for this machine is the sequence $r_1 r_2$ (i.e., reading the contents of the two cells). This follows directly from the definition of homing sequences. Thus reading the contents of both the cells uniquely identifies the final state of the machine regardless of the initial state (which is the same as the final state in our case, as read operations cause no state transitions).

The minimum length **preset distinguishing sequence** for this machine is the sequence $r_1 r_2$. This also follows directly from the definition because for any two different states q_1 and q_2, the machine's response to the sequence $r_1 r_2$ will be different.

A minimum length **synchronizing sequence** for a 2-cell memory is any of the four sequences of the form $w_1^x w_2^y$, i.e., the input sequence on the memory that causes the cell-pair to be lodged into the state (x, y), where $x, y \in \{0,1\}$.

We can recall from the literature on fault detection of Mealy automata (see, for example, [76]) that there are two phases involved in a simple fault detection experiment: first, the machine has to be somehow taken to a prespecified state, which is the initial state for the second phase; then, the machine has to be taken through all possible transitions. The first phase is usually done via an *adaptive* experiment and the second one via a *preset* experiment. However, in the case of a k-cell memory ($k = 2$ in our simple example), we may easily circumvent the adaptive experiment and can perform a purely preset experiment. This is because of the simple observation that *this* Mealy automaton has a synchronizing sequence. Hence any fault detection experiment can begin by writing a set of known initial values x and y onto the memory cells.

The second part of the experiment can be performed in various ways, but any scheme chosen must check all the transitions. The important state transitions for our purpose are the transition writes, since other kinds of state transitions (namely, reads and non-transition writes) are assumed to be fault-free. A traditional approach is to make the machine undergo a state transition and then immediately apply a *distinguishing sequence* to check the identity of the new state (since every state responds uniquely to the distinguishing sequence, this is a sufficient test); and then reach new states along *already verified transitions* and check each of these until there are no more left.

In our case, the complexity of the above fault detection experiment for an n-cell memory is n (for initialization) $+ (n + n^2) \cdot 2^n = O(n^2 \cdot 2^n)$, with each cell being read $n^2 \cdot 2^n / n = n \cdot 2^n$ times (once after every transition write). This experiment is therefore too complex to be of practical use. However, some fault modeling techniques may help us to find a much simpler approach. For example, if we assume a coupling fault model comprising only two-cell interactions, we may ignore the 2^{n-2} distinct combinations of values in the remaining cells and regard them alike. This would substantially reduce the size of the test sequence. Another way to reduce the test application time is to minimize the number of single-cell operations. No satisfactory method for reducing the number of read operations has been found so far. Hence, only the number of transition write operations on single cells can be reduced. To minimize the number of such write operations, the number of bit positions at which two successive test patterns differ must be minimal. For example, if a test for a two-cell memory requires the memory to be taken to each state in the set $\{(00), (01), (11)\}$ of states, then the write sequence $00 \rightarrow 01 \rightarrow 11$ involves only two single-cell transition write operations, whereas the write sequence $00 \rightarrow 11 \rightarrow 01$ involves three such operations. This observation leads us to the choice of an *Eulerian cycle* around the state-output transition graph for a fault-free memory as the checking sequence.

By definition, an Eulerian cycle [13] traverses each arc exactly once, and comes back to the starting node. In the process, it reads each cell once in every state, for a total of 2^n times. For a 2-cell memory, an Eulerian cycle traversal involves a total of at least 16 single-cell operations (corresponding to the 16 arcs shown in Figure 3.2(iv)). Thus the total number of single cell operations needed for checking a 2-cell memory is *at least*: 2 (for the first phase) + 16 (for the second phase) = 18.

An example of a checking experiment for a 2-cell memory is given in Table 3.1. The first two steps denote the initialization of the memory with a synchronizing sequence (two write operations), and the remaining steps denote the manner in which state transitions are verified. This experiment is discussed in [34]. For an n-cell memory, such an exhaustive experiment would have a total complexity of $n + 2n.2^n$, since there are a total of 2^n states with $2n$ arcs associated (n reads and n transition writes) associated with each state.

3.2.4 Fault classification

We notice from the above discussion that a Mealy machine modeling an n-cell memory contains 2^n states, and treats every functional fault as a (unique) modification of the transition-output graph. It thereby produces an impractically long test. Hayes [53, 54] restricted the fault model so that the contents of any cell can be affected only by read and write operations in its immediate physical neighborhood. Such realistic assumptions about the nature of memory faults are a necessity for designing efficient tests. Making such assumptions constitutes *fault modeling*. A fault model is a set of faults that are realistically assumed to be sufficient to describe the behavior of all or a large proportion of the expected defective circuits. Often, it may be possible to test for various kinds of faults simultaneously. For the purpose of efficient utilization of the tester's time, it is advantageous to design tests that have a high fault coverage. Fault models allow us to study the relationship between different fault types and design tests that have a high fault coverage for the most probable faults.

Fault models are built in various ways. The basic approach is to construct a set of logical faults and then compute tests for these faults. Modeling logical faults is a non-trivial task and involves examining the physical design and layout defects in the memory cell array and associated circuitry. Once these physical defect mechanisms are understood, simplified fault models are designed to correspond to these defects. In order to reduce the test length, it is often helpful to design tests which can be re-used for multiple fault models.

Functional Fault Modeling and Testing 85

This, in turn, becomes easy if we establish a fault classification. For example, tests that detect all two-cell coupling faults cause transition writes on each cell and thereby also detect all transition faults; hence, coupling faults dominate transition faults. Likewise, pattern-sensitive faults usually involve more than three cells and therefore dominate 3-coupling faults, k-coupling faults dominate 2-coupling faults, and so on.

An early fault model to represent the behavior of common failure mechanisms in both SRAMs and DRAMs is found in [22] (1975). This fault model considers defects in the memory cell array and the address decoder, and classifies the faults into four groups:

1. **Shorts and opens in the memory cell array**: These defects cause "stuck" bits or fixed interactions among adjacent bits.

2. **Shorts and opens in the address decoder**: These defects produce cells that cannot be accessed, or cells that can be accessed by multiple addresses, or addresses that access multiple memory locations.

3. **Access time failures**: These defects are produced by very fast or very slow transitions in the address decoder input, and get manifested as functional faults when worst-case address sequences are applied.

4. **Disturb sensitivities**: These defects produce what later became known as *coupling* faults. A coupling fault causes a single-cell transition write to disturb the contents of a neighboring cell as a result of stray capacitances between the two cells.

Cocking [22] has proposed various tests to detect faults of each type. In particular, a *march test* of length $10n$ is shown to detect all singly-occuring disturb sensitivities. He also notes that multiple disturb sensitivities can be detected by adding write operations to this test. He has proposed specialized tests to detect each of these fault types individually. He does not present a functional fault model that combines multiple failure mechanisms.

Thatte and Abraham [163] (1977) have described the first functional fault model for RAMs that includes some generic fault types based on specific failure mechanisms in various parts of the RAM circuitry. First, we shall describe this traditional fault model and then extend this model to obtain a more useful one. As mentioned before, a feasible test procedure can be developed only if we restrict ourselves to a subset of faults that are the most likely to occur. Exhaustive testing of all possible combinations of faults would result in computational explosion.

Before we introduce the fault models, a few definitions are in order.

Definition 1 *A defect of a memory device is an actual physical condition of one or more hardware components that might produce erroneous circuit behavior; such a physical condition may be* process-related *or* circuit design-related.

Definition 2 *A fault of a memory device is the erroneous circuit behavior that a defect produces.*

Definition 3 *A fault model is a set of faults sufficient to describe the erroneous behavior of a large proportion of potentially defective circuits.*

Thatte and Abraham observe that, although physical defects arise randomly all over the RAM chip, including the address decoders and the read/write logic, all the resulting erroneous behaviors can be represented as combinations of the following three faults in the memory cell array alone:

1. **Stuck-at faults (SAF or SF):** the value stored in a cell is permanently 0 or 1.

2. **Transition faults (TF):** a cell can be written from 0 to 1 (1 to 0), but not from 1 to 0 (0 to 1).

3. **Coupling faults (CF):** a write operation to a cell i, which changes the contents of cell i, forces the contents of another cell j to 0 or 1. However, changes in the contents of cell j need not disturb those of cell i. Such faults are called *dynamic coupling faults* (DCF). An alternative scenario is when the state of cell i forces a particular state on cell j; this is called a *state coupling fault* (SCF).

A read and a *non-transition* write (i.e., a write operation on a cell that keeps its contents intact) are assumed to have no effect on the state of either good or faulty memories.

The model described here is derived from the above, but considers other kinds of faults for the sake of simplicity, even though such faults may possibly be derived

Functional Fault Modeling and Testing

from some of the above types of faults. For example, a pattern-sensitive fault can be regarded as a special case of a k-coupling fault, but we have regarded it as a separate fault type for ease of analysis. Also, these models are relevant mostly to static RAMs, because we assume that reads and non-transition writes are fault-free, and the fault is always triggered by a transition write operation. In DRAMs, however, this assumption is not realistic, because read cycles are followed by refresh cycles during which the data is written back. Hence fault modeling for DRAMs is more complex than that for SRAMs. Some of the SRAM functional test *algorithms*, however, achieve good fault coverage for DRAMs also.

The fault model, therefore, consists of the following:

1. *Stuck-at and Transition Faults*: These faults are caused by shorts and opens in the memory cell array. In this model, one or more cells in the memory are assumed to remain unchanged by a write operation. There are two possibilities: a faulty cell may be permanently forced to a fixed value, either 0 or 1, and its state cannot be modified by any means whatsoever — this situation is known as a stuck-at-zero or stuck-at-one fault respectively (SAF). The faulty cell fails to undergo a write transition when a 1 is written at its 0 state or vice versa (a *transition write*), however, a coupling fault with another memory cell might change its state — this being described as a transition fault (TF). These faults may be caused by various reasons and may affect different components of the memory circuitry. This fault model is shown in Figure 3.3.

2. *Coupling Faults*: These faults are typically produced either by address decoder faults (known as AF), causing multiple-access, or by capacitive coupling between neighboring storage cells. Various kinds of coupling are possible between two or more cells. In this fault model, the state of one cell or a transition in one cell may affect the contents of another cell. For example, a transition in one cell might force another cell to a fixed value — this is called an *idempotent coupling fault* (CFid) [165], or may cause a spurious transition in another cell, called an *inversion coupling fault* (CFin) [165]. These two kinds of coupling faults are known as *dynamic coupling faults* because they are triggered by a *transition* within a cell. Static coupling faults are triggered by a logic level instead of a transition. Two such faults are: bridging faults (BFs), also called shorts [165], and state-coupling faults (SCFs). A bridging fault, also called a *short* is caused by a spurious galvanic connection (called 'bridge') between two or more cells or lines. It is a bidirectional fault and is of two types — AND and OR. In an AND bridging fault (ABF), the logic value of the bridge is the AND

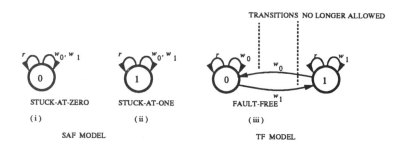

Figure 3.3 Stuck-at fault (SAF) and transition fault (TF) model

of the logic values of the interacting cells or lines; in an OR bridging fault (OBF), the logic value of the bridge is the OR of those of the interacting cells or lines. Another type of coupling fault is noticed when the coupled cell is forced to a value x only if the coupling cell j is in state $\bar{x}$, and vice versa, this being called a *state* or a *static coupling fault* (SCF). Also, coupling faults could be symmetric or asymmetric; that is, the coupling between two cells could be one-way or mutual. This fault model is shown in Figure 3.4.

3. *Pattern Sensitivity Fault Model*: A fault that affects the state of a memory cell as a result of the presence of a certain pattern of 0s, 1s, and/or $0 \to 1$ or $1 \to 0$ transitions is called a pattern-sensitive fault (PSF). Such faults are caused mostly by leakage currents whose strengths are often data-dependent. In [97], we examine the following leakage currents due to variations in process parameters: the *weak inversion* current I_W, from the storage node to the bit line, the *field inversion* current I_F between two adjacent cells, the *gate leakage* current I_G due to pin-hole defects in the gate oxide, and the *dark* current I_B between the storage node and the substrate.

Pattern-sensitive faults can cause either a spurious alteration of the state of a cell or a read/write operation on the cell to fail. PSFs are of two types — *unrestricted* PSFs (UPSFs) and *restricted*, or *neighborhood* PSFs (NPSFs). Testing for UPSFs is infeasible in practice as proved by Hayes in 1975 [53]. However, NPSFs belong to a useful fault model and are amenable to efficient testing algorithms. This model is shown in Figure 3.5.

Based on the above fault-modeling techniques, we find that the most common functional fault types are: stuck-at fault (SAF), of which there are two types,

Functional Fault Modeling and Testing

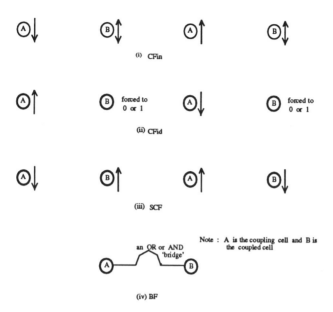

Figure 3.4 Coupling fault (CF) model: (i) CFin – B changes state when A undergoes a transition; (ii) CFid – B is forced to 0 or 1 when A undergoes a transition; (iii) SCF – B is forced to a value opposite to that of A; (iv) BF – an 'OR' or 'AND' bridge is formed between A and B, so that both of them are forced, respectively, to the OR or AND of their fault-free states

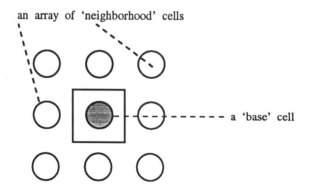

Figure 3.5 Pattern-sensitive fault (PSF) model

namely stuck-at-zero (SA0) and stuck-at-one (SA1); inversion coupling fault (CFin), in which a transition in one cell causes an inversion is another; idempotent coupling fault (CFid), in which a transition in one cell forces a fixed value in another; state coupling fault (SCF), in which the state x of one cell forces another cell to have a state $\bar{x}$; address decoder fault (AF); transition fault (TF) of which there are two kinds, one in which a $0 \rightarrow 1$ transition fails when the cell is at state 0, denoted as $\uparrow$ /0, and the other one in which a $1 \rightarrow 0$ transition fails, denoted as $\downarrow$ /1, and neighborhood pattern-sensitive fault (NPSF).

Unless otherwise stated, we shall consider a memory array consisting of single cells, that is, the memory word size is 1. Most of the concepts can be clearly understood and generalized even with this simplified approach. For example, storing a 0 in a single cell would be analogous to storing the all-0 word in a memory location with more than one bit. Likewise, storing a 1 in a single cell would correspond to storing the all-1 word.

Definition 4 *A test is said to* detect *a fault if it produces a yes/no answer corresponding to whether or not the fault is present.*

Definition 5 *A test is said to* locate *a fault if it detects the fault and also finds out which cells are involved in the fault.*

This book focuses on testing and testable design and hence most of the algorithms discussed are detection algorithms. Since the entire memory device is a single chip, fault location is deemed important only if the memory array has redundant rows and columns that can be switched in to replace the faulty rows and columns. This is done either externally through laser personalization or by using self-repair circuitry to replace faulty cells by fault-free ones. Another possible objective of fault or physical defect diagnosis is to refine the steps in the fabrication process. Defects caused by phenomena such as misaligned masks or overetching can be diagnosed by fault location algorithms.

3.2.5 Deterministic tests for detecting reduced functional faults

In memory testing, one common problem faced is the fact that multiple faults of the same or different types can be linked together. A *fault linkage* corresponds

Functional Fault Modeling and Testing 91

to a situation in which a memory cell c has two or more reduced functional faults associated with it. For example, two CFs may be linked such that a cell is coupled to more than one cells at the same time. Since SAFs and TFs are single-cell faults, they can each be linked only with other types of faults. *Active* and *passive* NPSFs can also be linked together. Also, SAFs can be linked with CFs and NPSFs.

It may be recalled that for testing SRAMs, read operations can realistically be assumed to be fault-free — that is, they retrieve the correct data and do not have undesirable side effects on unread cells, changing the state of the memory. This assumption is not realistic for DRAMs because DRAM read cycles are followed by refresh cycles in which the data is written back. Hence, read operations are as fault-prone as write operations for DRAMs. It should also be emphasized again at this point that the simplifying assumptions used in fault modeling must correspond to the physical nature of the device being tested, otherwise, the fault model itself will be unrealistic and tests based on this model may be inefficient or even useless.

3.2.6 Algorithms for detecting SAFs, TFs, AFs and CFs

The traditional deterministic algorithms for functional fault detection can be broadly divided into two classes: non-march tests and march tests. First we shall discuss some simple, traditional tests for fault detection assuming stuck-at faults and coupling faults. Discussion on functional testing for PSFs will be taken up later. We shall deal with some non-march tests which are easy to understand, and then we shall discuss march tests. March tests, for which built-in self-test (BIST) hardware can be easily designed, have been known to detect various combinations of linked and unlinked faults such as SAFs, AFs, TFs and CFs. NPSFs cannot be detected or located by march tests, for the following simple reason: a march test treats all the memory cells in exactly the same fashion and performs the same finite set of operations on each; in contrast, a test for detecting and/or locating NPSFs must distinguish a *base cell* from its *neighborhood*.

Before starting this discussion, we introduce some definitions that would enable us to understand these algorithms.

Definition 6 *A test is said to be* complete *if it achieves full fault coverage (i.e., 100 %) with regard to the fault model chosen before the design of the algorithm.*

Definition 7 *A test is said to be* irredundant *if it is complete and removal of any operation from the test results in a test that is not complete. (The concept of irredundancy is of particular relevance for march tests.)*

In the following paragraphs, we discuss each of the two test approaches with examples that illustrate fault coverage and running time.

3.2.7 Non-march tests

A number of memory tests exist that do not necessarily perform the same set of read and write operations on each memory location. Such tests are called **non-march** tests. Examples of such tests are: Checkerboard, GALPAT (GALCOL and GALROW), Walking 1/0, Sliding Diagonal and Butterfly. Such test algorithms are discussed in [165, 72]. Some more sophisticated ones, for example, Thatte and Abraham's divide-and-conquer approach, to be discussed later, consider a reduced functional model.

We shall describe and give examples of each of these algorithms and compare their fault coverages and running times.

1. A *checkerboard* test divides the memory array into two groups of cells forming a checkerboard pattern (see Figure 3.6), writes 1 in one group and 0 in the other group, reads all cells, and repeats the process with complementary data. This simple test can detect a coupling fault between two adjacent cells, given that there is no AF. Its time complexity is $4n$, n being the number of memory cells.

    It is easily seen that the checkerboard test won't detect all CFs. Consider two cells i and j that belong to the same group; that is, the checkerboard test writes 0 (1) to both i and j in the first (second) pass. If a CF is such that it can be detected only by setting cells i and j to opposite values, then the CF will not be detected by this checkerboard test. SAFs will of course all be detected if it is assumed that AFs don't exist, because each cell is written with a 0 and a 1. AFs that cause cells of the same group to be coupled will go undetected, whereas those that cause cells of different groups to be coupled will be detected. TFs will not all be detected because each cell undergoes either a $0 \rightarrow 1$ or a $1 \rightarrow 0$ transition, but not both.

Functional Fault Modeling and Testing

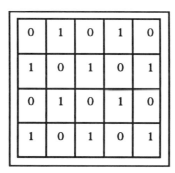

Figure 3.6 A checkerboard pattern

2. *GALPAT* (GALloping PATtern) (also called *ping-pong*), [14] *GALCOL* (GALloping pattern with COLumn read), *GALROW* (GALloping pattern with ROW read), *Walking 1/0* and *Butterfly* (also known as *bit surround disturb*) are based upon modifications of the same basic algorithm [165]. This basic algorithm is shown below and each modification that leads to one of the above algorithms is also shown.

(1) *for* $d := 0$ *to* 1 *do*
begin
    for $i := 0$ *to* $n - 1$ *do*
        $A[i] := d$;
    for base-cell $:= 0$ *to* $n - 1$ *do*
    begin
(2) $A[\textit{base-cell}] := \bar{d}$;
        perform READ ACTION;
(3) $A[\textit{base-cell}] := d$;
    end;
end;

READ ACTION for GALPAT:
begin
    for cell $:= 0$ *to* $n - 1$ *do*
    begin
        if (cell $\neq$ base-cell) *then*
        begin
(4)   *if* ($A[\textit{cell}] \neq d$) *then* *print*('Error at cell', *cell*);
(5)   *if* ($A[\textit{base-cell}] \neq \bar{d}$) *then* *print*('Error at cell', *base-cell*);

```
        end;
    end;
end;
```

READ ACTION for GALROW/GALCOL:
{Read cells on the same row for GALROW, same column for GALCOL}
```
begin
    for cell := 0 to √n − 1 do
    {These are cells located on the same row(for GALROW)
    or column(for GALCOL) as the base-cell.}
        begin
            if (cell ≠ base-cell) then
            begin
(4)             if (A[cell] ≠ d) then print('Error at cell',cell);
(5)             if (A[base-cell] ≠ d̄) then print('Error at cell',base-cell);
            end;
        end;
end;
```

READ ACTION for Walking 1/0:
```
begin
    for cell := 0 to n − 1 do
    begin
        if (cell ≠ base-cell) then
        begin
(4)         if (A[cell] ≠ d) then print('Error at cell',cell);
        end;
    end;
(5) if (A[base-cell] ≠ d̄) then print('Error at cell',base-cell);
end;
```

READ ACTION for Butterfly:
```
begin
    distance := 1;
    while distance ≤ maxdist do
    begin
        read cell at distance north from base-cell;
        read cell at distance east from base-cell;
```

           read cell at *distance south* from *base-cell*;
           read cell at *distance west* from *base-cell*;
           read base-cell;
           distance := *distance* * 2;
    end,
end;

The idea behind these tests is to initialize the memory with the all-0 pattern (or all-1 pattern) with the exception of the *base cell*, which is set to the complementary value; a possible CFin between the base cell and another cell is detected by reading each cell in the memory, and then making the base cell walk through the memory. The motivation for doing this is to verify all possible pairwise couplings between memory cells. The difference between GALPAT and Walking 1/0 is in reading the base cell — in GALPAT, the base cell is read after reading each cell whereas in Walking 1/0, all the other cells are read and the base cell is read last, in other words, there is a *walking read sequence*. GALPAT thereby has the following two properties — it can detect slow transition in the address decoder, and can also locate mutually coupled faults between cell pairs which can possibly be overlooked by Walking 1/0. For example, if the base cell A triggers a CFin in B and B in turn triggers one to A immediately after, then GALPAT will detect and locate this, while Walking 1/0 will detect but may not locate the base cell error. This happens if this error is subsequently masked by a CF involving some other cell coupled to the base cell. Thus GALPAT has a higher fault location coverage than Walking 1/0.

It is quite easy to see that GALPAT and Walking 1/0 both have a time-complexity of $O(n^2)$. Even though both these algorithms have a high fault coverage (all AFs, SAFs, TFs and most CFs) yet the quadratic running time is rather prohibitive. GALROW and GALCOL achieve a faster running time but a smaller fault coverage by performing the read operation on only the rows or the columns of the memory cell array, thereby reducing the complexity to $O(n^{1.5})$. In Butterfly, the read operation is done in a somewhat more complex manner, but fewer cells are read than any of the other four. In each pass through the loop, only those cells whose addresses are at a fixed Hamming distance from the base cell are read — this is done in order to reduce the running time to $O(n \lg n)$, at the cost of missing a number of possible coupling faults and also some AFs.

3. *The Sliding Diagonal* method is a shorter alternative to GALPAT, using a diagonal of base cells instead of a single base cell and then shifting the diagonal one step at a time north-east and south-west. Since in a square

memory array of n cells, the diagonal has $\sqrt{n}$ elements, the complexity of this algorithm is given by $n(\sqrt{n} + \sum_{i=1}^{\sqrt{n}} 2(\sqrt{n} - i))$, which is of $O(n^{1.5})$.

It is seen quite clearly that the Sliding Diagonal method will not detect all CFs; for example, a CFin with A coupled to B where A and B belong to the same diagonal may not be detected, if this fault is masked by another CFin between A and a third cell on the same diagonal. The sliding diagonal method detects and locates all SAFs and TFs. Not all AFs are detected because AFs on cells on the same diagonal can mask each other.

4. *The divide-and-conquer approach of Thatte and Abraham*: The divide-and-conquer approach [163] (1977) assumes the reduced functional fault model as explained earlier. This approach has been inspired by GALPAT but is a major improvement as far as time-complexity is concerned. The algorithm is described in [163]. It partitions the memory recursively into k equal parts, and then performs GALPAT, considering each part as a *super-base cell*, and each of the other parts as (possibly) coupled to this super-cell. This algorithm has a reduced complexity to $O(n \lg n)$, as proved below. The basic steps of the algorithm are made quite clear below.

The memory array is first initialized with the all-0 pattern. It is then partitioned into k equal parts. Then a part is chosen and its cells are made to undergo transition writes, keeping the remaining $k-1$ parts unchanged. The data belonging to the remaining $k-1$ parts are then read out. This step is followed by making the cells of the chosen part revert to their original state (by applying the reverse transition writes), and reading out the cells of the remaining $k-1$ parts. These steps are repeated (k-1) more times for k possible initial choices of the part undergoing transition writes. Finally, the entire procedure is repeated with the all-1 pattern used to initialize the memory.

The running time $T(n)$ of Thatte and Abraham's algorithm is $T(n) = \frac{4k+2}{\lg k} n \lg n$. This is obtained by solving the following recurrence relation for the running time.

The recurrence relation for the running time of Thatte and Abraham's algorithm is as follows:

$$T(n) = T_{init}(n) + T_{r/w}(n) + kT(\frac{n}{k}),$$

$$T(2) = a \text{ constant},$$

where $T_{init}(n)$ denotes the time for initialization of the memory with all-0 in one case and all-1 in the other; clearly then, $T_{init}(n) = 2n$, $T_{r/w}(n)$ denotes the time for reading and writing — it comprises making each of the k equal parts undergo two transitions ($0 \to 1$ and $1 \to 0$), reading data present in the other $k-1$ parts for each of the two transitions. The above process has to be performed twice, first with a background of 0s and then with a background of 1s. Hence,

$$T_{r/w}(n) = 4k(\frac{n}{k} + (n - \frac{n}{k}))$$

Finally, since the formulation is recursive in nature, the algorithm has to be run on each of the k equal parts and the total running time of these k parts, which is $kT(\frac{n}{k})$, must be added to the above sum.

3.2.8 Tests for detecting k-coupling faults

Nair, Thatte and Abraham (1978) [121] propose two algorithms to detect k-coupling faults in RAMs. Their first algorithm, to be discussed later, is a test for CFids (2-coupling faults). Their second algorithm is designed to detect 3-coupling faults and has a complexity of $n+32n \log n$. Papachristou and Sahgal (1985)[133] propose a test for detecting 3-coupling faults that has a length of $36n + 24n \lg n$. Cockburn (1994) [25] has proposed an algorithm called SVCTEST which can be adapted for various values of k (or V, the symbol used by Cockburn to denote the number of cells involved in a coupling fault). For $k = 3$, his test is known as an S3CTEST and has a complexity of $4n \lg n + 18n$.

Nair, Thatte and Abraham's second algorithm, called Algorithm B [121] deserves mention as one of the earliest tests for k-coupling faults. It detects all 2-coupled faults and also restricted 3-coupled faults. The algorithm is described in the section on march tests because it involves two sets of march operations on the memory array iteratively partitioned into two halves.

In Papachristou and Sahgal's algorithm, the total number of addresses in memory are split into two halves: a top half (T) and a bottom half (B). When a coupling cell j is in T, a coupled cell i is in B, and vice versa — this assumes that the faults are restricted 3-coupling faults, with cells j that are in one half of the memory making transitions followed by cells i in the other half being tested for possible k-coupling. This is done in accordance with the state diagram of the finite state machine shown in Figure 3.7 that describes the transition between the 4 possible states of the memory: each state corresponding to a value (0 or 1) stored in all the cells of T and a value (0 or 1) stored in all the cells of B. A minimum test sequence, in terms of the fewest possible write operations for the state diagram, can be achieved by following an *Eulerian sequence* through the state transition graph. By following around this graph, it is noticed that

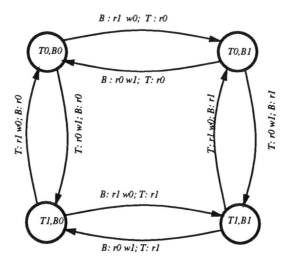

Figure 3.7 State-transition graph for testing 3-coupling faults; courtesy [133]

the total number of operations performed by an Eulerian sequence on each half (T or B) of the memory is 12, hence the complexity of the Eulerian sequence is $24 \cdot n/2 = 12n$ operations. The test is then repeated with the address orders reversed, increasing the number of operations to $24n$. The entire process above is repeated for $\lg n = N$ iterations, where N is the total number of address bits; at each iteration i, the address incrementing and decrementing within the T and B blocks is performed in steps of $2^i (0 \leq i \leq N)$. This ensures that all possible 3-coupling faults are detected.

The complexity of the algorithm, including the time taken for initialization, has been shown to be $36n + 24n \lg n$ [133], (1985).

Cockburn [24] (1994) discovered a relationship between single k-coupling fault detection and a family of binary codes called (n, θ)-exhaustive codes. He proved that if a test t detects all single k-coupling faults excitable by a transition write operation on a cell i, then a set of vectors can be constructed to produce an $(n-1, k-1)$-exhaustive code.

Suppose a code of length n consists of codewords of length μ. We construct an $n \times \mu$ matrix M whose rows are the codewords. A $p \times q$ binary matrix A with $p > 0$ and $q > 2^p - 1$ is said to be exhaustive if all 2^p distinct binary vectors are present among the columns of A. The code is said to be (n, θ)-exhaustive if every θ-submatrix of M is exhaustive. For a detailed

Functional Fault Modeling and Testing

mathematical derivation of Cockburn's results and discussion of the code, see [24]. The SVCTEST test algorithm is informally described below:

(1) Generate a 'short' $(n, k-1)$-exhaustive code G. Each codeword is a binary tuple that corresponds to the values to be written into a particular cell over a period of time. The columns of the matrix M correspond to snapshots of the state of the n cells taken at various time instants.

(2) Write the first n-bit-long column of M into the n memory cells, each bit to its corresponding cell. This is the first background.

(3) For each bit in the memory, read its stored data, compare this data with the appropriate entry in M, write back its complement, and then write back the original content. In this manner, a *read-write-read* sequence of operations is "walked" through the memory.

(4) Steps (2) and (3) are repeated for each column of M, to generate each background pattern specified by the code. After each background pattern is set up, the cells are read and compared with the expected values indicated by the matrix M. The values obtained from M are then written back in the cell.

(5) Finally, each cell is scanned one last time and its contents compared with the expected ones obtained from M.

The algorithm can be adapted for various values of k (by generating an appropriate $(n, k-1)$-exhaustive code, and is proved to have a high fault coverage with regard to coupling faults. In particular, Cockburn [24] obtained two different tests to detect single 3-coupling, by using two different choices for the $(n,2)$ exhaustive code in step (1). One of them is a $(10,6)$-exhaustive code and the other one is a $(8,7)$-exhaustive code, which Cockburn found to be easier to implement by hardware. These two tests are known respectively as S3CTEST and S3CTEST2.

Table 3.2 [24] displays various tests for coupling faults and a comparison of their test lengths and typical testing times.

3.2.9 March tests

March tests, defined and described in [72] and [165], involve the application of a finite number of read and write operations to each memory cell before proceeding to the next cell, and traversing the memory only along a certain predetermined address sequence or its reverse; thereby treating all the memory cells identically. March algorithms are described in the literature using a special notation [165] outlined below:

Test	Length	256K	1M	4M	16M
Nair, et al.	$32n \lg n + n$	15.1s	1m 7.2s	4m 55.7s	21m 30.2s
Papach.	$24n \lg n + n$	11.4s	50.4s	3m 41.9s	16m 8.0s
S2CTEST	$10n$	0.3s	1.0s	4.2s	16.8s
S3CTEST	$4n \lg n + 18n$	2.4s	10.3s	44.5s	3m 11.3s
S3CTEST2	$4n \lg n$ $4n \lg n + 4n \lceil \lg(1 + \lg n) \rceil + 9n$	2.6s	11.4s	49.1s	3m 11.3s

Table 3.2 A comparison of testing times for coupling fault tests; courtesy [24]

(a) A march algorithm is a set of *march elements*. The order of these elements is important.

(b) Each element is a finite tuple preceded by exactly one of these three symbols - $\Uparrow$, $\Downarrow$, or $\Updownarrow$; for example, $\Uparrow (r0, w1, r1)$.

(c) Each component of a march element can be any of the following only — $r0$, $r1$, $w0$, or $w1$; they denote, respectively, a 0 is expected to be read from a fault-free cell, a 1 is expected to be read from a fault-free cell, a 0 is written and a 1 is written.

Each march element denotes a finite set of (read/write) operations to be performed on each memory location before moving on to the next location. The symbol $\Uparrow$ denotes any arbitrarily chosen sequence of memory locations to be accessed (the 'forward sequence'); the symbol $\Downarrow$ denotes the reverse of the forward sequence chosen (the 'reverse sequence'); and $\Updownarrow$ indicates that the direction of traversal is unimportant and either the forward or the reverse sequence may be chosen. Often, the increasing address sequence as produced by a binary up-counter is chosen as the forward sequence in a march test.

The above description of a march test indicates that it always has a time complexity of $O(n)$, n being the number of cells in the memory. This linear time complexity is one of the most attractive features of these algorithms. Another important advantage of march tests is that their regularity and symmetry (i.e., treating all cells identically) allow built-in self-testing hardware to be designed quite easily; besides, self-test circuitry for march tests can be implemented with a minimum of silicon area.

Examples of march tests

Two small examples of the use of march algorithms, using the notation for march elements given in [165] are presented below. The first example is with regard to unlinked AFs, and the second one is for an AF linked with another fault, such as a CF.

Suppose we want to detect an unlinked AF in a memory consisting of just 3 cells A, B, and C. Without loss of generality, suppose that the march algorithm we are about to design looks at either the sequence A-B-C or the reverse one C-B-A. Our AF causes cell C to be wrongly accessed when the address decoder inputs are set up to select A. Then the march sequence $\{\Uparrow (wx), \Uparrow (rx, wx'), \Downarrow (rx', wx)\}$ will detect the fault. The write operation wx in the second march element will trigger the fault because it causes a transition in exactly one cell at a time in the fault-free case and possibly multiple cells in case of an AF. If a cell is never accessed because of the AF, then the read operations in the second and third march elements will detect a discrepancy between actual and expected values. Moreover, this set of march elements is irredundant, as can be easily verified (removal of any march element destroys the full AF detection capability of the algorithm). This simple observation for this example can be easily extended to determine the necessary and sufficient conditions for detecting unlinked AFs. As proved in [165], a necessary and sufficient condition for detecting unlinked AFs is to have at least two march elements, one of the form $\Uparrow (rx, \ldots, wx')$, and another of the form $\Downarrow (rx', \ldots, wx)$. It should be noted that we need to march through the cells of the memory array in an ascending followed by a descending sequence (or vice versa, or any other sequence followed by its reverse sequence) to exhaust all possible coupled cell pairs produced by AFs. However, it is sometimes possible to relax this constraint if the behavior of the read operation on cells coupled due to an AF is known. If the technology used for designing the memory is known (i.e., whether the memory device returns the AND or the OR function when an AF causes multiple cells to be read), we can simplify these necessary and sufficient conditions to yield the following: for an OR technology, we need to include the march sequences $\Updownarrow (\ldots, w0); (r0, \ldots, w1)$ and $\Updownarrow (\ldots, w1); (r1, \ldots)$, for an AND technology, we need to include the march sequences $\Updownarrow (\ldots, w0); (r0, \ldots)$ and $\Updownarrow (\ldots, w1); (r1, \ldots, w0)$.

A second example will also be useful here. Suppose now we have a linked fault, namely, an AF linked with a TF. Suppose, we again have a 3-cell memory as before, with the above address fault linked to the TF ($\uparrow /0$) in C. Any march test that detects both the faults must make a cell undergo an up-transition, and then read its contents. This requires modifying the above sequence to $\{\Uparrow (wx), \Uparrow (rx, wx'), \Downarrow (rx', wx, rx)\}$.

March tests for stuck-at, transition and coupling faults

Two simple and well-known march algorithms, developed in the mid-70s, will first be presented. The first one is called Zero-One or MSCAN (Memory Scan), and the second one is called ATS (Algorithmic Test Sequence). ATS (as originally proposed by Knaizuk and Hartmann [73, 74]) is *not* a march test but a combination of three march tests. It partitions the memory array into three parts as explained below and performs a march test on each partition.

The MSCAN (or Zero-One) Test Procedure (Breuer and Friedman, 1976 [14]) is extremely simple. The algorithm visits each (addressable) cell of the memory and writes a 0. It then re-visits each cell and reads out its contents, and then repeats the above process for 1. In terms of the march notation above, it can be represented as: $\{\Uparrow (w0), \Uparrow (r0), \Uparrow (w1), \Uparrow (r1)\}$. This test has a very limited fault coverage because in the worst case, it will always visit one cell of the memory (for an AF that causes only one memory cell to be accessed always) and verify the correct functionality of that cell.

The ATS procedure of Knaizuk and Hartmann [73, 74] (1977) is an $O(n)$ test called algorithmic test sequence that detects any combination of SAFs (i.e., single and multiple). In their first paper [73], they introduce the algorithm and prove that it detects any single stuck-at fault in a RAM. In their second paper [74], they extend their proof to multiple stuck-at fault detection. We shall briefly describe ATS in this section.

The memory contains n words, with addresses ranging from 0 to $n-1$, and the ith word is denoted as W_i. These words are grouped into three partitions G_0, G_1, and G_2, such that $G_j = \{W_i | i = j \pmod 3\}$, $0 \le j \le 2$. The address decoder is assumed to be of a *non-creative* nature, that is, a single fault within the decoder does not spuriously access a memory cell without also accessing the programmed address. Also, SAFs are assumed to occur only at decoder inputs, memory cells array, and the MDR. The procedure is described below:

(1) Write 0 in all cells of $G_1 \cup G_2$.
(2) Write 1 in all cells of G_0.
(3) Read cells of G_1. (Expected value $= 0$)
(4) Write 1 in all cells of G_1.
(5) Read cells of G_2. (Expected value $= 0$)
(6) Read cells of $G_0 \cup G_1$. (Expected value $= 1$)
(7) Write 0 in all cells of G_0, and then read all cells of G_0.
(8) Write 1 in all cells of G_2, and then read all cells of G_2.

If we examine this algorithm closely, we notice that it can detect any fault that causes an erroneous access of a cell in a partition *different from* the correct one. Any single SAF in the address decoder input (note that the address decoder is an l-input, 2^l-output device) will cause the decoder to select a memory cell whose address differs from the correct one by a power of 2. For example, the 6-bit address 100011 is converted into the address 101011 by an SAF in bit no. 3, and the difference between these two addresses in decimal notation is 8. Now, no two numbers that differ by a power of 2 can belong to the same residue class modulo 3, hence a single SAF would be detected.

Let us now examine what happens for a multiple SAF. Once again, we consider a simple example: suppose the 6-bit address 100100 is erroneously converted into 001110, by 3 SAFs (one s-a-0 and 2 s-a-1 faults), at bit positions 1 (MSB), 3 and 5 respectively. Let us consider the cell W_2 in the partition G_2. This cell has the address 000010 in the fault-free case. Because of these three SAFs, however, its address will be converted into 001010. Now there exists at least one cell with its Hamming distance from W_2 equal to exactly 1 (and hence it is not in the same partition as W_2), which also gets erroneously mapped to this wrong address. An example of such a cell is W_{34} (with 100010 as its address). Now, W_{34} and W_2 are in different partitions but point to the same data in memory, because both their addresses would cause an erroneous access to W_{10} (corresponding to the address 001010). This is detected in steps (4) and (5) of the algorithm. Figure 3.8 illustrates this example.

It is quite easy to argue that ATS will detect any stuck-at fault in the memory array or the memory data register.

The complexity of ATS is $4n$, because there are a total of 12 operations each of which is performed on a group of $n/3$ cells belonging to one of the three partitions.

From the reduced functional fault classification described earlier, the following families of march tests have been designed. Names such as *Test-US*, *Test-UT*, and others, are given by Van de Goor and have been borrowed from his book [165].

(a) **Test-US**: *MATS, MATS+* are tests for unlinked SAFs which also cover AFs. MATS (*M*odified *A*lgorithmic *T*est *S*equence) is the shortest march test for SAFs. The Algorithmic Test Sequence (ATS) algorithm first presented by Knaizuk and Hartmann [74] (1977) has been improved by Nair [122] (1979) and developed into a full-fledged march test called MATS. One important difference between ATS and MATS is that MATS treats the memory as being divided into n partitions instead of only 3, each cell itself being a partition. Since a key feature of ATS is that a single SAF in the decoder input will fail

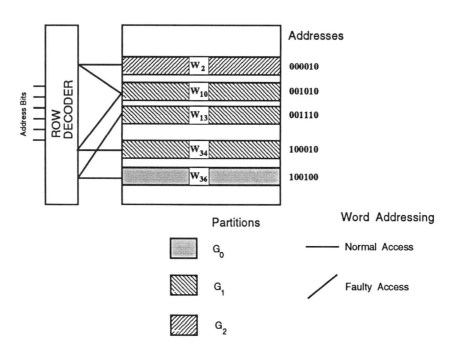

Figure 3.8 An example of the ATS procedure

to map an address in one partition into another address within the same partition, the number of partitions is not very important provided this condition is satisfied. This condition is trivially satisfied if each partition has exactly one cell. The march notation for MATS is $\{\updownarrow (w0), \updownarrow (r0, w1), \updownarrow (r1)\}$. This test detects all SAFs, since a 0 and a 1 is written into and read from each cell, all AFs, since the necessary and sufficient conditions for an OR technology are satisfied, as explained earlier. One advantage of MATS over ATS is that all SAFs are detected by MATS, independent of the decoder design, as proved in Nair's original paper [122]. This algorithm is superior to zero-one and checkerboard algorithms (discussed later) because it achieves a higher fault coverage with an equal number of operations (namely, $4n$). When the used memory technology is not known, a special version of MATS named MATS+ [1] (1983) [1] can be used. This algorithm has the march notation $\{\updownarrow (w0), \Uparrow (r0, w1), \Downarrow (r1, w0)\}$. This test detects all SAFs, since a 0 and a 1 are both written in each cell, AFs causing a cell-write to affect another at a higher address (by the first march element), AFs linking a cell with another at a lower address (by the second march element) — hence all AFs and SAFs.

(b) **Test-UT**: *Marching 1/0, MATS++* are tests for unlinked TFs, also covering AFs and SAFs. Marching 1/0 is a simple, traditional test, described by the march notation $\{\Uparrow (w0), \Uparrow (r0, w1, r1), \Downarrow (r1, w0, r0), \Uparrow (w1), \Uparrow (r1, w0, r0), \Downarrow (r0, w1, r1)\}$ — it detects all AFs, SAFs and TFs, and most 2-coupling CFs. The fact that it detects all AFs, SAFs and TFs can be verified by comparing it with the necessary and sufficient conditions detailed earlier. However, Marching 1/0 is a *redundant* test. It is seen quite clearly that the last three march elements can be deleted, because they cause the same write transitions to be performed and tested as in case of the first three march elements. Also, the $r1$ in the second march element is redundant, as it is followed immediately by an $r1$ in the third march element.

This algorithm has 14 read/write operations on each cell of the memory, consisting of n cells. Therefore, the complexity of the Marching 1/0 algorithm is $14n$. The following CFs are not detected by this test [165]:

- $\langle\uparrow;0\rangle, \langle\downarrow;1\rangle, \langle\uparrow;\downarrow\rangle, \langle\downarrow;\uparrow\rangle$ between coupling cell C_i and coupled cell C_j where the address of C_i is less than that of C_j.
- $\langle\downarrow;0\rangle, \langle\uparrow;1\rangle, \langle\downarrow;\downarrow\rangle, \langle\uparrow;\uparrow\rangle$ between coupling cell C_i and coupled cell C_j where the address of C_i is greater than that of C_j.
- Linked faults.

MATS++ optimizes Marching 1/0 scheme by deleting the redundant operations listed above, reducing the complexity to $6n$, to produce the march sequence $\{\updownarrow (w0), \Uparrow (r0, w1), \Downarrow (r1, w0, r0)\}$.

(c) **Test-UCin**: March X is a test for unlinked CFins, covering AFs, SAFs, and TFs not linked with CFins. The march notation for this test is as follows — $\{\updownarrow (w0), \Uparrow (r0, w1), \Downarrow (r1, w0), \updownarrow (r0)\}$. Again we may verify, by looking at the necessary and sufficient conditions for detecting faults, that AFs, SAFs, unlinked CFins, and TFs not linked with CFins can be covered by March X. The complexity of this test is $6n$.

(d) **Test-UCid**: March C- tests for unlinked CFids and covers AFs, SAFs, and TFs and CFins not linked with CFids. It is also known as Marinescu's algorithm [83] (1982), and is denoted by the march notation $\{\updownarrow (w0), \Uparrow (r0, w1), \Uparrow (r1, w0), \updownarrow (r0), \Downarrow (r0, w1), \Downarrow (r1, w0), \updownarrow (r0)\}$. March C- satisfies the necessary and sufficient conditions for AFs as described earlier; it also detects all SAFs and TFs because each cell undergoes both up and down transitions, followed by read operations. March C- also detects unlinked CFins and CFids. It has a complexity of $11n$.

(e) **Test-LCid**: March A [159] is a test for linked CFids, covering AFs, SAFs, TFs not linked with CFins, and some CFins linked with CFids. Its march notation is $\{\updownarrow (w0), \Uparrow (r0, w1, w0, w1), \Uparrow (r1, w0, w1), \Downarrow (r1, w0, w1, w0), \Downarrow (r0, w1, w0)\}$. It has a complexity of $15n$. March A is complete and irredundant.

(f) **Test-LTCin**: March Y is a test for CFins linked with TFs, and covers AFs and SAFs. Its march notation is $\updownarrow (w0), \Uparrow (r0, w1, r1), \Downarrow (r1, w0, r0), \updownarrow (r0)\}$. It has a complexity of $8n$. March Y is complete and irredundant.

(g) **Test-LTCid**: March B [159] tests for linked CFids, covering TFs linked with CFids or CFins, certain CFins linked with CFids, AFs and SAFs. Its march notation is $\{\updownarrow (w0), \Uparrow (r0, w1, r1, w0, r0, w1), \Uparrow (r1, w0, r0), \Downarrow (r1, w0, w1, w0), \Downarrow (r0, w1, w0)\}$. It has a complexity of $17n$. March B is complete and irredundant.

(h) **Tests for single and multiple k-coupling faults**:

Definition 8 *A fault is said to be a* single k-coupling fault *if a transition on one of k cells triggers a certain fault in another of these k cells given some fixed pattern written on the background consisting of $k-2$ cells, and appears only once (i.e., produces only one set of k-coupled cells) in the memory array.*

Definition 9 *A fault is said to be a* multiple k-coupling fault *if a transition on one of k cells triggers a certain fault in another of these k cells given some fixed pattern written on the background consisting of $k-2$ cells.*

Definition 10 *A fault is said to be a* restricted k-coupling fault *if it is a multiple k-coupling fault and the sets of k-coupled cells are disjoint.*

Cockburn [24] (1994) proved that any test to detect single 3-coupling must have a length of at least $2n \lg n + 11n$. This immediately rules out any march test for detecting 3-coupling faults, because march tests are always $O(n)$. However, *march elements* are often used as part of other (non-march) tests, such as Algorithm B of Nair, et al. [122] which will be discussed shortly.

An adaptation of the SVCTEST proposed by Cockburn also (for V-coupling faults) uses a short march sequence, as we saw earlier. For detecting 3-coupling faults, the test S3CTEST uses a march sequence of the form read-write-write, where the read operation compares the value stored in the storage cell with the entry in the matrix M corresponding to the code, and the two writes are transition writes. This test is proved by Cockburn to have an expected length (or complexity) of $4n \lg n + 18n$, a value that is fairly close to the theoretically-obtained lower bound on the test length [24].

A march algorithm with a complexity of $30n$, proposed by Nair, Thatte and Abraham [122] (1978), detects multiple 2-coupling faults (CFids), SAFs and AFs. They have also given a second algorithm for detecting restricted 3-coupling faults. Their second algorithm is a non-march test that uses march elements. Their algorithms are briefly described below.

Algorithm A [122] is a march test that can be described by the march notation:

$\{\Uparrow (w0), \Uparrow (r0, w1), \Downarrow (r1), \Uparrow (r1, w0), \Downarrow (r0), \Downarrow (r0, w1), \Uparrow (r1),$

$\Downarrow (r1, w0), \Uparrow (r0), \Uparrow (r0, w1, w0), \Downarrow (r0), \Downarrow (r0, w1, w0), \Downarrow (r0),$

$\Uparrow (w1), \Uparrow (r1, w0, w1), \Downarrow (r1), \Downarrow (r1, w0, w1), \Uparrow (r1)\}$

Algorithm B [122] is a test for detecting all 2-coupling faults and restricted 3-coupling faults. In this test, the memory is iteratively partitioned into two halves corresponding to the i-th address bit value, with i ranging from 1 to $\lg n$. In each iteration, the march sequences S_1 and S_2 are applied on the top and bottom halves respectively,

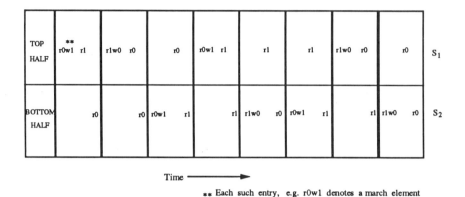

** Each such entry, e.g. r0w1 denotes a march element

Figure 3.9 March sequences used in Algorithm B of Nair, et al.(1978); courtesy [122]

first in the forward direction and then in the reverse direction of addressing. S_1 and S_2 are shown in Figure 3.9. These two sequences have a certain temporal relationship between them with regard to the time instants at which various operations are performed. This temporal relationship can be best expressed with a table as in Figure 3.9 instead of by a march notation. The algorithm takes $n + 32n \lg n$ operations.

A march algorithm for multiple non-interacting 2-coupling faults is described by Papachristou and Sahgal [133] (1985). Its march notation is:

$\{\Uparrow (w0), \Uparrow (r0, w1, r1), \Uparrow (r1, w0, r0), \Uparrow (r0, w1, w0), \Uparrow (r0, w1),$

$\Uparrow (r1, w0, w1), \Uparrow (r1, w0), \Uparrow (r0, w1, w0), \Downarrow (r0, w1), \Downarrow (r0, w0),$

$\Downarrow (r0, w1, w0), \Downarrow (r0, w1), \Downarrow (r1, w0, w1), \Downarrow (r1, w0), \Downarrow (r0, w1, w0)\}$

It has a complexity of $37n$. This algorithm is found to detect the simultaneous presence of transition faults, stuck-at faults, decoder faults, and memory read-write logic faults.

Marinescu [83] has discovered a lower bound of $9n-2$ as the test length for any test that detects single 2-coupling. The shortest algorithm known for testing single 2-coupling faults is Cockburn's S2CTEST [24] whose generic version, SVCTEST, is described earlier. It is not a march test but uses a short march sequence of the form read-write-write, the read operation being used to detect if the readout data

Functional Fault Modeling and Testing

matches the appropriate entry in the matrix M for the chosen (n, θ)-exhaustive code, and the two writes being transition writes. It has a complexity of $10n$.

Other march tests

Moving Inversions (MOVI) [32] (1976) is a test which has been designed to serve as both a functional test and an AC parametric test. The functional test has to ensure that no memory bit is disturbed by a read or write operation on another bit, and the AC parametric test allows best and worst access times together with address changes causing these to be determined. The functional test part of MOVI is *not* a true march algorithm as per the definition of march algorithms but can be expressed quite conveniently by a march notation. The name MOVI is derived from the way the data patterns are written into memory; initially, the memory is written with the all-0 pattern, thereafter, 0s are successively inverted to become 1s, and vice versa. MOVI has been designed to act as a shorter alternative to GALPAT. In this section, we shall outline only the functional test part of MOVI, its parametric test aspects will be examined later.

The functional aspect of MOVI can be denoted by the march notation $\{\Downarrow (w0), \Uparrow (r0, w1, r1), \Uparrow (r1, w0, r0), \Downarrow (r0, w1, r1), \Downarrow (r1, w0, r0)\}$, in which the first march element is used for initialization and the remaining ones are repeated N times, where N is the number of address bits in the memory cell array, (i.e., $N = O(\lg n)$). The algorithm therefore has a complexity of $O(n \lg n)$.

The basic idea in the functional part of this test is to sandwich each write operation (except the first one) by two read operations to verify that no memory bit i is modified by an operation on another unrelated bit j. It can be easily shown that this sandwiching of writes with reads achieves AF, SAF and TF detectability.

March tests for DRAMs

Functional testing of SRAMs and DRAMs differ in a few respects. First, a DRAM read operation is followed by a write-back refresh cycle; hence, reads and transition writes are both likely to trigger faults in DRAMs. In SRAMs, however, read operations can be assumed to be fault-free. Also, defect mechanisms in DRAMs and SRAMs are quite different. Since a DRAM cell stores charge on a capacitor, it is more susceptible to functional faults produced by various charge leakage mechanisms such as voltage imbalance on the bit lines (described in Chapter 5). An SRAM cell has a 'solid' connection to V_{CC} and is usually more robust to leakage currents.

Oberle, Maue and Muhmenthaler have described several march tests to detect and locate faulty cells in 4Mb DRAMs manufactured by Siemens [130, 131]. These tests have lengths ranging from $6n$ to $10n$ and varying fault coverages that can be evaluated via fault simulation. Shen and Cockburn [149] (1993) have reported fault modeling for the DRAM manufactured by Siemens and have proposed a march test which is provably optimal for *diagnosing* faulty cells in the DRAM. They analytically derive a lower bound of $8n$ on the length of any march test to locate all faulty cells and then propose a march algorithm whose length matches the lower bound.

March tests for single-order addressed memories

Van de Goor and Zorian [168] have proposed march tests for functional faults in memory devices such as FIFOs where the addressing sequence is unique. Such memory devices may also be used where the BIST area overhead and/or speed penalty for dual or multiple order addressing are not acceptable. March tests can be adapted to such memory devices much better than non-march tests, because a march test has at most two memory addressing sequences. In [168], single-order addressed (SOA) versions of MATS+ and March C- tests are proposed for detecting SAFs, AFs, TFs and CFs. These are the basic march tests with the simple modification that the same addressing order is used for all the march elements in a test. An SOA-March B- test that detects some state coupling faults is also discussed in [168]. An example of an SOA test is the following, called SOA-CFst [168]:

$$\{\Uparrow(w0), \Uparrow(r0, w1, r1), \Uparrow(r1, w0, r0)\},$$

which can detect SCFs (state coupling faults) between pairs of memory cells.

Fault coverage of march algorithms

March algorithms are somewhat constrained by the fact that each memory cell must be treated in exactly the same manner; this constraint creates difficulty in detecting coupled faults which mask one another. Suppose a fault triggered by a write operation in a march element when cell i is accessed gets masked by another fault triggered by a write operation in the same march element when cell j is accessed subsequently. This will cause the failure of a march test which accesses the cells sequentially to detect the faults. Hence, a pair of coupling inversion faults (CFins) at arbitrarily chosen memory locations which mask one another, or a CFin (with A as the coupling cell and B as the coupled cell) and an active NPSF (to be

Functional Fault Modeling and Testing

discussed later), which inverts the contents of B, may not be detected by a march algorithm. However, march tests can detect all AFs, SAFs and TFs, and certain CFs. Such tests, for example, Marching 1/0, MATS++, and others, have been discussed in [1] and [165]. Some simple march algorithms, such as Zero-One, which has the representation $\{\Uparrow (w0), \Uparrow (r0), \Uparrow (w1), \Uparrow (r1)\}$ have a very low fault coverage. The necessary and sufficient conditions for AFs and CFs indicate that this simple test would not be able to detect all TFs, CFs and AFs. It is only somewhat useful for detecting SAFs, particularly if there is prior knowledge that no AF is present.

In general, linked CFins cannot all be detected by march tests. The proof of this is very simple — suppose we consider three cells, i, j and k ($i < j < k$), with cell k being $\langle \uparrow; \updownarrow \rangle$ coupled to cell i and to cell j, and both i and j being visited either before or after k is visited by a march element. Then any march element or combination can traverse these 3 cells in either of the following ways:

(a) using the sequence $i \rightarrow j \rightarrow k$ and/or its reverse, in which case (i) the two CFins will mask each other for any march element marching 'up', and (ii) neither will be triggered for an element marching 'down'; thereby failing to detect the linked CFins.

(b) using the sequence $k \rightarrow i \rightarrow j$ and/or its reverse, in which case any 'up' or 'down' march would mask the two CFins.

3.2.10 Algorithms for unrestricted PSFs (UPSFs)

Hayes [53, 54] (1975,1980) studied unrestricted pattern sensitive faults. He modeled a memory cell array as a sequential machine of the Mealy type with 2^n possible states (corresponding to all possible n-bit patterns that can be stored in n cells) and $3n$ inputs (3 for each cell — read, write 0 and write 1). With respect to this model, an unrestricted PSF can be regarded as a phenomenon that modifies the state diagram of the sequential machine in an arbitrary manner. To test such a PSF with a largely unknown behavior, Hayes formulated a checking sequence comprising all possible 2^n patterns and having a total length of $(3n^2+2n)2^n$. It is quite clearly seen that this is computationally infeasible and hence testing unrestricted PSF is not a particularly tractable problem.

3.2.11 Algorithms for row/column PSFs

As defined earlier, row/column PSFs fall somewhere in between UPSFs and NPSFs. In UPSFs, the cell contents are influenced by all n cells in the memory and in NPSFs, the cell contents are influenced by a *constant* number of neighborhood cells known *a priori*. In row/column PSFs, however, the cell contents are influenced by all the cells located in the same row or column. For a square memory array, therefore, each cell is influenced by $\sqrt{n}$ cells. In the following discussion, we shall, for simplicity, consider the memory as an $\sqrt{n} \times \sqrt{n}$ square array, and describe a fault model for row/column PSFs based upon this organization. It should be noted that in practice, memories are typically divided into square subarrays.

Franklin, et al. [43] (1989) gave a recursive parallel algorithm for detecting row/column PSFs., Their algorithm is briefly described below [72]:

(1) Test the four corner cells of the RAM simultaneously, for cell-value 0.

(2) Test the cells around the periphery — that is, all cells on the top row, the rightmost column, the bottom row and the leftmost column for cell-value 0.

(3) Test the two middle rows and the two middle columns for cell-value 0.

(The above step partitions the array into four parts, with the periphery of each partition completely tested for cell-value 0).

(4) Perform steps 1 through 3 recursively for each of the four subarrays produced.

(5) Repeat steps 1 through 4 for cell-value 1.

The algorithm they presented consists of two procedures: A and B. Procedure A is used to test the four corner cells simultaneously, and procedure B is used to test the remaining cells.

3.2.12 Algorithms for NPSFs

Functional tests for neighborhood pattern-sensitive faults are described in this section.

Functional Fault Modeling and Testing 113

1. **Types of NPSFs:** Let us recall the definition of NPSFs and introduce some more terminology that will help us in describing and characterizing tests for these faults:

   Definition 11 *A neighborhood pattern-sensitive fault (NPSF) is a reduced functional fault that causes the contents of a cell, or the ability of the cell to change its contents, to be influenced by the contents of k cells forming a certain* neighborhood *in the memory, for some positive integer $k > 0$, (i.e., including the cell itself).*

   Definition 12 *The* deleted neighborhood *of an NPSF is the finite set of cells that influence the contents of the base cell, excluding the base cell itself.*

   With these notions, we can now define active, passive and static NPSFs (known as ANPSF, PNPSF and SNPSF, respectively) as follows.

   Definition 13 *An* active neighborhood pattern-sensitive fault *(ANPSF) [158], also called a* dynamic neighborhood pattern-sensitive fault *[140] is one in which the base cell can change its contents only due to a change in the deleted neighborhood pattern; a non-transition write in the deleted neighborhood pattern being assumed to have no effect on the base cell. A useful notation for an ANPSF is as follows [165]: $C_{i,j}\langle d_0, d_1, d_2, d_3; b\rangle$; where $C_{i,j}$ is the location of the base cell; for example, an ANPSF in which the base cell becomes a 1 when d_2 makes a $\uparrow$ transition, and the other cells d_0, d_1, and d_3 contain the value 101, is denoted as $C_{i,j}\langle 1, 0, \uparrow, 1; 1\rangle$; an ANPSF that causes the base cell to be inverted can be written as $C_{i,j}\langle 1, 0, \uparrow, 1; \updownarrow\rangle$.*

   Definition 14 *A* passive neighborhood pattern-sensitive fault *(PNPSF) [158], is one in which the contents of the base cell can fail to undergo a transition, when a certain pattern is present in the deleted neighborhood; following the same notation as in the case of ANPSFs, we denote the fault in which the neighborhood pattern 1011 prevents a low to high transition as $C_{i,j}\langle 1, 0, 1, 1; \uparrow/0\rangle$.*

   Definition 15 *A* static neighborhood pattern-sensitive fault *(SNPSF) [140] is one in which a base cell can be forced to a certain value due to a certain deleted neighborhood pattern; for example, $C_{i,j}\langle 0, 1, 1, 1; -/1\rangle$ denotes an SNPSF in which the base cell is forced to the value 1.*

Since ANPSFs and PNPSFs both cause transitions to occur in some neighborhood cell, they are often grouped together as APNPSFs (active and passive NPSFs) for testing purposes.

2. **Necessary/sufficient conditions for detecting and/or locating different kinds of NPSFs:**

   Before discussing these, we make an important assumption — read operations of memory cells are fault-free. This assumption, which is reasonable for SRAMs, simplifies the design of detection and location algorithms [140, 158].

   To detect and locate ANPSFs, each base cell must be read in state 0 and state 1, for all possible transitions in the deleted neighborhood pattern [165]. There are k cells in the neighborhood and we allow only one deleted neighborhood cell to undergo a transition. Hence we have two different values to be chosen for the base cell (namely, 0 and 1), $(k-1)$ ways of choosing the deleted neighborhood cell which has to undergo one of two possible transitions (up and down), and 2^{k-2} possible choices for the contents of the remaining neighborhood cells. Therefore, the total number of ANPs (active neighborhood patterns) to be applied for detecting and/or locating all ANPSFs is $2(k-1)2 \times 2^{k-2} = (k-1)2^k$. Table 3.3 gives the active neighborhood patterns (ANPs) necessary to detect/locate all ANPSFs.

   To detect and locate PNPSFs, for each of the 2^{k-1} deleted neighborhood patterns, the two possible transitions ($0 \to 1$ and $1 \to 0$) must be verified. Therefore, the total number of PNPs is $2.2^{k-1} = 2^k$, as shown in Table 3.3.

   The total number of patterns for detecting and locating APNPSFs (which included active and passive NPSFs) is, therefore, $(k-1)2^k + 2^k = k2^k$.

   For detecting and locating SNPSFs, we merely need to apply the 2^k combinations of 0s and 1s to the k-cell neighborhood, and verify by reading each of them that each can be stored.

3. **Mechanisms for optimal write sequence for testing NPSFs:**

   If a certain predetermined number of patterns has to be applied to a neighborhood, a sequence for these patterns must be chosen. Suppose we have a memory consisting of k-cell neighborhoods on which we wish to apply p test patterns. For NPSF testing, it is imperative to read each base cell of the memory for every test pattern applied. Hence, for every neighborhood, the number of read operations necessary is p, and this is a constant independent of the sequence in which the patterns are applied. However, we can minimize the number of write operations. This can be done at two

Functional Fault Modeling and Testing 115

Table 3.3 ANPs and PNPs; courtesy [165]

	Cell	Value in each pattern
ANPs	b	00000000000000001111111111111111
	d_1	↑↑↑↑↑↑↑↑↓↓↓↓↓↓↓↓↑↑↑↑↑↑↑↑↓↓↓↓↓↓↓↓
	d_2	00001111000011110000111100001111
	d_3	00110011001100110011001100110011
	d_4	01010101010101010101010101010101
	b	00000000000000001111111111111111
	d_1	00001111000011110000111100001111
	d_2	↑↑↑↑↑↑↑↑↓↓↓↓↓↓↓↓↑↑↑↑↑↑↑↑↓↓↓↓↓↓↓↓
	d_3	00110011001100110011001100110011
	d_4	01010101010101010101010101010101
	b	00000000000000001111111111111111
	d_1	00001111000011110000111100001111
	d_2	00110011001100110011001100110011
	d_3	↑↑↑↑↑↑↑↑↓↓↓↓↓↓↓↓↑↑↑↑↑↑↑↑↓↓↓↓↓↓↓↓
	d_4	01010101010101010101010101010101
	b	00000000000000001111111111111111
	d_1	00001111000011110000111100001111
	d_2	00110011001100110011001100110011
	d_3	01010101010101010101010101010101
	d_4	↑↑↑↑↑↑↑↑↓↓↓↓↓↓↓↓↑↑↑↑↑↑↑↑↓↓↓↓↓↓↓↓
PNPs	b	↑↑↑↑↑↑↑↑↑↑↑↑↑↑↑↑↓↓↓↓↓↓↓↓↓↓↓↓↓↓↓↓
	d_1	00000000111111110000000011111111
	d_2	00001111000011110000111100001111
	d_3	00110011001100110011001100110011
	d_4	01010101010101010101010101010101

levels: (a) at the k-cell neighborhood level, by hitting upon a scheme for generating a sequence of all p neighborhood patterns such that each pattern differs at as few component positions as possible from its successor; (b) at the entire memory cell array level, by an optimal *partitioning* of the memory. The former is achieved by using a *Hamiltonian sequence* for SNPs and an *Eulerian sequence* for APNPs, and the latter using the concepts of the *tiling neighborhood* and the *two-cell* methods, to be discussed later. First the graph-theoretic model will be introduced, and then we shall discuss how an optimal write sequence will be implemented at both levels of abstraction.

Optimal write sequence at the neighborhood level: Let us start with a few definitions.

Definition 16 *A* neighborhood pattern-transition graph *(NPTG) is a directed graph (V, E) where each vertex v in V represents a k-tuple (of 0s and 1s) denoting a state of the k-cell neighborhood, and each directed edge (v_i, v_j) denotes a transition from the state denoted by v_i to the state denoted by v_j by a single bit change; hence (i)$|V| = 2^k$; (ii) (v_i, v_j) is an edge if and only if (a) the Hamming distance between the states denoted by v_i and v_j is exactly 1 and (b) (v_j, v_i) is an edge; (i.e., a NPTG is always symmetric) — such a graph is called an* Eulerian graph.

From our previous discussion on necessary and sufficient conditions for detecting ANPFs, PNPFs, and SNPFs, it is easy to see that a vertex of a NPTG corresponds to an SNP whereas an edge corresponds to an ANP or an PNP (or an APNP), because ANPs, PNPs and APNPs have transitions associated with them. A *traversal sequence* through a graph $G = (V, E)$ is defined as a finite sequence of the form $\langle v_0, e_0, v_1, e_1, \ldots, v_{j-1}, e_{j-1}, v_j \rangle$, where $v_i, 0 \leq i \leq j$ denotes a vertex ($v_i \in V$) and e_i denotes the edge (v_i, v_{i+1}), ($e_i \in E$).

An *Eulerian sequence* is a traversal sequence through an Eulerian graph such as a NPTG, that visits each edge exactly once; on the other hand, a *Hamiltonian sequence* is a sequence through a graph (not necessarily Eulerian) that visits each vertex exactly once. From elementary graph theory texts such as [13], it is a well-known fact that every simple Eulerian graph has a Hamiltonian sequence in it.

Therefore, it is quite easy to see that a Hamiltonian sequence through a NPTG will optimally generate the set of patterns necessary to test for SNPFs. If a Hamiltonian sequence is used, after one SNP is generated, a single bit change can generate the next one; therefore, after p bit changes, all p patterns will be applied successively to a k-cell neighborhood of the

(1) Construct a reflected k-bit Gray code sequence G_k with initial state $S_0 = 00\ldots0$.

(2) Let $X = x_0 x_1 \ldots x_{k-1}$ denote an element of G_k. We define a transformation $T^1 : G_k \to G_k$ as follows:

$$T^1(x_0 x_1 \ldots x_{k-1}) = \overline{x_{k-1}} x_0 x_1 \ldots \overline{x_{k-2}}$$

(3) Let T^i denote i consecutive applications of T^1. Then the sequence of length $k2^k + 1$ formed by concatenating G_k, $T^1(G_k)$, ..., $T^{k-1}(G_k)$ is an Eulerian sequence.

Figure 3.10 Algorithm for constructing an Eulerian sequence for a k-cube

memory. Therefore, if $p = 2^k$, the total number of write operations on cells of a neighborhood is $k + 2^k - 1$, k initial write operations for the first pattern, followed by $2^k - 1$ single-cell changes to generate the remaining patterns. An Eulerian sequence through a NPTG will optimally generate the set of patterns necessary to test all APNPSFs, because it will generate all possible single cell transitions keeping the other $k - 1$ cells unchanged and cause each transition to be generated exactly once. Hence the total number of write operations performed for detecting APNPSFs is $k + k2^k$.

Hayes [54] gave an algorithm to generate an Eulerian sequence in a k-cell neighborhood. His algorithm is given in Figure 3.10.

Optimal write sequence at the memory cell array level: For a $(k, 1)$ NPSF (refer to the notation in Section 1), each memory cell belongs to at most k different neighborhoods, playing the role of a base cell in one and that of a deleted neighborhood cell in each of the others. As such, testing the entire memory for an NPSF would involve applying each pattern to each neighborhood separately. If we follow a naive approach for doing this, the complexity of write operations for testing all SNPSFs will be about $n2^k$, because, there are n possible choices for the base cell (i.e., any cell of the memory can be chosen), and for each such choice there are 2^k possible SNPs that can be applied to the k-cell neighborhood with the chosen cell at the base. To reduce the complexity of the write operation, two approaches are adopted — (a) *the tiling method* and (b) *the two-group method* [165].

In the tiling method, the memory is tiled by a group of (non-overlapping) neighborhoods. There are two possible choices for such neighborhoods — a *Type-1* neighborhood, which is depicted in Figure 3.11, and a *Type-2*

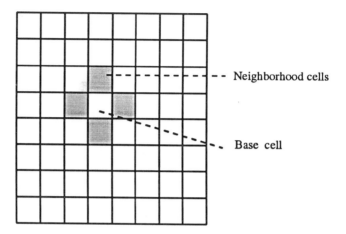

Figure 3.11 Type-1 neighborhood

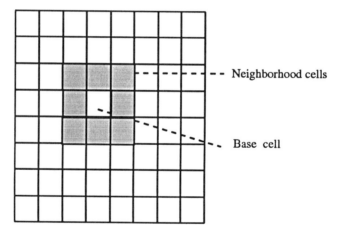

Figure 3.12 Type-2 neighborhood

neighborhood, shown in Figure 3.12. It should be noted that in a Type-1 neighborhood, all deleted neighborhood cells are treated alike whereas in a Type-2 neighborhood, there are 2 kinds of deleted neighborhood cells — corner cells and middle cells (cells that are in the middle of a row or column).

Let us examine the Type-1 tiling shown in Figure 3.11. The tiling method for applying test patterns applies each pattern simultaneously on all tiling neighborhoods. Hence for testing SNPFs on this 5-cell neighborhood, each of the $2^5 = 32$ patterns would be simultaneously applied on all the tiles whose base cell is 2. This would automatically cause each of these patterns to be also applied to any *other* possible Type-1 tiling arrangement, involving any of the other numbered cells as the base cell. Thus, this scheme would result in all the SNPs being applied to all possible 5-cell neighborhoods of this memory array, with a total of $\frac{n}{5}2^5$ write operations, because there is a total of $\frac{n}{5}$ tiling neighborhoods. Thus, the complexity of write operations is reduced by a factor of 5. In general, for a k cell Type-1 neighborhood, the number of write operations will be reduced from $n2^k$ to $\frac{n}{k}2^k$. An exact similar argument is also valid for a Type-2 neighborhood (Figure 3.12). We can extend the same argument to APNPFs also, because with APNPs, all the cells are treated identically, an arbitrarily chosen cell being made to undergo a transition in any given APNP. However, the argument is *not* valid for ANPs or PNPs separately because with these patterns the k neighborhood cells are not treated identically.

The *two-group* method [140, 158] is based on the duality of cells — it is based upon treating a Type-1 neighborhood cell as a base cell in one case and a deleted neighborhood cell in the other. This is shown in Figure 3.13. The cells are divided into two groups with the base cells of one group acting as the deleted neighborhood cells of the other group. The deleted neighborhood forms part of a Type-1 neighborhood and consists of 4 kinds of labeled cells, namely A, B, C and D, as shown in Figure 3.13, and each of these is a member of 4 neighborhoods. Each group consists of $\frac{n}{2}$ base cells and $\frac{n}{2}$ deleted neighborhood cells each with one of 4 possible labels. Each labeled group of deleted neighborhood cells is called a *subgroup*, thus there are $\frac{n}{8}$ cells in each subgroup (and they all have the same label).

A write operation in the two-group method consists of choosing a subgroup and writing into all $n/8$ cells of this subgroup, each subgroup being chosen exactly once; this reduces the number of write operations by a factor of 4. This would cause all p patterns to be applied to all possible k-cell neighborhoods. This method assumes duality for each cell and also the fact that the cells are all identical, hence this method cannot be used with a Type-2 neighborhood because there are two kinds of deleted neighborhood cells for such a neighborhood, as observed earlier.

4. **NPSF detection and location algorithms:**

The techniques for minimizing the number of transition-write operations in NPSF testing are of special importance in parallel testing. Built-in self-testing (BIST) architecture can be designed to exploit test parallelism as much as possible. Efficient parallel algorithms suitable for use in a BIST

120   CHAPTER 3

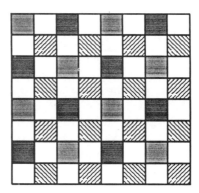

Cells of Group 1

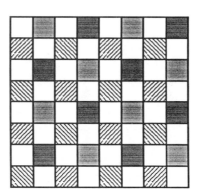

Cells of Group 2

Figure 3.13   The two-group method

(1) *Initialize* all base cells with 0;
(2) *loop*

>   apply a pattern; {this pattern might cause the base cell to flip}
>   read base cell;
>   *endloop*;

(3) *write* 1 in all the base cells;

(4) *loop*

>   apply a pattern; {this pattern might cause the base cell to flip}
>   *endloop*;
>   read base cell.

Figure 3.14 Basic detection algorithm for NPSFs

environment have been designed by us for NPSF detection in [101]. These will be discussed in Chapter 5.

The basic paradigms for detection and location of NPSFs are described in Figures 3.14 and 3.15 [165]. The two algorithms are virtually similar except for the second loop in which the base cell is not read after each pattern application when it is at state 1. This suffices for detection because when the second loop is executed, any forced transition in the base cell is guaranteed not to be masked by another forced transition. This is true because the first loop verifies that no $0 \rightarrow 1$ transition occurred in the base cell for any possible deleted neighborhood pattern. This scheme is not suitable for detection or location of PNPSFs or APNPSFs because patterns for testing these faults require transitions to be made in the base cell. There are three well-known algorithms for locating active and passive NPSFs (APNPSFs) — TLAPNPSF1G (*Test to Locate APNPSFs*, using a Type-1 neighborhood and the Two-Group method; TLAPNPSF2T, having a similar objective as the previous one, but using a Type-2 neighborhood and the Tiling method; and TLAPNPSF1T which uses a Type-1 neighborhood and the Tiling Method). The well-known algorithms for testing SNPSFs are called TLSNPSF1G, TLSNPSF1T, TLSNPSF2T, and TDSNPSF1G (this last one does purely detection). The names of these tests have been designed by Van de Goor and are borrowed from his book [165].

(1) *Initialize* all base cells with 0;

(2) *loop*

   apply a pattern; {this pattern might cause the base cell to flip}
   read base cell;

   endloop;

(3) *write* 1 in all the base cells;

(4) *loop*

   apply a pattern; {this pattern might cause the base cell to flip}
   read base cell;

   endloop.

Figure 3.15   Basic location algorithm for NPSFs

Figures 3.16 and 3.17 show the algorithms TLAPNPSF1G and TLAPNPSF2T. TLAPNPSF1T is basically the same algorithm as TLAPNPSF2T but applied to a Type-1 neighborhood instead of a Type-2 neighborhood, and hence has not been shown separately. A brief explanation of the algorithms follows.

TLAPNPSF1G uses the two-group method, writing onto cells in group-1 or group-2 with the same label. Table 3.4 shows 64 operations corresponding to all write operations to create the Deleted Active Neighborhood Patterns (DANPs) for a base cell — these are ANPs minus the base cell values, forming an Eulerian sequence of a 4-bit Eulerian graph. The algorithm makes two passes through the memory; in one pass it initializes base cells of groups 1 and 2 with 1s and 0s respectively in a checkerboard fashion and in the other pass it initializes them with all 0s. As shown later, these actions are sufficient to test the whole array.

At step 1, the memory is initialized with 1s in all the base cells of group-1 (called cells-1) and 0s in all the base cells of group-2, (called cells-2), followed by reading all cells to verify the initialization. At step 2, two different Eulerian sequences are applied, one starting with all zeroes and the other starting with all ones; this is necessary since all cells-1 are initialized to 1 and all cells-2 are initialized to 0. The term $Write(i, j), j \in \{1, 2\}$ denotes a change in the contents of the deleted neighborhoods of all cells-j; the cell A_i

(1) *Initialize* all cells-1 with 1 and cells-2 with 0;
   L1: *read* them out;
(2) $p := 0; q := 0;$
   for $i := 1$ *to* 64 *do*
   begin
      if $i = 13$ *then* $p := 1, q := 0$; {All cells-2 are 0, $j = 64$}
      if $i = 33$ *then* $p := 0, q := 1$; {All cells-2 are 1, $j = 52$}
      if $i = 53$ *then* $p := 1, q := 1$; {All cells-2 are 0, $j = 32$}
      $write(i,1)$;{Write deleted neighborhood cells of group-1}
   L2 : *read* all cells written by this write operation;
           {Test group-2 for PNPSFs}
   L3 : *read* all cells-1;{Test group-1 for ANPSFs}
      $j = (i + 52 + p * 33 + q * 58) \ mod \ 65;$
      $write(j,2)$; {This is the inverse of write(i,1)}
   L4 : *read* all cells written by this write operation;
           {Test group-1 for PNPSFs}
   L5 : *read* all cells-2;{Test group-2 for ANPSFs}
   end;

(3) *write* 0 in all cells-1;
   read all cells-1 and all cells-2;

(4) {NPSF test with initial value of base cells : cells-1 = 0 and cells-2= 0}

   for $i := 1$ *to* 64 *do*
   begin
      $write(i,1)$;{Write deleted neighborhood cells of group-1}
   L6 : *read* all cells written by this write operation;
           {Test group-2 for PNPSFs}
   L7 : *read* all cells-1; {Test group-1 for ANPSFs}
      $write(i,2)$; {This is the inverse of write(i,1)}
   L8 : *read* all cells written by this write operation;
           {Test group-1 for PNPSFs}
   L9 : *read* all cells-2;{Test group-2 for ANPSFs}
   end;

Figure 3.16 TLAPNPSF1G algorithm; courtesy [165]

to be changed is determined by the position of the ↑ or ↓ symbol in entry i of Table 3.4, therefore, a little reflection tells us that the inverse of $Write(i,j)$ for cells of the other group $(3-j)$ is $Write((i+52+33p+58q) \mod 65, 3-j)$ where $p=0, q=0$ for $1 \leq i \leq 12$; $p=1, q=0$ for $13 \leq i \leq 32$; $p=0, q=1$ for $33 \leq i \leq 52$; and $p=1, q=1$ for $53 \leq i \leq 64$. The read operations in step 2 check different kinds of NPSFs — the one at label L2 verifies the cells written by $Write(i,1)$, hence it checks for PNPSFs of group-2 (because deleted neighborhood cells for group-1 are base-cells for group-2). At label L3, the base cells of group-1 are tested — this corresponds to a test for ANPSFs. At labels L4 and L5, checks are made for PNPSFs of group-1 and ANPSFs of group-2 respectively.

Since each of the two sequences of write operations is Eulerian, all cells-1 have a value of 1 and all cells-2 have a value of 0 again after step 2 is completed. At step 3, the base cells of group 1 (cells-1) are initialized again with 0s and read along with cells-2. At step 4, a single Eulerian sequence is traced through the memory array — this sequence involves $Write(i,1)$, followed by 4 successive read operations, which test PNPSFs, ANPSFs, PNPSFs and ANPSFs for group-2, group-1, group-1 and group-2, respectively. Hence, during steps 1 and 2, all possible APNPSFs with an initial value of 1 in all base cells of group-1 are tested; during steps 3 and 4, all possible APNPSFs with an initial value of 0 in all base cells of group-1 are tested.

The time complexity of TLAPNPSF1G is, as shown in [165] $195.5n$ (obtained as follows - $2n$ (Step 1) $+2 \times 64(\frac{n}{8} + \frac{n}{8} + \frac{n}{2} + \frac{n}{8} + \frac{n}{8} + \frac{n}{2})$ (Steps 2 and 4) $+(\frac{n}{2} + n)$ (Step 3)), and the fault coverage is very impressive — all NPSFs. This follows from the fact that the algorithm generates all ANPs and PNPs for each neighborhood, and tests any cell whose contents are changed (because the cell happens to be a base-cell undergoing a transition) or affected by a change in some other cell (due to its presence in the deleted neighborhood of a base cell) before and after the changes are made.

TLAPNPSF2T uses a Type-2 tiling neighborhood, the cells being identified with labels $A_0, A_1, \ldots, A_{k-1}$ and correspond with the nodes of the Eulerian graph for k-tuples. In the algorithm displayed in Figure 3.17, $Write(j), 1 \leq j \leq k2^k$, associated with the j-th element of the k-bit Eulerian sequence EP_k, causes the contents of cells with symbol A_i to change, where i is the position that changed in the j-th element of EP_k. This algorithm locates all ANPSFs and PNPSFs because each execution of $Write(j)$ causes one ANP or PNP to be created for each neighborhood at step 2; the Eulerian write sequence ensures that all APNPs are generated optimally (i.e., by single bit changes each time without repeating previously gener-

Functional Fault Modeling and Testing

Table 3.4 Optimal write sequence to generate DANPs for TLAPNPSF1G; courtesy [165]

Oper	DANP $A_1A_2A_3A_4$	Oper	DANP $A_1A_2A_3A_4$	Oper	DANP $A_1A_2A_3A_4$	Oper	DANP $A_1A_2A_3A_4$
1	000↑	17	0↑00	33	00↑0	49	1 0↓0
2	00↑1	18	↑100	34	001↑	50	1↑00
3	0↑11	19	110↑	35	↑011	51	11↑0
4	01↓1	20	11↑1	36	1↑11	52	111↑
5	010↓	21	↓111	37	1↓11	53	111↓
6	01↑0	22	011↓	38	↓011	54	11↓0
7	↑110	23	0↓10	39	001↓	55	1↓00
8	1↓10	24	↑010	40	00↓0	56	10↑0
9	↓010	25	1↑10	41	↑000	57	101↑
10	0↑10	26	↓110	42	100↑	58	10↓1
11	011↑	27	01↓0	43	1↑01	59	↓001
12	↑111	28	010↑	44	↓101	60	0↑01
13	11↓1	29	01↑1	45	0↓01	61	↑101
14	110↓	30	0↓11	46	↑001	62	1↓01
15	↓100	31	00↓1	47	10↑1	63	100↓
16	0↓00	32	000↓	48	101↑	64	↓000

---

(1) *Initialize* all cells with 0; *read* 0 from all cells;

(2) **for** $j := 1$ **to** $k2^k$ **do**
    begin
        $write(j)$;
        read all cells;
    end;

Figure 3.17 TLAPNPSF2T algorithm; courtesy [165]

ated patterns); and all cells of the memory are read before and after each write operation. This algorithm performs $5122n$ operations for a neighborhood of 3 cells × 3 cells, (for a $k \times k$ cell neighborhood, its complexity is $n(2 + (k+1)2^k)$ [165]).

TLAPNPSF1T is a more efficient version of TLAPNPSF2T that uses a Type-1 neighborhood and the tiling method instead of a Type-2 neighborhood. This results in a decrease in the number of neighborhood cells, and thus the test time, which becomes $194n$ (the expression for the running time is exactly the same as for TLAPNPSF1T, but k has the value 5 instead of 9).

If only ANPSFs have to be tested, the above algorithms can be simplified by dropping all the read operations corresponding to detection of PNPSFs — for example, from the algorithm, TLAPNPSF1G, dropping the statements with labels L2, L4, L6, and L8 results in an algorithm to test only ANPSFs. This algorithm is known as TLANPSF1G, and requires fewer operations than its parent algorithm TLAPNPSF1G. A march test for ANPSF detection using a Type-1 neighborhood and the two-group method (TDANPSF1G) is proposed by Suk and Reddy [159] (1981). It has a complexity of $99.5n$.

TLSNPSF1G [140] is a simple algorithm for testing static NPSFs; Saluja and Kinoshita [140] (1985) present an improved version of this algorithm. The original version and Saluja's version (known as advanced TLSNPSF1G) are shown in Figures 3.18 and 3.19 respectively. The original version as shown in Figure 3.18 is self-explanatory, and has been subsequently modified to produce a more efficient algorithm (not shown here) in which a Hamiltonian sequence is used for writing the 15 remaining patterns on the deleted neighborhood cells after writing the first pattern — hence the first pattern involves $k - 1 = 4$ bit (cell) write operations for each $(k-1)$-cell deleted neighborhood followed by single-bit (cell) changes for each of the other 15 patterns. This algorithm is seen to have a complexity of $4(\frac{n}{2} + \frac{n}{2} + 15\frac{n}{8})$ (for writes) $+32n$ (for reads) $= 43.5n$. We observe that the Hamiltonian sequence is useful for cutting the complexity of the *remaining* write operations after the first one has been performed; hence trying to reduce the complexity of the first write operation (i.e., writing the *seed* pattern) can motivate a better algorithm, as discovered by Saluja and Kinoshita in 1985 [140].

In the advanced version, three different Hamiltonian sequences (denoted as H_1, H_2 and H_3), shown in Table 3.5 are used to reduce the complexity of the write operation for the starting pattern. This algorithm is shown in Figure 3.19. The basic idea is to write the seed pattern on only $\frac{n}{8}$ deleted neighborhood cells in such a manner that we need only three additional

Functional Fault Modeling and Testing

(1) {Locate SNPSFs in group-j cells, $j = 1,2$}
    for $j := 1$ *to* 2 *do*
    begin
        for data $:= 0$ *to* 1 *do*
        begin
            write data in all cells-j;{Write base cells of group-j}
            for $i := 1$ *to* 16 *do*
            begin
                write pattern i in the deleted neighborhood cells;
                read all cells-j;
            end;
        end;
    end;

Figure 3.18 TLSNPSF1G (original version)

writes to change the last deleted neighborhood pattern of a Hamiltonian sequence into the first pattern of the next sequence, the sequence followed is as shown in Table 3.6. This table has 67 entries, formed by the sequences H_1, H_2, H_2, and H_3.

The complexity of this algorithm is seen to be: $8\frac{7}{8}n$ (for writes)$+32n$ (for reads) $= 40.875n$, a slight improvement over the simple one, described above.

TLSNPSF1T (Figure 3.20) is an algorithm presented by de Jong (1988) [31] to locate SNPSFs, using the tiling method and a 5-bit Hamiltonian sequence ($k = 5$). The algorithm performs n initial cell write operations for the seed pattern and then follows a Hamiltonian sequence in each of the $\frac{n}{k}$ tiling neighborhoods to perform $(2^k - 1)\frac{n}{k}$ cell writes for the remaining $2^k - 1$ successive patterns. This results in a total of $7.2n$ write operations and $32n$ cell read operations, giving a total complexity of $39.2n$. This algorithm is equally applicable to Type-2 neighborhoods (except that we need a Hamiltonian sequence of more than $k = 5$ bits, for example, for a 3 cell × 3 cell neighborhood, $k = 9$). The Type-2 neighborhood version of this algorithm is called *TLSNPSF2T*. This has a much higher complexity, the total number of operations (read and write) turns out to be equal to $569\frac{7}{9}n$.

(1) *Initialize* all cells-1 with 0;
    read cells-1; {Read base cells of group-1}

(2) {Perform write operations of sequence H_1 on cells-2}
    for $i := 1$ *to* 16 *do* {Perform Entry 1 through 16 of Table 3.6 }

   begin
       $write(H_1,i,2)$;
       read cells-1; {Test group-1 for SNPSFs}
   end;

(3) *initialize* all the D group of cells of group-2 to 0; {Perform Entry 17 of Table 3.6 }

(4) *read* cells-2; {After base cells have been changed from cells-1 to cells-2}

(5) {Perform write operations of sequence H_2 on cells-1}
    for $i := 1$ *to* 16 *do* {Perform Entry 18 through 33 of Table 3.6}

   begin
       $write(H_2,i,1)$;
       read cells-2; {Test group-2 for SNPSFs}
   end;

(6) *initialize* all the C group of cells of group-1 to 1; {Perform Entry 34 of Table 3.6 }

(7) *read* cells-1; {After base-cells have been changed from cells-2 to cells-1}

(8) {Perform write operations of sequence H_2 on cells-2}
    for $i := 1$ *to* 16 *do* {Perform Entry 35 through 50 of Table 3.6}

   begin
       $write(H_2,i,2)$;
   L1: *read* cells-1; {Test group-1 for SNPSFs}
   end;

(9) *initialize* all the C group of cells of group-2 to 1; {Perform Entry 51 of Table 3.6 }

(10) *read* cells-2; {After base-cells have been changed from cells-1 to cells-2}

(11) {Perform write operations of sequence H_3 on cells-1}
    for $i := 1$ *to* 16 *do* {Perform Entry 52 through 67 of Table 3.6}

   begin
       $write(H_3,i,1)$;
   L2: *read* cells-2; {Test group-2 for SNPSFs}
   end;

Figure 3.19 TLSNPSF1G (advanced version)[140]

Table 3.5 Hamiltonian sequences used by advanced TLSNPSF1G; courtesy [165]

Entry No.	H_1 ABCD	H_2 ABCD	H_3 ABCD
1	0000	0000	1111
2	0100	0001	1110
3	1100	0011	1010
4	1000	0010	1011
5	1001	0110	1001
6	1011	0111	1000
7	1010	0101	1100
8	1110	0100	1101
9	1111	1100	0101
10	1101	1000	0111
11	0101	1001	0110
12	0111	1011	0010
13	0110	1010	0011
14	0111	1110	0001
15	0011	1111	0000
16	0001	1101	0100

(1) *Initialize* all cells with 0; *read* 0 from all cells;

(2) *for* $j := 1$ *to* $2^k - 1$ *do*
    begin
        write pattern no. j of the chosen Hamiltonian sequence;
        {This is done by flipping a bit (i.e., cell) of pattern no. $j - 1$}
        read all cells;
    end;

Figure 3.20 TLSNPSF1T

Table 3.6 Test sequence to locate SNPSFs for advanced TLSNPSF1G; courtesy [140]

Entry No.	Sequence No.	Base cell	ABCD	Group	Entry No.	Sequence No.	Base Cell	ABCD	Group
1	H_1	0	0000	1	35	H_2	1	0000	1
2	H_1	0	0100	1	36	H_2	1	0001	1
3	H_1	0	1100	1	37	H_2	1	0011	1
4	H_1	0	1000	1	38	H_2	1	0010	1
5	H_1	0	1001	1	39	H_2	1	0110	1
6	H_1	0	1011	1	40	H_2	1	0111	1
7	H_1	0	1010	1	41	H_2	1	0101	1
8	H_1	0	1110	1	42	H_2	1	0100	1
9	H_1	0	1111	1	43	H_2	1	1100	1
10	H_1	0	1101	1	44	H_2	1	1000	1
11	H_1	0	0101	1	45	H_2	1	1001	1
12	H_1	0	0111	1	46	H_2	1	1011	1
13	H_1	0	0110	1	47	H_2	1	1010	1
14	H_1	0	0111	1	48	H_2	1	1110	1
15	H_1	0	0011	1	49	H_2	1	1111	1
16	H_1	0	0001	1	50	H_2	1	1101	1
17		0	0000	1	51		1	1111	1
18	H_2	0	0000	2	52	H_3	1	1111	2
19	H_2	0	0001	2	53	H_3	1	1110	2
20	H_2	0	0011	2	54	H_3	1	1010	2
21	H_2	0	0010	2	55	H_3	1	1011	2
22	H_2	0	0110	2	56	H_3	1	1001	2
23	H_2	0	0111	2	57	H_3	1	1000	2
24	H_2	0	0101	2	58	H_3	1	1100	2
25	H_2	0	0100	2	59	H_3	1	1101	2
26	H_2	0	1100	2	60	H_3	1	0101	2
27	H_2	0	1000	2	61	H_3	1	0111	2
28	H_2	0	1001	2	62	H_3	1	0110	2
29	H_2	0	1011	2	63	H_3	1	0010	2
30	H_2	0	1010	2	64	H_3	1	0011	2
31	H_2	0	1110	2	65	H_3	1	0001	2
32	H_2	0	1111	2	66	H_3	1	0000	2
33	H_2	0	1101	2	67	H_3	1	0100	2
34		0	1111	2					

Functional Fault Modeling and Testing   131

Table 3.7 A comparison of testing times for NPSFs; courtesy [24]

Author & Test	Length	256K	1M	4M	16M
Suk & Reddy	$99.5n$	2.6s	10.4s	41.7s	2m 46.9s
Franklin & Saluja	$64n(\lceil \lg_3 \sqrt{n} \rceil)^4$	36m 14.3s	4h 28.5m	17h 54.2m	5d 2.2h
Cockburn's S4CTEST	$8.6n(\lg n)^{1.585}$	22.9s	1m 54.2s	8m 20.4s	33m 21.4s
Cockburn's S5CTEST	$9.6n(\lg n)^{2.322}$	3m 16.7s	19m 14.3s	1h 17m	5h 7.8m

TDSNPSF1G is a simplified version of *TLSNPSF1G*, tailored to perform merely *detection* of SNPSFs by taking the 'read' statements at labels L1 and L2 in Figure 3.19 outside of their respective loops; the idea being derived from the same basic simplification approach that resulted in simplifying the generic NPSF location algorithm to a detection algorithm. This simplification causes a total of $36.125n$ operations.

A comparison of the test times of various NPSF algorithms is shown in the table in Table 3.7 [24].

3.3 TECHNIQUES DEPENDENT ON LAYOUT AND CIRCUIT DESIGN FOR FUNCTIONAL TEST OF RAMS

All the above algorithms and fault modeling techniques are of a traditional nature; that is, they do not recognize the fact that layout variations can lead to varying fault coverages and test effectiveness. Several of the fault models described above may be constructed from realistic defect models but the tests themselves have little or no built-in knowledge of the physical layout and the cell geometry. As a result, implementing some of the above tests may result in some practical overheads for the tester. For example, after a memory chip is reconfigured, physically adjacent cells may no longer have consecutive addresses [30, 69]; moreover, the logical to physical address mapping is also lost. This mapping is known at the time of repair but storing it for later use can be very costly [69]. After packaging, a number of functional tests have to be performed — before shipping (production testing), after delivery (incoming inspection) and during operation (maintenance testing). Test algorithms used at these stages for the detection of functional faults have to consider layout-related issues, for example, the fact that logical and physical neighborhoods are not identical and the address mapping is not available. For various reasons [69], storing the address mapping on-chip or deriving it with special circuitry may not be an acceptable solution.

Moreover, layout-independent fault models may achieve a high fault coverage, but they often produce a low *defect* coverage. Physical defects in the memory chip layout may or may not produce faults, but necessitate replacement or repair. We are currently in the age of multi-megabit memories with line widths and geometries shrinking at a rapid rate. The technology is pushing itself down to 0.3 μ at a steady rate, thanks to new processing techniques and changes in the structure of the basic storage cell. These, in turn, are associated with new defect types which adversely affect the yield of memory chips. Moreover, more complex built-in self-testing (BIST) circuits often result in an increase in the area of the chip, causing a greater probability of chip defects. Testing for physical defects has, therefore, become more important nowadays than testing for mere faults.

In this chapter, we shall examine functional test algorithms that recognize that the chip layout and circuit techniques play an important role in testing. These parameters of a chip are (realistically) assumed here to impact the fault coverage of a test algorithm. The algorithms described in the remainder of this chapter may or may not themselves perform fault modeling at the layout level, however, the tests proposed are robust enough to deal with layout-related issues such as *address scrambling* from the logical to the physical address space.

3.3.1 Operational model for RAMs

An **operational** model for RAM is similar to a functional model with the added capability that it can verify whether a sequence of operations can be performed correctly. Such a model called **Paragon** has been proposed by Winegarden and Pannell [176]. This model verifies that a sequence of operation pairs is executed correctly for every address combination, data combination and operation. This approach can thereby be used to perform a write recovery test which checks

Functional Fault Modeling and Testing

the timing of the address decoder, and as a sense amplifier recovery test which checks the response time of the sense amplifier. The algorithm is as follows:

(1) Write a background pattern of 0s in the memory.

(2) For every pair of choices for the base cell and the test cell, repeat the following four pairs of operations:
   (a) read the test cell and then read the base cell;
   (b) write 1 into the test cell and 0 into the base cell;
   (c) read the test cell and then read the base cell;
   (d) write 0 into the test cell and then write 0 into the base cell.

(3) Read out all the cells, (expected data being the originally written data).

(4) Repeat steps (2) and (3) with complementary data (that is, interchanging 0s with 1s and vice versa).

This algorithm has a time complexity of $16n^2 - 12n$.

3.3.2 Functional fault modeling of megabit SRAMs at the layout level

The fault modeling proposed by Dekker [33, 34, 36] is based on physical defects as observed in the layout of an SRAM memory array, the Philips 8 kB chip. The defects have been modeled at the electrical level resulting in logical faults for which functional tests, as described before, can be performed.

The defects at the layout level have been modeled using the *Inductive Fault Analysis* (IFA) technique [148, 84] which assumes only a single defect per memory cell. IFA is a simple procedure comprising the following steps [120]:

1. **Generation of defect characteristics:** This involves computation of the physical characteristics of a defect which could occur in the actual process (for example, size, location and layer of the memory structure in which the defect is found).

2. **Defect placement:** This involves actually placing the defect at an appropriate layer in the memory layout. In other words, the nominal layout itself is modified and a new layout generated.

3. **Extraction of schematic and electrical parameters of the defective cell.**

4. **Evaluation of the results of testing the defective memory:** This can be performed using an approach for deducing the testing results, or if not possible, through Monte Carlo simulation, also known as the *dot-throwing* approach, of the circuit.

Defects can be generated in a layout and Monte Carlo simulation performed using a package such as VLASIC [172].

Two classes of such defects have been distinguished — global and local. Global defects affect a wide area of a wafer and often spread over multiple ICs. Examples of such defects are too thick gate oxide, mask misalignments and line dislocations — they manifest themselves as *dynamic* faults, described later. Local defects affect only a small area of the chip and are modeled as spots of extra, missing or improper material — such defects are named *spot defects* and are the main cause of *functional* faults [148, 84].

At the layout level, our memory functional fault model consists of spot defects modeled as (*a*) broken wires, (*b*) shorts between wires, (*c*) missing contacts, (*d*) extra contacts, and (*e*) newly created transistors. Dekker [33] has analyzed 60 spot defects in detail, starting with a two-stage translation — from the layout to the electrical level, and then from the electrical to the logical level, and came up with six functional fault classes. They are: (*a*) SAF in a memory cell; (*b*) a *stuck-open fault* (SOF) in a memory cell, this means that the cell cannot be accessed, for example, due to an open word line; (*c*) a TF in a cell; (*d*) a state coupling fault between two cells such that the coupled cell is forced to a certain value only if the coupling cell is in a given state; (*e*) a CFid between two cells; (*f*) a *data-retention fault*, in which a cell loses its contents after some time, this is caused by a broken pull-up device, and the retention time is determined by the leakage current from the cell with the broken pull-up device to the substrate. The exact nature of these faults has been explained in [33] and [165].

For testing the six faults mentioned above, a test called *Inductive Fault Analysis-9* (IFA-9, also described popularly as the $9N$-Test) has been proposed by Dekker. In this test, the chip layout is analyzed using IFA, and then nine march operations are used to detect all faults except the data retention faults. The march notation for this test is $\{\Uparrow (w0), \Uparrow (r0, w1), \Uparrow (r1, w0), \Downarrow (r0, w1) \Downarrow (r1, w0), \text{Delay}, \Uparrow (r0, w1), \text{Delay}, \Uparrow (r1)\}$. These march elements are labeled M_0 through M_6. March elements M_0 through M_5 satisfy the necessary and sufficient conditions for detecting AFs, SAFs, TFs, unlinked CFins and unlinked

Functional Fault Modeling and Testing 135

CFids. In addition, it can be shown that all pairwise state coupling faults will be detected by these march elements if the four states of any two cells i and j can be reached — M_0 through M_4 make this happen. March elements M_4 through M_6 with intermediate delays, detect data retention faults — the delays (periods for which the RAM is disabled) are typically about 100 ms each, as per the specifications for the Philips 8 KB chip tested by Dekker in [33]. The state table and corresponding state diagrams for detecting SCFs are described in [34]. This test works for a purely combinational read/write logic. The part of the test before the delay consists of 9 march operations, and has a complexity of $9N$. The remaining part of the test is the *Data Retention Test*.

The extension of the above algorithm for a sequential read/write logic resulted in the algorithm IFA-13 (also described popularly as the $13N$-test). This algorithm is IFA-9 plus 4 more read operations that expect to read 1 and 0 alternately. The purpose of this modification is to detect stuck-open faults. If the IFA-9 algorithm is not modified, the read/write logic may always pass the latest proper value read to the output pin during a read operation, thereby preventing detection of a stuck-open fault on a cell. Thus IFA-13 can be denoted by the march notation $\{\Updownarrow (w0), \Uparrow (r0, w1, r1), \Uparrow (r1, w0, r0), \Downarrow (r0, w1, r1), \Downarrow (r1, w0, r0), \text{Delay}, \Uparrow (r0, w1), \text{Delay}, \Uparrow (r1)\}$.

The fault model has been validated via photographs taken using a scanning electron microscope, and the test algorithm has been validated using a metric called the *score number*. Twenty three different test algorithms, including IFA-9 and IFA-13, have been compared in the following manner:

(a) for each of 1192 devices, each test algorithm is assigned a signature bit with value 0, 1, or 2; this bit being a function of the number of faulty cells detected by the test algorithm, called the *bit error number b*, as follows: if b is less than $(m - r/5)$, m being the mean value and r being the standard deviation of the bit error numbers for one device, then the signature bit is 0; if b is between $(m - r/5)$ and $(m + r/5)$, the signature bit is 1, and if b is greater than $(m + r/5)$, the signature bit has the value 2 — thus for each device, the greater the number of faulty cells detected by the test algorithm, the greater the value of the signature bit;

(b) the score number of a given test algorithm is computed as the mean value of all the signature bits from all the devices for a given test algorithm — obviously this is a number between 0 and 2, the larger its value, the greater the fault coverage of the corresponding algorithm. IFA-9 is seen to have a score number of 1.81 and IFA-13 with data-retention test, 1.83.

3.3.3 Test algorithms for double-buffered RAMs and pointer addressed memories (DBMs and PAMs)

Van Sas, et al. [169] (1993) have explored the implementation of test algorithms for *double-buffered* and *pointer-addressed* memories (DBMs). Figure 3.21 shows an SRAM, a single-buffered memory (SBM), and a double-buffered memory (DBM). Unlike a conventional SRAM, an SBM or a DBM cell can be read/written simultaneously as there is a dedicated read line and a dedicated write line. The design of test algorithms for single-buffered memories and content addressable memories has been discussed in [93].

For many practical applications that involve making good interfaces with fast microprocessors, a single-buffered memory (SBM) has not enough bandwidth available over a single port. Using several SBMs in parallel will often not alleviate this problem because of extra area overhead for additional hardware. A DBM, on the other hand, can overcome this problem by having two parts — a master and a slave, with a *conditional* buffering in between controlled by a global *TRANSFER* signal. DBMs can therefore be used for on-chip interprocessor communication and as an ideal basis for I/O interfaces.

A pointer-addressed memory (PAM) is a FIFO (first-in-first-out, like a queue) memory that is accessed using a cheaper mechanism than address decoding and subsequent word select. In a FIFO memory, for both read and write select, a shift register can be used with the memory array. A 'running one' points to a certain memory location and is shifted further after each memory access. Hence, a set of words would be accessed consecutively in increasing order of addressing. The principle of a running one can be extended to allow shifting over dedicated routing tracks, instead of through adjacent cells. These are illustrated in Figure 3.22.

Fault modeling and test strategy for DBMs

The mechanism of testing DBMs is complicated by the fact that each memory cell is now associated with two values instead of one — one stored in the master and the other stored in the slave. Conventional functional testing and parametric fault algorithms deal with single buffered or simple SRAMs that are capable of storing one value per cell, and have therefore to be modified for DBMs. In [169], a realistic fault model based on *actual* defects that are noticed in the chip layout is developed. The approach is based on an inductive fault

Functional Fault Modeling and Testing

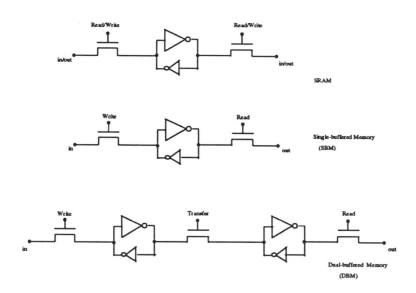

Figure 3.21 An SRAM, a single-buffered memory (SBM), and a double-buffered memory (DBM); courtesy [169]

analysis (IFA) procedure, as opposed to abstract ones that employ heuristics. Leakage, device anomalies, and parasitic effects between adjacent memory cells of a DBM are used to derive the fault model.

Unlike conventional functional test generation in which decoder and address generation logic faults could readily be translated to faults in the memory array alone, a DBM consists of three structurally different components — the memory array, the Read/Write logic and the address generation or decoder logic. It is somewhat difficult to develop a unified fault model for all three components, and hence they are modeled separately. There are two reasons for this difficulty - firstly, the fact that not one, but *two* address generation circuits are used; and secondly, pointer addressing and exclusively combinational decoding or a mixture of both may be used.

The above IFA procedure systematically detects all faults that are extremely likely, assuming that only *single spot* defects per memory cell are present. Typical defects are caused by dust particles on the chip or mask, scratches or gate oxide pinholes, oxide defects, conducting layer defects (shorts/opens) and interconnect defects, and defects caused by extra or missing material resulting in complete shorts and breaks. Only physically close nodes are considered

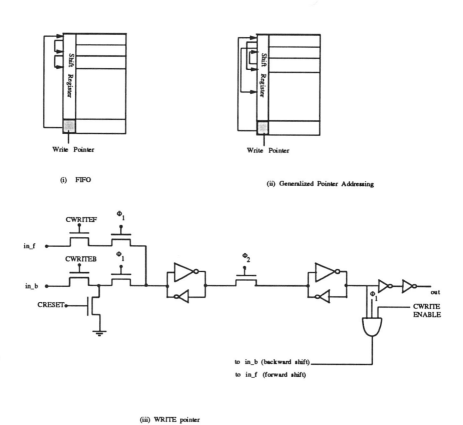

Figure 3.22 Pointer-addressed memory cell; courtesy [169]

Functional Fault Modeling and Testing 139

for dealing with shorts. Furthermore, faults due to leakage, device anomalies and parasitic effects between adjacent memory cells are considered. The next section lists the fault classes and the test algorithm to detect these faults.

Fault classes for the DBM

Twelve fault classes for the DBM memory array based upon the circuit-level characteristics of the observed defects are proposed in [169]. These classes are listed below:

1. **IFA.1**: One or more master or slave parts have an SAF.

2. **IFA.2**: One or more master or slave parts are stuck-open.

3. **IFA.3**: One or more master or slave parts have a TF.

4. **IFA.4**: One or more master or slave parts are permanently written or read.

5. **IFA.5**: One or more pairs of *adjacent* master-slave parts of the same or different DBM cells are coupled by a bridge (BF), with one memory part being coupled to at most one other memory part; therefore, the final state of the two bridged parts is either the logical AND or the logical OR of its values.

6. **IFA.6**: Destructive read operation in one or more slave parts because of precharging error.

7. **IFA.7**: A BF between input and output data lines of a DBM.

8. **IFA.8**: A BF between master and slave part of a DBM cell.

9. **IFA.9**: A BF between one or more master or slave parts to an input or output data line.

10. **IFA.10**: Data retention fault in one or more master or slave parts.

11. **CCF.1**: A master part is capacitively coupled with a slave part of the same or of a different DBM cell.

12. **CCF.2**: A master part is capacitively coupled with another master part.

The following three fault classes are identified for the read/write logic:

1. **R/W.1**: One or more lines of the sense amplifier logic or write driver logic have an SAF.
2. **R/W.2**: One or more data input/output lines have shorts or hard coupling faults between them.
3. **R/W.3**: One or more output cells of the sense amplifier logic or write driver logic are stuck-open.

The fault model proposed for the address generation logic consists of two faults, and assumes that the decoder is not converted into a sequential circuit as a result of the faults:

1. **AG.1**: The corresponding word is not accessed. In addition, a wrong word may be accessed.
2. **AG.2**: Multiple words are accessed, with or without the addressed word — with the output block performing either the logical AND or logical OR of the multiple accessed words.

Test algorithms for the array and the surrounding logic

1. **Test algorithm for the DBM array:** A modified march test has been proposed to detect all faults of the fault model. The modified march test has the following notation (with & between two operations denoting the fact that they have to be performed simultaneously, and tr denoting transfer of data from master cell to slave cell):

    $\{\Uparrow (w0), tr, \Uparrow (r0\&w1), tr, \Uparrow (r1\&w1), tr, \Uparrow (r1\&w0), tr, \Uparrow (r0\&w0),$
    $tr, \Uparrow (r0\&w1, r0, tr), \Uparrow (r1\&w0, tr, w1, tr), \Uparrow (r1\&w0, w1, r1, tr),$
    $\Uparrow (r1\&w0, r1, tr), \Uparrow (r0\&w1, tr, w0, tr), \Uparrow (r0\&w1, w0, r0, tr),$
    $\Uparrow (r0\&w1), tr, \Uparrow (r1\&w1), tr, \Uparrow (r1\&w0), tr, \Uparrow (r0\&w0),$
    $tr, \Uparrow (r0\&w1, r0, tr), \Uparrow (r1\&w0, tr, w1, tr), \Uparrow (r1\&w0, w1, r1, tr),$
    $\Uparrow (r1\&w0, r1, tr), \Uparrow (r0\&w1, tr, w0, tr), \Uparrow (r0\&w1, w0, r0, tr)\}$

    This algorithm can be proved to detect all the faults of IFA.1, IFA.3, IFA.4, IFA.5, IFA.6, IFA.7, IFA.8, IFA.9, CCF.1 and CCF.2 fault classes. The algorithm has a complexity of $53n$, n being the number of words in the memory.

Functional Fault Modeling and Testing 141

2. **Test algorithm for the read/write logic:** For purposes of testing, the fault class R/W.1 is found to be equivalent to IFA.1, R/W.2 to IFA.5, and R/W.3 to IFA.2. Thus read/write faults can be detected by the above modified march algorithm.

3. **Test algorithm for the address generation logic:** The decoder logic is found to be testable by the following algorithm: $\{\Uparrow (w0), tr, \Uparrow (r0\&w1, tr), \Uparrow (r1\&w0, tr), \Downarrow (r0\&w1, tr), \Downarrow (r1\&w0, tr)\}$. This test has a complexity of $9n$.

Test algorithm for a double-buffered PAM

The above algorithms require some modification if read and write signals are pointer-generated, because read and write addressing sequences are often quite different. The following three conditions are required for the pointer addressing scheme, to enable the correct application of the first algorithm:

1. Both pointers must be able to generate the same sequence, so that the read and write addressing sequences are the same.

2. Each pointer should be able to shift both forwards and backwards.

3. During normal operation, the 'running 1' of a pointer is shifted to the next location over a dedicated route after one clock period. However, to apply some of the march elements (for example, the eleventh one in the first algorithm), it is also necessary to address the same word during multiple clock periods. To achieve this, an enable signal has to be added to the pointer logic.

Since one pointer has to be modified to perform the same addressing sequence as the other, the first algorithm may not cover all defects in this modified pointer. This difficulty can, however, be circumvented if the following test algorithm for the pointer logic is performed in addition to the original algorithm for the memory array testing.

1. Write a unique test vector, known as an *identification test vector* while running through the sequence of the write pointer.

2. Transfer (data from master to slave parts).

3. Read along the sequence of the read pointer.

This algorithm guarantees the correct read/write sequence for each word. This method is feasible only if $n < 2^b$, where n is the number of words, and b is the number of bits per word. Otherwise, faults in the read or write logic can be detected by writing an all-1 vector in one location, and an all-0 in all other locations, similar to GALPAT.

3.3.4 Physical (or scrambled) neighborhood pattern-sensitive faults

The algorithms we have already discussed for detecting or locating NPSFs mostly assume that the memory cells are arranged in a rectangular grid and the logical and physical addresses of the cells match. However, this is often not a realistic assumption for two reasons [165]: IC designers often scramble the natural order of the bit and/or word lines to obtain better placement of the decoders. Moreover, state-of-the-art memory chips are designed with spare rows and spare columns for reconfiguration purposes. Reconfiguration often causes physically adjacent cells to have non-consecutive logical addresses. Hence, either the tester hardware or the NPSF detection algorithm itself has to be adapted to incorporate knowledge regarding physical neighborhood of cells. Several solutions have been proposed to the problem of testing memory chips after reconfiguration or repair. Furnweger [48] proposed the *roll-call* feature, which uses a decoder, called the *roll-call decoder* to identify the columns that have been replaced after repair. During the roll-call, all column addresses are applied to the chip by an external tester to identify the replaced columns. The exact address mapping can be mathematically derived using the knowledge of the repairing strategy. Though this scheme correctly identifies repaired chips, there are several disadvantages — excessive area overhead due to the use of the roll-call decoder, requirement for an external tester, and the need for a large test program for the external tester.

Kantz, et al. [69] implemented the roll-call feature for both rows and columns. They also developed a test procedure consisting of the following three steps — (*a*) a signature test to determine if the chip has been repaired or not; (*b*) a roll-call test that uses an external tester to identify the rows and columns that have been replaced; and (*c*) a modification of *address descrambling* based upon the replacement strategy and the information obtained from the roll-call test — this is also done by the tester which modifies the test algorithm to incorporate the correct address mapping.

Functional Fault Modeling and Testing 143

The above approaches are based upon direct acquisition of knowledge regarding the physical to logical address mapping. Daehn and Gross [30] developed test algorithms that *do not* depend upon the ordering of rows and columns. They developed an $O(n^{1.5})$ test procedure called *GALPROCO* (*GAL*loPing *R*ows and *C*olumns) to test reconfigurable memories and memories with distinct logical and physical addresses for SAFs, AFs, CFs within a row, and so on. Unfortunately, their test does not detect physical NPSFs with larger neighborhoods.

Franklin and Saluja [46] modeled the problem of detecting physical NPSFs as follows: testing a memory array without finding its address mapping can be achieved if each base cell is tested for *all possible* address mappings, for example, all possible 5-cell neighborhoods that could arise as a result of scrambling row and column addresses. This at first glance seems like a combinatorial explosion, but a little insight tells us that even after scrambling both row and column addresses, the physical north and south neighbors of the base cell are in the same logical column as the base cell, though they may no longer remain immediate north and south neighbors; likewise the physical east and west neighbors of the base cell are in the same logical row as the base cell, though they may no longer remain immediate east and west neighbors.

With this observation, the next step is to identify the n^2 5-cell neighbors efficiently. Franklin and Saluja [46] achieved this by dividing the address into row address and column address, and considering each 5-cell row-column neighborhood to be the union of two 3-cell neighborhoods — a 3 cell row-neighborhood and a 3 cell column-neighborhood, having a common base cell. The problem thereby reduces to identifying all possible triplets of rows and columns to detect 3-cell neighborhood PSFs in linear arrays, instead of identifying quintets that form 5-cell row/column neighborhoods.

Identification of these triplets is performed using a *3-coloring approach*, as follows: each row (and column) is divided into blocks of size 3 (one base cell and two neighbors), and all these blocks are colored in the same way using 3 different colors. Thus we have, for example, an $RBGRBGRBGRBG...$ coloring where R denotes cells colored red, B denotes cells colored blue, and G denotes cells colored green. Each cell acts as a base cell in one block and a deleted neighborhood cell in two other blocks, which overlap with the block in which it is a base cell. Hence, by applying 8 different patterns to the boolean variables B, R, and G we can simultaneously test each 3-cell row (or column) neighborhood. For achieving this, we need to ensure that each possible 3-cell neighborhood is colored trichromatically (i.e., by 3 different colors) at least once. The efficiency with which this can be achieved obviously depends on the efficiency of the coloring method. Franklin and Saluja [46] have presented the *trichromatic triplet 3-coloring algorithm* to identify the triplets efficiently.

The algorithm models the problem of efficient triplet identification as follows: given n objects and 3 colors, find a minimal set of 3-colorings such that every possible triplet of objects is trichromatically colored in at least one of the 3-colorings. This problem can, in turn, be abstracted as a graph-theoretic problem involving hypergraph coloring:

"Given $H = (X, E)$, a complete uniform hypergraph of rank 3 on n vertices, [12] find the minimum number of strongly 3-colorable partial hypergraphs $H_i = (X_i, E_i)$ such that $\bigcup_i E_i = E$."

The algorithm is applied on an $m \times n$ memory array as follows:

(1) Initialize all memory cells to 0.

(2) Denote the 3 colors as 0,1,and 2.

(3) Encode each of the n cells in a row in ternary by expressing its number i in base 3, forming a codeword, say $a_{\lceil \log_3 n - 1 \rceil} \ldots a_1 a_0$, where each a_i belongs to the set $\{0,1,2\}$.

(4) for $j := 0$ to $\lceil \log_3 n - 1 \rceil$ do
    for $k := 0$ to $m - 1$ do
    {for each row}
        assign color a_j to each cell of row r_k, where the cell
        is encoded as $a_{\lceil \log_3 n-1 \rceil} \ldots a_1 a_0$,
        color_columns_and_test();

(5) for $j := 0$ to $\lceil \log_3 n - 1 \rceil$ do
    for $k := 0$ to $m - 1$ do
        for $l := 0$ to $j - 1$ do
        {for each row}
            assign color $a_j + a_l$ to each cell of row r_k, where the cell
            is encoded as $a_{\lceil \log_3 n-1 \rceil} \ldots a_1 a_0$,
            color_columns_and_test();

(6) for $j := 0$ to $\lceil \log_3 n - 1 \rceil$ do
    for $k := 0$ to $m - 1$ do
        for $l := 0$ to $j - 1$ do

Functional Fault Modeling and Testing   145

> {*for each row*}
> assign color $2a_j + a_l$ to each cell of row r_k, where the cell
> is encoded as $a_{\lceil \log_3 n - 1 \rceil} \ldots a_1 a_0$,
> color_columns_and_test();

color_columns_and_test() has an exact similar code as above, with rows replaced by columns, and the procedure call replaced by a pair of calls to two procedures known as *5-cell test()* and *9-cell test()*. These procedures are outlined below:

5-cell test()

> for $i := 0$ to 2 do
>> for $j := 0$ to 2 do
>>> let all (i,j)-cells be the B-cells (base cells)
>>> let all $((i+1) \bmod 3, j)$-cells be the S-cells
>>> let all $((i-1) \bmod 3, j)$-cells be the N-cells
>>> let all $(i,(j-1) \bmod 3)$-cells be the W-cells
>>> let all $(i,(j+1) \bmod 3)$-cells be the E-cells
>> Apply all 32 0/1 patterns to the memory cells such that
>> all cells of one type (B,N,S,W or E) have the same data
>> stored in them; thus all B cells have the same value,
>> all N cells have the same value, and so on.

9-cell test()

> for $i := 0$ to 2 do
>> for $j := 0$ to 2 do
>>> for i : 0, 1, 2 correspond to N, don't care, S respectively,
>>> for j : 0, 1, 2 correspond to W, don't care, E respectively,
>>> $i = j = 1$ corresponds to B.
>>> (for example, (0,2) corresponds to NE cells, (1,2) corresponds to E
>>> cells, and others. A don't care is ignored when it occurs in
>>> combination with a non-don't care in an (i,j) pair).
>> Apply all 512 0/1 patterns to the 9 types of memory cells such that
>> all cells of one type (from among the types B,N,S,W or
>> E) have the same data stored in them. Thus all B cells have
>> the same value, all NE cells have the same value, and so on.

It has been proved by Franklin and Saluja [46] that the test length for this algorithm is in $O(N\lceil \log_3 N\rceil^4)$. One nice property of this algorithm is that in a self-reconfigurable memory array, the spare rows and columns can be introduced in any order in the array, because this algorithm allows all possible permutations of the row and column addresses. This algorithm applies all required patterns to detect all possible 5-cell and 9-cell physical NPSFs, regardless of the scrambling of address lines. This algorithm also detects SAFs, and can be modified to detect all 3-coupling faults within rows, all 3-coupling faults within columns, all 3-coupling faults between any triplet of cells containing two cells in the same row or column. This modification involves a read before each write operation.

Cockburn [24] proposed algorithm S5CTEST, whose generic version SVCTEST has already been described. This algorithm has a complexity of $9.6n(\lg n)^{2.322}$. He proved that it detects all single scrambled physical NPSFs. Sometimes cell positions in the array layout are such that simpler algorithms can be used, for example, for the 4 Mb layout examined by Oberle, et al. [130]. For this layout, the cell positions are based on a hexagonal grid and it is more reasonable to assume that each cell has three nearest neighbors intead of four. Hence, S4CTEST can be applied for this device. This test is barely 11 times longer than Suk and Reddy's TDANPSF1G, whereas S5CTEST is about 111 times longer.

3.4 TRANSPARENT TEST ALGORITHMS FOR RAMS

A challenge during maintenance testing of RAMs, especially those that are embedded in larger circuits such as microprocessors and digital communication ICs, is that it is often necessary to test the components of a system periodically. In such cases, the test algorithms must not destroy the contents of the RAM. To deal with this problem, Nicolaidis [125, 126] proposed a method of transforming any RAM test algorithm into one that is *transparent*. A test is said to be *transparent* if prior to the test, there is no need for backing up the data stored in the RAM and the test is such that it automatically restores the initial contents of the RAM if the RAM is fault-free. RAMs that are equipped with BIST circuitry running these transparent algorithms can thereby be tested periodically in a running system without the requirement that the stored data be first copied elsewhere during the test and then written back once the test has completed.

Functional Fault Modeling and Testing

The basic idea behind Nicolaidis' method [125] is that a RAM test can be made transparent if we ensure that the data in each cell is complemented an even number of times during the application of the test. The transparent test, therefore, consists of two phases: in the first phase, only read operations are applied to the RAM and a data-dependent signature is computed; in the second phase, both read and write operations are performed while another signature is computed. The RAM is regarded as fault-free if the two signatures match. The detailed description of this approach together with a comprehensive discussion of BIST hardware for various transparent tests, is given in Chapter 5. Nicolaidis [125, 126] proved analytically that if the fault universe satisfies a certain symmetry property, then the fault coverage of the transparent test will be the same as that of the original, non-transparent version.

3.5 FAULT SIMULATION OF RAMS BASED ON MEMORY CELL LAYOUT

Many researchers [60, 131] have studied ways to perform functional fault simulation of RAM test patterns. The basic objective for fault simulation is to observe the fault coverage of well-known tests without requiring large amounts of data to be stored and processed.

Oberle and Muhmenthaler [131] have reported a fault simulator for DRAMs. Their fault models are based on a realistic analysis of DRAM failures and include complex pattern sensitivity mechanisms. Also, the simulator can simulate several defects simultaneously and can be easily adapted to SRAMs and other storage devices.

This approach is based upon studying the distribution of faults occurring during the manufacturing process of ULSI circuits. Faults can be particle inclusions, faulty structuring or residues of photoresist resulting in physical defects like line interruptions and short circuits.

As we know, the architecture of a DRAM consists of two major components: control and interface circuits, and the cell array. The control and interface circuits cause less faults because they are of larger size than the cell array and can be easily detected. However, faults within the cell array need special consideration.

Table 3.8 Defect classes in DRAMs; courtesy [131]

Faults resulting in Cells Mutually Interfering	Faults that don't result in Interfering Cells
Short-circuit between adjacent word lines	Short-circuit between word line and capacitor plate
Short-circuit between adjacent bit lines	Short-circuit between bit line and capacitor plate
Short-circuit between word and bit lines	Interrupted word line
Short-circuit between two or more cells	No connection between bit line and cell
Interrupted bit line	No connection between word line and transfer gate
	Leakage current through cell capacitor

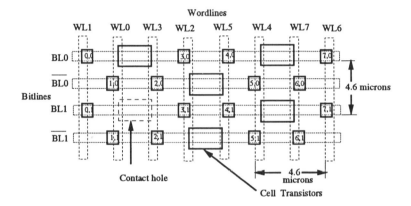

Figure 3.23 A DRAM cell matrix array (4 Mbit); courtesy [131]

3.5.1 Faults and defects modeled by Oberle and Muhmenthaler's simulator

Oberle and Muhmenthaler consider two categories of faults. The first category includes faults that result in mutual interference between cells, the second one includes the remaining faults. These faults are shown in Table 3.8 [131].

These faults are often manifested as single and double-bit errors and faults on bit and word lines. The ones shown on the left hand column are expressed as coupling faults between cells. The ones on the right result regardless of the test pattern used and can be regarded as stuck-at faults for special test conditions. There are well-known test patterns for detecting SAFs.

A simple DRAM cell array layout is shown in Figure 3.23 [131]. A memory cell is bounded by three immediate neighbors, such that the addresses of adjacent word lines differ by 1 or 3, as shown.

Functional Fault Modeling and Testing

It should be noted that in a physical cell array, it is not always the case that the logic value 1 is internally stored as 5 V and the logic value 0 as 0 V. However, the voltage values applied at the input are 5 V for 1 and 0 V for 0. Depending on the location in the cell field, the level applied at the memory input can be internally stored as either a 0 V level or a 5 V level. This mapping from voltage level to logic value is known as *data topology*.

3.5.2 Characteristics of Oberle and Muhmenthaler's fault simulator

The fault simulator described by Oberle and Muhmenthaler performs several different functions. It can simulate all built-in faults in parallel with one test pattern run. Therefore, only a small sector of the cell field is simulated. During the simulation, the individual memory cells are written and the simulator responds to any fault occurrence by activating further cells in accordance with the fault mechanism being modeled — 1 being high and 0 being low.

The simulator outputs the following pieces of information to the user:

- *failbitmap* is a two-dimensional matrix that contains the values Fail-0, Fail-1, and Fail-0&1 per read cycle and the final result. Fail-0, Fail-1 and Fail-0&1 correspond respectively to the following: a 0 could not be stored in a cell, a 1 could not be stored in a cell, and at least one Fail-1 and one Fail-0 are found in relation to the cell

- *fault coverage* per read cycle according to the above statistics

- *fault coverage* according to defect types

- *minimum fault coverage* for 100% defect detection

The simulation performed by the DRAM simulator has the following three independent components:

1. **Model of the cell array with description of the cell array layout:** The memory cell array has a regular structure and consists of several identical or similar blocks. The regular structure is built up from a small basic element, for example, 4 × 8 cells. The configuration is repeated after every 4 or 8 cells in the horizontal or vertical direction. All faults are

simulated by a relatively small cell array for example, 32 × 32 cells. All faults are simultaneously simulated at their representative locations. The size of the cell array depends on the number of simultaneously simulated defects. Faulty cells with different defects must not intersect.

2. **Description of fault models for specified layout, circuit and induced faults:** The simulation considers faults at all possible locations in the cell array. Since the same configurations are repeated regularly we need to look at the different forms in which a fault can occur. The failbit map gives the locations for the following categories of defects: (*a*) pass cell, (*b*) short between cells, (*c*) short between two word lines, fail-0 and fail-1, (*d*) short between two word lines, fail-1 only, (*e*) short between word and bit lines in the bit line section, (*f*) short between word and bit lines in the word line section, and (*g*) interrupted bit lines.

3. **Test Pattern Programming:** The test pattern, which is programmed as an algorithm (note: this algorithm simulates the mechanism of test application by an external tester), requires a computing time proportional to the number of cells and the number of read and write operations per cell. As with the tester, some frequently used basic settings for the cell array (for example, all-0 or all-1, chessboard pattern, and others) can be selected and the test pattern sequence can be simulated for logic values or voltage level (taking into account the data topology). Different kinds of test pattern, some of which are march tests, have been examined in [131].

The main advantages of simulation are: no tester time or silicon real estate is necessary, test patterns can be easily modified and resimulated, and all defects can be simulated in parallel with one test run. Besides, the procedure is pretty fast (taking only a few minutes for a 4 Mb DRAM). The number of fault models can be increased to include further defects which occur less frequently.

3.5.3 Defective hashing

Fault simulation of embedded memories is often found to require large amounts of storage. Huisman [60] discovered a technique that drastically reduces the storage required for fault simulation. The proposed technique allows the amount of storage to be decided either at compile time or at load time, and can almost always be scaled down to the available storage at the expense of only a small decrease in the fault detection (or *exposure*) probability.

Functional Fault Modeling and Testing

For a memory with separate read and write ports (i.e., a multi-port memory) a fault that affects the write port may be exposed (i.e., detected) if the contents of the memory become incorrect and are subsequently read. Since a fault simulator has no prior knowledge of the test algorithm, it should maintain a complete copy of the memory for the faulty circuit, in addition to a complete copy of the memory for the fault-free circuit. When only the read port is affected by a fault, incorrect data may appear in the read buffer but the contents of the memory will not be affected. Therefore, no separate copy of the memory is required for such a fault.

Faults can affect the memory in many different ways. When a fault affects the read port, the memory is simulated as if it were fault-free, but with a different enable input or address vector than in the fault-free design. The manner in which faults impact the write operation for multi-port memories is more complex. To simplify the discussion, Huisman named the fault-free design as the *g.machine* and the faulty design as the *f.machine*. In an embedded situation, both the address and the data inputs may be faulty. Depending on whether the write port of the *g.machine* and/or the write port of the *f.machine* are enabled or not, and whether the address and/or data are good or faulty, we have 16 possible fault effects on the write port of a memory. These 16 possibilities are shown in [60]. For example, if we have the 0/1 combination for the write enable of the *g.machine* and the *f.machine*, that is, the write enable of the *f.machine* is inadvertently turned on, data will be written into an address that may not be the same as the *g.machine*'s address, though data should not have been written at all. The 16 possibilities can be encoded in a variable in the simulation program. For each of these 16 possibilities, the program may try to determine the fault effect by examining the corresponding write operation.

Defective hashing is a process of efficiently recording the addresses of the faulty circuit using a hash table-based storage technique. Each record has a key consisting of an address, a memory name, a fault name, and a body containing the data stored at that address. In many applications, fault simulation is done to establish that faults are exposed in a sequence of tests, that is, by at least one of the patterns in the sequence. In such cases, it is not necessary to identify all possible ways of exposing faults, just one will suffice. Moreover, it is often not even necessary to ascertain that all faults have been exposed — we only need to expose a sufficiently large fraction. It is therefore helpful to relax the fault simulation requirements and require a fault to be exposed as soon as the simulator identifies a pattern that exposes the fault. Such a *pessimistic* approach eliminates the problems with collisions and hash-table overflows. The general strategy is simply to overwrite on a collision, that is, to use a new item to replace an item that is already present. This has the effect of missing certain possible exposures, resulting in a decrease in the fault exposure probability. However, it causes efficient retrieval of defective addresses, and as a result, many memories can be handled simultaneously, usually within the constraint of existing storage. The disadvantage of this approach is a possible decrease in the calculated test coverage. However, this decrease can be adjusted by the user, as specified in [60].

3.6 CONCLUSION

We have presented both conventional and modern functional testing techniques for RAMs. The basic difference between them is their fault modeling approach. Conventional algorithms use simplified fault models such as stuck-at and transition fault models. These fault models do not take into consideration the physical layout of the memory chip. The recent approaches are more powerful in that they consider realistic fault and defect models based on a close examination of the cell layout. The latter is thereby more relevant to modern-day RAM test engineers, as new processing and device types are associated with new defect mechanisms. Defects in the cell array or the peripheral circuitry may or may not produce faults but require some corrective measures to be taken, such as cell repair or replacement. A test that achieves a very high defect coverage would also achieve a very high fault coverage, but not vice versa.

Conventional fault models and tests often cause a low fault or defect coverage because they do not use knowledge regarding the physical layout of the chip. They are useful for classifying faults in a more generic sense, but the present generation of high-density, multi-megabit RAMs built with various new and innovative technology, require more comprehensive layout-based testing approaches.

Functional Fault Modeling and Testing 153

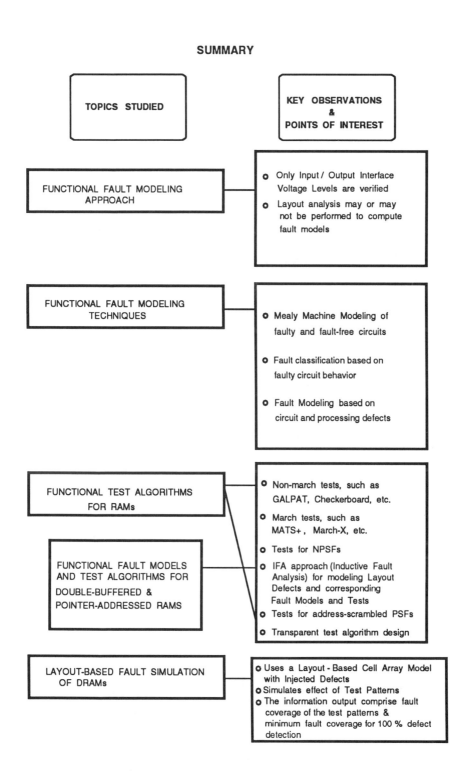

Table 3.9 Relationship between functional faults and reduced functional faults

Functional fault	Reduced Functional fault
Cell stuck	SAF
Driver stuck	SAF
Read/write line stuck	SAF
Chip-select line stuck	SAF
Data line stuck	SAF
Open in data line	SAF
Short between data lines	CF
Crosstalk between data lines	CF
Address line stuck	AF
Open in address line	AF
Short between address lines	AF
Open decoder	AF
Wrong access	AF
Multiple access	AF
Cell can be set to 1 but not to 0	TF
Pattern-sensitive interaction between cells	NPSF

3.7 PROBLEMS

3.7.1 Worked-out Example

1. Explain the relation between functional faults and **reduced** functional faults, with the help of a simple table.

    (**Sample Answer Sketch:**) A functional fault reduction is performed by mapping various functional fault types into some generic fault classes. Table 3.9 gives an example.

3.7.2 Exercises

1. Prove that March A is a complete and irredundant test for linked idempotent coupling faults. Prove that Butterfly is an $O(n \lg n)$ test.

2. Display, in a tabular fashion, the fault coverages (with respect to AF, SAF, TF, CF, sense amplifier recovery, write recovery, read access time, and operational faults), complexities and actual running times (in seconds, minutes, etc.) of the following tests when applied to a 1 Mb SRAM:

    Zero-One, Checkerboard, Walking 1/0, GALPAT, GALROW, GALCOL, Sliding Diagonal, Butterfly, MATS, MATS+, Marching 1/0, MATS++, March X, March X, March C-, March A, March Y, March B, MOVI, 3-coupling and Paragon. For illustrating fault coverages, use the letters d, l, ds and ls to denote detection, location, detection of some faults, and location of some faults, respectively.

3. Design a MATS+ test for a 4-bit wide memory (that is, word size = 4 bits), and analyze its fault coverage. What is the advantage of having an initial checkerboard data background ?

4. What is the difference between an operational model and a functional model ? Describe the operational model used by Paragon. Explain the **Paragon sequence**. How can Paragon be adapted for a word-oriented memory ? Give the complexities of a bit-oriented and a word-oriented Paragon-based approach.

5. Design a complete and irredundant march test to detect AFs, SAFs and TFs that disallows down transitions in single memory cells. Prove that your algorithm is complete and irredundant.

6. A technique often employed in functional testing of RAMs is to use the binary value of the address of a memory word as the test data. Derive formal fault models and tests based on this approach, and analyze the fault coverage. See [141] for a hint.

7. Prove that a test which detects any fault from the fault model consisting of every toggling (transition) fault to any cell, contains at least $2n^2 - 2n$ read operations and $3n$ write operations. (Note: such a fault model is known as a general toggling fault model).

8. Prove that a test which detects any fault from the fault model consisting of every coupling fault to any cell, contains at least $2n^2 - 2n$ read operations and $5n$ write operations. (Note: such a fault model is known as a general coupling fault model).

9. Propose algorithms for general coupling and general toggling to approach their lower bound complexities as closely as possible.

10. Using the theory of finite state automata, design a systematic approach to **verify** a functional test for RAMs, that is, given the test algorithm and a target set of faults (claimed to be detectable by the algorithm), design a RAM model to simulate the test, verify whether the desired set of faults is detected or not by the test; if not, compute the fault coverage. Describe, using pseudocode, the verification strategy.

11. How can you perform a guided probe-based test for troubleshooting a memory board comprising multiple memory chips? Design fault models and functional tests for memory boards. Describe the probe hardware and software. Describe how the board information is conveyed to the guided probe test software, and describe, using pseudocode, how pattern generation and signature analysis are performed. For a hint, see [111].

4
TECHNOLOGY AND LAYOUT-RELATED TESTING

Chapter At a Glance

- Motivation for Technology and Layout-Oriented (Parametric) Testing

- Faults in Bipolar Multi-Port RAMs and their Testing

- Faults in MOS SRAMs and their Testing
  -- SDD, Access Delay, Defect Signature, Current Testing

- Faults in GaAs HEMT SRAMs and their Testing
  -- Resistive Leakage Currents, Read and Write Errors

- Faults in MOS DRAMs and their Testing
  -- Sleeping Sickness, Bit-Line Voltage Imbalance, Bit-Line to Word-Line Coupling, Single-Ended Write, Multiple Selection

4.1 INTRODUCTION

In the literature on functional [1] and electrical testing [165], no distinction is made between fault models that study only the behavior at the I/O interface of the memory device, and those that also perform a more detailed examination of the internal layout, circuit design, and the processing technology of the memory device. This chapter is an attempt to highlight such fault modeling and testing techniques.

Functional and electrical tests for RAMs are typically designed from simple fault models, such as stuck-at, coupling, and slow-access faults. These models describe faulty behavior at the I/O interface without a thorough analysis of the chip layout and technology. The reason these simplified fault models are popular is because the tests for these faults are found to have a satisfactory *fault coverage*. However, with the advent of deep submicron processing technologies, there exist peculiar *technology and layout-related defects* (also called *parametric* defects) that require a more comprehensive testing approach because such defects may or may not give rise to functional or electrical faults. However, these defects cause unreliable future operation of the memory device and should be detected by a memory test. Parametric defects cannot be tested by purely functional or purely electrical testing because they often cause peculiar circuit behavior that requires a combination of electrical and functional testing, and both voltage and current measurements. Examples of faults that may be produced by such defects are: (*a*) static data loss, (*b*) abnormally high currents drawn from the power supply, (*c*) bit-line voltage imbalance in DRAMs, and others. Fault modeling and test development for such faults require a thorough investigation of the processing technology and circuit techniques. In this chapter, we shall examine ways to model and test such faults [18] by first dividing such faults into three categories based on manufacturing technology — bipolar, MOS and GaAs.

4.2 MOTIVATION FOR TECHNOLOGY AND LAYOUT-ORIENTED TECHNIQUES

During the last two decades, the increasing complexity and speed of RAMs has necessitated rapid advances in testing techniques. The complexity of testing and cost of test equipment have also increased considerably. The major driving

Technology and Layout-Related Testing   159

force for manufacturers choosing a particular processing technology has been high manufacturing yield, reliability and low processing cost. In the early days of memory testing, only simple electrical measurements were performed at the I/O interface. As speeds of devices increased, the need was felt for rigorous speed tests such as access time test. These tests came to be classified as AC parametric tests. With increasing complexity, it became necessary to perform algorithmic functional testing as well as electrical testing. Electrical and functional behavior are often correlated and their respective testing techniques are often applied together. Both these techniques use simple fault models, such as stuck-at, coupling, and delay faults. The tests designed using these simple fault models have been found to achieve a good fault coverage but not necesarily a good *physical defect* coverage. As time passed, the need was felt to devise new tests for defects and faults by examining the circuit layout and the processing technology. It has been found that a large number of defects that might translate into faults are technology-specific. A Gallium Arsenide (GaAs) SRAM, for example, has a different set of canonical fault types than a MOS SRAM, because of some unique processing and device characteristics. Layout defects due to process variations must be thoroughly analyzed for a proper understanding of testing requirements.

There are three popular techniques used in layout-based fault modeling and testing. Often these techniques are used together. They include:

- **Simulating the effect of layout defects**: Defects such as broken pullup transistors and floating gates in the RAM cell are analyzed and modeled accurately. A common approach is the **Inductive Fault Analysis** (IFA) technique that models shorts and opens as spots of missing metal or extra metal respectively, and injects them into the layout. The functional aspects of IFA, with special emphasis on the design of the popular '$9N$' and '$13N$' tests, are described in the previous chapter; in this chapter, we shall perform a more rigorous examination of this technique. The equivalent circuits for the device layout with and without defects are simulated using Monte Carlo or SPICE simulation to obtain the electrical characteristics of the defects, and the fault model, or the *fault-defect vocabulary* is derived. Functional and electrical tests are then designed to target the faults included in this fault model.

- **Current measurements**: Physical measurements of the quiescent and dynamic current drawn from the power supply (called I_{DDQ} and I_{DD}, respectively) is also a very popular technique. Defective RAM cells are typically associated with an abnormally large value of these currents.

- **Simulating the effect of varying circuit parameters**: This is done by designing a simplified equivalent circuit model to predict fault effects of varying circuit parameters (due to process variations or design-related errors) and verifying these using analog circuit simulators such as SPICE. This study leads to the design of functional and electrical tests for parametric defects. This technique is very useful in measuring the limits of correct operation of RAMs.

We shall now describe some past work in this area on testing bipolar and MOS RAMs and then give a detailed account of our own work in the area of fault modeling and testing of Gallium Arsenide (GaAs) SRAMs. The tests that are discussed here use layout and circuit-related information. Some of these algorithms are modified functional algorithms, that is, they verify the logic (0/1) level of each memory cell or bit/word line under conditions that try to maximize the likelihood of occurrence of the fault(s) (for example, the parametric checkerboard test for MOS DRAMs). Others are electrical tests that measure the actual values of circuit parameters (for example, current tests and delays), and some are combinations of various testing approaches.

4.3 FAULTS IN BIPOLAR (MULTI-PORT) RAMS AND THEIR TESTING

Some possible 'point' defects (i.e., physical process defects) for bipolar RAMs are examined in this section. These defects can be divided into two groups: those that affect the manufactured transistor, and those that affect inter-transistor wiring [170]. The typical process defects for bipolar RAMs include:

- Isolation opens
- Emitter-base shorts
- Collector-isolation shorts
- Collector-emitter leakage
- Resistor opens
- Resistor shorts
- Schottky diode leakage

Technology and Layout-Related Testing 161

- Collector contact integrity
- Emitter contact integrity
- Base contact integrity
- Sidewall integrity

The typical personalization defects for bipolar RAMs consist of:

- First metal opens
- First metal shorts
- Second metal opens
- Second metal shorts
- Third metal opens
- Third metal shorts
- Fourth metal opens
- Fourth metal shorts
- First-second metal shorts
- Second-third metal shorts
- Third-fourth metal shorts
- Contact integrity

The faults corresponding to most of these defects are of two types: shorts between two nodes, and open contacts. Vida-Torku, et al. from IBM [170] have used the circuit simulator ASTAP to simulate these defects on a cross section of a three-port memory with two independent read ports and one write port. They used the technique of defect injection and simulation and then generated test algorithms for the functional faults caused by these defects. In their scheme, the defects are injected in a bottom-up approach, and comprise *component defects* and *intercomponent defects*. Most defects in the memory cell result in a stuck cell — the output of the cell driving the sense amplifier is stuck at one or at zero. There are two other fault types — a defective cell dominates the output, that is, the output of the memory is always a function of the defective cell; and a defect at a cell causes the memory output to be a function of the good cells on the same bit line. Decoder defects result in the three conventional decoder faults: no memory word is read or written, multiple words are selected, and an access is made to a wrong word in the memory. Multi-port memories have two or more independent ways to address on a single memory array. A port denotes an address group and a data-in/out group and the controls needed to manage the grouping. These ports are referred to as Read or Write ports.

For each of the defined failure modes, faults that affect one or more than one ports are considered. These fault types are described in the sections below.

4.3.1 Single-port fault models

The single-port fault models are **stuck cells, no select on write operation** (i.e., write decoder fault), **no select on read operation** (i.e., read decoder fault), **write multiple words, read multiple words,** and **defective cell dominating the output**. Their test requirements are briefly described below:

1. **Stuck cells**: To detect a stuck-at-1 (0), the cell must be initialized with the value 0(1), and this value must be read before the next write operation.

2. **No select on write**: This causes a memory word to not be selected during a write operation on that word. However, a read operation on the word will be normal. A necessary and sufficient test for this fault is to write a randomly generated pattern into the word, followed by reading the word. To overcome the tiny probability that the contents of a non-(write)-selected word could be exactly the same as the random test pattern applied to the selected word, the complement of this pattern is also written and read back.

3. **No select on read**: This is a read decoder failure whose detection depends on the technology and special design for testability (DFT) features present in the memory design. One requirement for detecting a no-select condition on read is that the sense amplifier design should discriminate between logic 0, logic 1 and *no-select*. Whether or not this condition is sufficient depends on the technology. In some bipolar current switch designs, the output of the sense amplifier is a function of the relative number of 0s and 1s. Hence, if there are more 1s (0s) than 0s (1s), the output is 1(0). The cells supply current equally down the bit line into the sense amplifier. If there are an equal number of 0s and 1s, the three-valued logic feature may be valid.

4. **Write multiple words**: Suppose i and f are the words being addressed and written to as a result of this failure. For detection:

   - f should contain the complement of the pattern stored in i.
   - A march test of the form $\{\Uparrow (w0), \Uparrow (r0, w1), \Uparrow (r1, w0)\}$ is needed, followed by another march test in descending sequence of addressing (i.e., $\Downarrow$).

5. **Read multiple words**: Detection of this fault is also dependent on the technology or special design of the circuit. Suppose i is a word being written and f is the word selected because of the fault. The necessary and sufficient conditions for detection are: words i and f should have complementary data, and the sense amplifier should be able to discriminate between reading 0 and 1 on the same bit line simultaneously and reading normal 0 or 1. The first condition thereby forces both logic 0 and logic 1 into the sense amplifier, and the second condition is needed to detect it. This condition is not valid in most differential sense amplifiers. Selecting multiple words for reading producing input values that are almost identical; resulting in a reduced noise margin. There are two possible design options: first, the differential amplifier can be design to favor one state over another, alternatively, it can monitor the word lines to force a known state on the output if multiple word lines are active.

   Vida-Torku, et al. [170] propose adding test circuits to every word select line and tying them together before gating them into the sense amplifier. The output of the test circuits must be 1 if one word line select is active, and 0 if more than one word line select are active. The output can then be used to control one of the data output pins of the RAM by gating the sense amplifier to it. These schemes transform a potential delay or noise-margin fault into an easy to detect stuck fault.

6. **Defective cell dominates output**: The output of a bit line depends on the value of the defective cell. The cell is always supplying current to the sense amplifier, whether or not it is accessed. This fault can be detected using tests that detect the read multiple words fault [170].

4.3.2 Multiple-port fault models

If there is a metallization short between decoder lines of two ports, then the resulting failure modes can be detected by multi-port testing. Most of the test conditions for detection are influenced by the technology. The principal multi-port faults observed are as follows:

1. **Shorts between two read ports**: Suppose that the short between two read port decoders places them both in read mode. Because of layout differences, it is assumed that a short between ports can only occur on decoder lines selecting the same word. Moreover, it is also assumed that during test, memory read is done one port at a time. In addition to the conditions for detecting a multiple read in a single port, we need to ensure that the two read ports do not both access a word f simultaneously during the test. If read port 1 points to a word f and read port 2 to a word g, and if $f \neq g$), then port 2 will be forced to have both words f and g chosen because of the fault, and this situation can be detected by the conditions of the *read multiple words* fault. Hence, both read ports must be independently controlled during the test.

2. **Other shorts**: Depending on the layout, it is possible to have read to write and write to write port shorts. These shorts can be transformed into single-port failure modes if we ensure, as before, that there is no memory contention between the ports.

4.3.3 Test generation for embedded multi-port memories

The various failure modes can be detected by running an alternating pattern P of 0s and 1s ($P = 0101...$) and then its complement on the address space consisting of n bits. First, the background pattern P is written in the address space from 0 to $n-1$. Then, each location is read to verify that P is still intact, and the bit-wise complement of P is written back. The process is repeated for all the memory words, and is followed by a *reverse ripple* in which the address space is traversed in the reverse sequence. Multiple ports are used to perform read and write operations for this test. For example, when a word is read and its complement written back, one read and one write ports work in combination. As long as there is no conflict between multiple read ports (trying to access the same location), multiple select faults and shorts are detected.

In a level-sensitive scan environment (LSSD), the scan latches may be used to test both the memory array and the surrounding combinational logic. The tests for multi-port RAMs described above have to be modified because they are defined in terms of the memory input, not in terms of the logic surrounding the memory. This modification involves signal propagation and backtracing through combinational logic, which are beyond the scope of the book.

4.4 FAULTS IN MOS SRAMS AND THEIR TESTING

MOS SRAMs also exhibit a variety of faults that are technology- and layout-dependent. A variety of approaches have been used to study and test these faults. Some of these approaches are based on modified functional test algorithms and others are based on a study of the actual electrical characteristics of faulty SRAM layouts. These faults and their tests are described below.

4.4.1 Testing for static data loss

SRAM devices can lose data stored in them because of leakage currents that can change the state of the cell when the pull-up device is defective or broken such that it forms an open circuit. In SRAMs, the function of the pull-up device (which is often a depletion mode NMOS with gate shorted to source, and should, theoretically, act as a resistance) is to supply the leakage current to hold the cell in its state. Static data loss in SRAMs is analogous to sleeping sickness for DRAMs.

Testing for static data losses in SRAMs is a difficult task. The difficulty arises because static data losses typically start to occur after long periods of time (of the order of seconds or even more) and that would result in impractical delays in testing. Testing them under worst case conditions (called 'burn-in') to accelerate failure is also not very reliable because it is not known precisely how chips age under burn-in conditions with regard to static data losses. Dekker (1988,1990) [34, 36] has described how to detect static data losses in SRAMs with 100 GΩ resistors as pull-up devices. He has reported that a broken pull-up resistor causes a static data loss in about 100 ms. This defect can be modeled using the IFA technique in which shorts (opens) are represented as spots of extra (missing) metal. Computer simulation or manual analysis using equivalent circuits can then be used to design a suitable fault model. Dekker has proposed the test IFA-9 based on equivalent circuit analysis. This test employs a march test plus two delays of 100 ms each.

A technique known as *soft-defect detection* (SDD), accomplished at room temperature is described by Kuo, et al. [78] — this provides a complete data retention test of a CMOS SRAM array. The SDD technique has two components — an *open circuit* test that checks for connectivity of the *p*-type load transistors, and a *cell-array* test that carefully monitors the standby array cur-

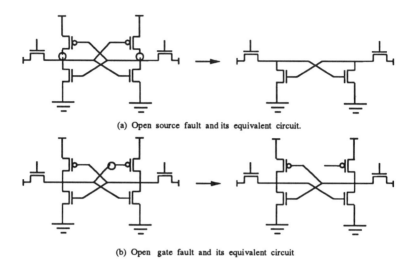

Figure 4.1 Fault model used by the SDD technique; courtesy [78]

rent to detect abnormally high leakage currents. The following sections describe this technique.

Fault model used by the SDD technique

An SRAM cell consists of a pair of cross-coupled inverters. In the SDD technique, each of the two cross-coupled inverters is tested independently. This model is shown in Figure 4.1.

The problem of treating two cross-coupled inverters independently is as follows: the parasitic effect of one inverter can clamp the input 1 level from reaching the switching point of the other inverter. The clamping can be overcome by reducing the ratio of the conductance between the driver and the access transistors. This can be achieved at the device fabrication level by decreasing the size ratio between the driver and access transistors, or at the circuit level by reducing the gate voltage for the driver transistor with respect to the access transistor.

Simulation studies indicate that lowering the array supply voltage to 2.9 V would cause an acceptable switching time for both input 0 and input 1 for a fault-free SRAM cell. Accordingly, the targeted layout and circuit parameters

Technology and Layout-Related Testing

used are: array supply voltage: 2.9 V, threshold for passing or failing the SDD open-circuit test: 150 ns. They used an effective channel length of 0.95μm for all cell transistors, and a gate voltage of 5 V for the access transistors.

The SDD test strategy

The SDD test strategy consists of (a) an open-circuit test, for testing the right as well as the left inverters; and (b) a cell-array current test, for fine-tuning the above test by performing a built-in or external test of the array current with all-0 and all-1 data backgrounds.

1. **SDD Open-Circuit Test:** In this test, each inverter is provided with a 0 (1) input and its output verified simultaneously for a 1 (0) level. This requires simultaneous read and write operations to the cell under test, which is done using a special circuit to be described shortly. An open circuit at the gate or the drain of the load might result in an abnormally long time for the output to respond to the input, if at all it responds. However, even if a cell passes this test, it does not necessarily follow that it is not defective; simulation studies indicate that a defective cell with a floating-gate P-channel load transistor can actually pass this test when the effective conducting P-channel current is between 50 and 290 μA for the prescribed 150 ns switching time. Therefore, the cell-array current test, described next, is used in conjunction with the open-circuit test.

2. **Cell-Array Current Test:** With a background of all-0 (all-1) for the left (right) inverters, a floating gate transistor of a faulty SRAM cell that passes the open-circuit test might produce an abnormally large standby current, which may be greater than the normal array leakage current. An internal current detection circuit is used for the SDD cell-array current test.

Circuit design

The circuit diagram for the SDD open-circuit test is shown in Figure 4.2. The circuit is described below.

This circuit can perform simultaneous read and write operations — this is necessary since the switching response of an inverter is verified by applying an input signal (write) and simultaneously detecting the output signal (read). This can be achieved if the right (left) output transmission gate and the left

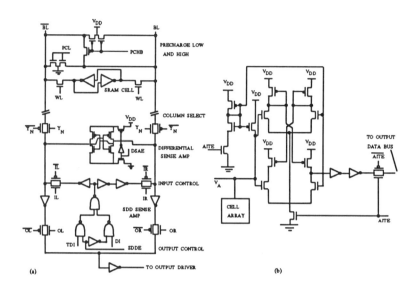

Figure 4.2 Circuit technique for (a) the SDD open-circuit test, and (b) the cell-array current detection test; courtesy [78]

(right) input transmission gate are simultaneously activated. One of the two input transmission gates ($IL/\overline{IL}$ or $IR/\overline{IR}$) and the corresponding output gate ($OR/\overline{OR}$ or $OL/\overline{OL}$) are activated using a mux and inverters as shown in the figure.

During the open-circuit test, the bit-line pair has to be precharged high and low (depending on whether the input is coming from the left or the right), so that switching times under both conditions are verified. This necessitates the incorporation of a special precharge circuit as seen in Figure 4.2.

It should be recalled that the open circuit test considers each inverter of a cross-coupled pair independently. Since during this test, we are not performing a normal read operation (i.e., precharging to high both the lines of the bit-line pair initially and then sensing differential voltage at the outputs of the cross-coupled inverters), the sense amplifier has to be completely isolated from the bit-line pair. This isolation is accomplished by gating a control signal called $DSAE$ to the sense-amplifier.

The array supply voltage is clamped to the prescribed 2.9 V from the usual 5 V using a special circuit, as shown in Figure 4.2. For the second part of

Technology and Layout-Related Testing

the test strategy (the cell array current test), a current-mirrored differential sense amplifier compares the cell-array current to a reference current. This current detection can be performed at high speed if the array power node V_A is precharged to the switching point of the differential sense amplifier, and then disconnected from the power supply V_{DD} during the array current detection.

Effectiveness of the SDD scheme

The SDD circuitry can be fully built-in with only a 2% increase in the module area for a 16K array, and this percentage decreases for a larger array size. The total SDD test time for the 16K module is about 1.5 ms, with most of the time being spent for current testing. Also, in the experimental 16K module, SDD is found to identify a greater percentage of defective cells than conventional test patterns such as marching 0s and 1s.

4.4.2 Delay time testing for bit-lines in megabit SRAMs

As the density of megabit SRAMs is increasing, the space between adjacent $bit/\overline{bit}$ lines is decreasing steadily. This has resulted in increased inter-line delays, caused by increased inter-line capacitance. One solution implemented for dynamic RAMs (DRAMs) has utilized the twisted bit-line technique to cancel the influence of coupling capacitance between bit-lines. For static RAMs (SRAMs), Kinoshita, et al. [70] have studied the characteristics of delay time due to parasitic coupling capacitances between adjacent bit-lines in several generations of megabit SRAMs. Such coupling capacitances disturb the sensing operation after the storage node data has been transferred to the bit line pair during read operation. Kinoshita, et al. [70] have measured the access time in a 1 Mb SRAM using three different test patterns which set up best case, worst case and average case conditions for delays. The amount of coupling noise depends on the memory cell data. The best case corresponds to a situation where physically neighboring bit (or $\overline{bit}$) lines from two different cells swing in the same direction, both either to 1 or 0, as this would cause little bit-line voltage difference, and consequently, little coupling noise between these neighboring lines from different bits. The worst case corresponds to a situation in which physically adjacent bit (or $\overline{bit}$) lines from two different cells swing to opposite value, as this would cause a large bit-line voltage difference between these lines. The average case is when one neighbor of a reference cell has the same value (as that of the bit-line for the reference cell) for the adjacent bit-line,

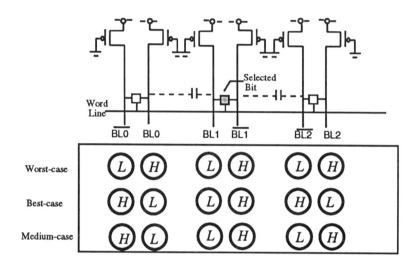

Figure 4.3 Test patterns for Layout-Based Delay Time Testing; courtesy [70]

and the other neighbor has the opposite value for its adjacent bit-line. These test patterns are illustrated in Figure 4.3.

The access time versus supply voltage (V_{CC}) of a 1 Mb SRAM is shown in [70]: supply voltage is varied and the corresponding variation in access time from the worst-case to the best-case patterns are recorded. For high supply voltages, the minimum access times possible are found to be lower than for smaller supply voltages. The ratio of propagation delay time to access time is found to vary between 0.8% in the worst case and 1.7% in the best case.

The experimentally observed differences between the best and worst case access times are confirmed using a distributed transmission-line simulation model. This model is illustrated in Figure 4.4.

Simulation results based on the data for the resistance and various capacitance values (namely, inter-metal capacitance for first metal (Al) C_L, first-metal to poly C_B) from three generations of processing technology, namely 0.8 μ (1 Mb), 0.5 μ (4 Mb) and 0.35 μ (16 Mb), are displayed in [70]. The conditions used in this simulation are summarized in Table 4.1. During simulation, the read-out voltage difference is modified to compare a wide bit-line swing (with the older technologies) with narrower bit-line swings characteristic of recent SRAMs. For a smaller bit-line swing, conductivity of the bit-line load is enhanced and for a larger bit-line swing, the conductivity is reduced. It should be noted that both

Technology and Layout-Related Testing

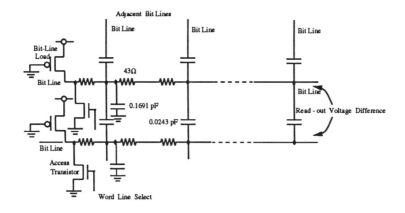

Figure 4.4 Bit-line model used in simulation; courtesy [70]

Table 4.1 Simulation parameters for three generations of process technology; courtesy [70]

Proc.						# Memory Cells	Cell Size	Bit-line Load L/w		Power Supply
			1024			2048	(μ^2)		(μ^2)	(V)
	CL (pf)	CB (pf)	$R(\Omega)$	CL (pf)	CB (pf)	$R(\Omega)$		$\Delta V =$ 200mV	$\Delta V =$ 30mv	
0.8-μ 1 Mb	0.194	1.353	683	0.388	2.705	1365	49.30	1.2/6.0	1.2/30	5.0
0.5-μ 4 Mb	0.187	0.853	638	0.374	1.718	1277	18.96	0.9/4.0	0.9/20	5.0
0.35-μ 4 Mb	0.238	0.412	497	0.476	0.824	995	7.29	0.6/2.6	0.6/13	3.3

the resistance and the total capacitance ($C_B + 2C_L$) of a bit-line reduce as the processing technology advances for the same number of cells connected to the bit line, since the cell size is scaled down.

4.4.3 I_{DDQ} testing using the Inductive Fault Analysis (IFA) approach

This testing paradigm also uses a test metric dependent on the detection of defects, rather than faults. This test monitors the quiescent power supply current (denoted by I_{DDQ} (for CMOS ICs), which should be very small under normal conditions) for each test vector, and detects any defect that causes abnormally high values of this current. For example, a short or an abnormal leakage causes a state dependent I_{DDQ} which is typically a few magnitudes larger than the steady-state quiescent value under normal conditions, the reason being that both the n- and the p-transistors will be turned on. The effectiveness of I_{DDQ} testing for SRAM chips especially for short circuit and open circuit defects has been demonstrated by Meershoek ([107, 108]), and electrical effects caused by stuck-open faults in CMOS ICs have been tested by Soden, et al. ([152]) using this paradigm.

Traditional I_{DDQ} tests on SRAMs

Shen, et al. (1985) ([148]) have performed a series of experiments that involve the introduction of spot defects in all layers of a memory layout (that of an SRAM chip of 8k 8-bit words, produced by Philips) and have analyzed effectiveness of the I_{DDQ} test method using the Inductive Fault Analysis (IFA) approach. In [107], the results of the analysis are presented in tabular form consisting of 33 groups of artificially injected defects. An I_{DDQ} test has been found to detect the defects of 27 of these groups. Examples of kinds of defects are gate of pull-up/pull-down transistor connected to its channel, and a bit-line shorted to a word-line.

Soden, et al. [152] (1989) have shown that I_{DDQ} measurements can detect stuck-open faults (SOF) in *some* designs, but not necessarily in all, based on experiments with a static CMOS ROM SA3002 (2K × 8). They analyze open metal defects in the drain interconnections of NMOS transistors located in the NOR gates of the address decoders of these ROMs. These open metal defects cause a high-impedance state at the output when certain test patterns are applied (see Figure 4.5). Some defects which have been examined are

Technology and Layout-Related Testing 173

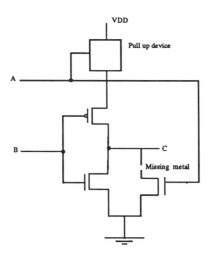

Figure 4.5 A defect causing a high-impedance output for a 2-input NOR gate; courtesy [152]

defective metal patterning (for example, missing metal in a long section on top of the polysilicon and intermediate oxide that ends abruptly at the edges of the polysilicon); this defect is believed to be the result of the failure of the photoresist to adhere in this region of the mask. The chip has been observed to flunk a ping-pong test in which the contents of each memory address are read before and after those of every other address — abnormally high I_{DDQ} values are associated with these failures.

It has been found that the measurement of I_{DDQ} is a very sensitive technique for the detection of stuck-open defects in CMOS ICs. Hence the time constant of the floating node is important for establishing the validity of the I_{DDQ} measurement technique for such defects; the time constant should be small compared to test vector clock periods, otherwise the elevated value of I_{DDQ} will be slow to be detected in the same vector that causes the high-impedance state.

The effectiveness of the I_{DDQ} fault model has been investigated by Meershoek, [108] (1990), who have run the following series of tests to a batch of 7072 devices:

1. **A continuity check**, to detect shorted input or output buffers and identify chips with contact problems;

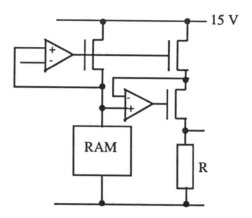

Figure 4.6 An I_{DDQ} monitor; courtesy [108]

2. **Functional tests** IFA-6 (having the march notation: $\{\updownarrow (w0), \Uparrow (r0, w1), \Downarrow (r1, w0, r0)\}$) and IFA-13;

3. **Current tests** for the standby static read and write currents, performed using a parametric measurement unit (PMU); the device is disabled and the quiescent current after a read and write cycle is measured, also a *dynamically* obtained average current through the device during reading or writing is measured. These tests are used to determine which devices need to be subjected to an I_{DDQ} test.

4. I_{DDQ} **test**, done using a special monitor whose circuit diagram is shown in Figure 4.6. The small I_{DDQ} current of the SRAM passes a field-effect transistor (FET), which forms part of a current mirror, controlled by an op amp to guarantee a constant V_{DD} for the SRAM. The current mirror measures the value of I_{DDQ} by dividing the voltage across the resistor R by the value of R. This monitor, built out of discrete components, has been found to have an accuracy of 5 μA and can operate at 200 kHz. The threshold of the detector is typically set at 20 μA. Each cell of the SRAM, containing a logic 1 or a logic 0, has to be subjected to the I_{DDQ} test. The small currents cause large settling times for the A/D converters and consequently, a slow measurement rate.

Table 4.2 [108] shows the results obtained by Meershoek (1990) for an actual I_{DDQ} test.

Technology and Layout-Related Testing 175

Table 4.2 An actual I_{DDQ} test; courtesy [108]

	Number of Devices	Percentage
Total tested	7072	100%
Failed continuity check	371	5.3%
Failed functional, current or I_{DDQ} test	3887	54.9%
Good devices	2814	39.8%
Functional, Current, or I_{DDQ} test failed	Number of Devices	Percentage
Functional only	2060	53%
Current only	0	0%
I_{DDQ} only	5	0.1%
Functional and Current	245	6.3%
Functional and I_{DDQ}	81	2.1%
Current and I_{DDQ}	87	2.2%
Functional and current and I_{DDQ}	1409	36.3%
Total	3887	100%

IFA testing with high-resolution defect vocabulary and I_{DDQ}

Naik, et al. [120] (1993) perform failure analysis of high density CMOS SRAMs using realistic defect modeling and I_{DDQ} testing. They have proposed the improvement of yield by the use of a methodology for rapid and accurate failure analysis, developed at Carnegie Mellon University [85, 86]. This methodology performs realistic defect modeling with the help of a realistic *high-resolution fault-defect vocabulary*. Such a vocabulary consists of multi-dimensional failure representations obtained by simulating both functional and I_{DDQ} testing. In [120], results obtained through the application of this methodology for a high-density SRAM fabricated in a 1.2 μ CMOS process at the Philips IC Advanced Development and Manufacturing Center, Eindhoven, the Netherlands, are summarized.

Introduction of redundant rows and columns (for example, during self-repair of embedded memories) has triggered interest in memory testing capable of indicating geometrical characteristics of defective parts of the memory. A widely used diagnostic approach is the use of bitmaps of faulty cells. A *bitmap* is an image of the memory core with marked defective cells. Use of a bitmap in failure analysis has many advantages and great potential for enhancements which may greatly improve the resolution of defect diagnosability. The traditional bitmap approach has a very serious limitation, namely, poor resolution of the mapping from defects (or signatures thereof) to bitmaps, as explained below.

A traditional approach to failure analysis of memories is to consider a set of easy to predict defects and convert each into a pattern of failing bits in a bitmap, called a *signature*. A typical defect-signature relation may be provided by analyzing a spot of extra metal shorting the bit and $\overline{bit}$ lines of a single column. Thus, an entire column of cells will be marked as faulty. When such a pattern appears in a bitmap, it may be concluded that the probable cause is a spot of extra metal with diameter smaller than 2λ of metal lines plus 3λ spacing between these lines, and larger than one spacing.

Now the spectrum of distinct signatures in the defect to bitmap vocabulary is usually small and incomplete, meaning that not every signature maps to a unique bitmap and actual bitmaps generated during testing may not be found in the defect-signature vocabulary. Hence, ad-hoc methods for builting the defect-signature vocabulary cannot support a high resolution of defect diagnosis.

Naik, et al. in their paper [120] (1993) have built a high-resolution vocabulary by using an Inductive Fault Analysis (IFA) approach. The last step of an IFA approach, namely, evaluation of the result of testing of the defective memory, is the most expensive and the most important in improving resolution of diagnosis. So instead of producing 'black-and-white' bitmaps consisting of the mere identification of faulty and fault-free cells, 'colored' bitmaps that contain information pertaining to full electrical characterization of affected cells are necessary for higher resolution. This study has been conducted on the Philips 64 Kb full CMOS SRAM chip, fabricated in a 1.2 μ twin-tub CMOS process with a p epitaxial layer on a $p+$ substrate using single polysilicon and double metal interconnect. This cell has a standard structure with 4 NMOS transistors, 2 PMOS transistors and 2 polysilicon diodes, as shown in Figure 4.7. The bit and $\overline{bit}$ are formed using second metal while the global rows, power and ground lines are horizontally running first layer metal. The word-line is made of polysilicon and runs horizontally. The cross-coupled connections within each cell are made with buried contacts to both $n+$ and $p+$ diffusions using n and p type polysilicon gate material respectively.

In the horizontal direction, the memory array has 16 blocks that are 32 bits (or 4 bytes) wide; in the vertical direction, there are 128 rows. Addressing of a byte involves (*a*) appropriate row selection (1 row out of 128), (*b*) appropriate block selection (1 block out of 16), followed by (*c*) appropriate byte selection (1 byte out of 4). Each memory cell can be independently accessed. Each block

Technology and Layout-Related Testing 177

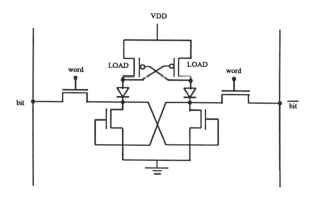

Figure 4.7 One cell of the Philips 64 Kb SRAM; courtesy [120]

has 32 vertical $bit\text{-}\overline{bit}$ line pairs. Read and write operations are performed in the usual manner.

1. **Generation of the Defect-Signature Vocabulary**: This is done using Monte Carlo simulation on the minimum repeatable section of the layout (an 8 × 2 cell portion of the core). The defect simulator, VLASIC has been used to randomly place 150,000 spot defects on the layout and determine what circuit faults, if any, occur. There are two operations which the defect simulator performs: (*a*) extraction of defects that are too small to cause a change in the electrical diagram of the circuit; (*b*) grouping of defects causing exactly the same modification of the circuit (for example, all defects shorting the same two nodes in the circuit are grouped together).

   The vocabulary is formed by considering one representative from each group formed in step (*b*). Defects that have easily predictable signatures on the bitmap are first identified, followed by those for which signatures have to be obtained by extensive circuit simulation. It has been reported [120] that only about 3000 of the 150,000 defects are major ones that have to go through step (*b*) of the two-step operation above, and only about 300 of these have to be analyzed to form the vocabulary. Approximately 10% of these cases have been evaluated by inspection and the rest have been simulated, using the Philips inhouse circuit simulator, described below.

   For simulating electrical testing, a neighborhood of 5 × 5 cells surrounding the defective cell is used in order to mimic all the load effects. Instead of performing the extremely expensive task of applying all 2^{25} possible combinations to the neighborhood pattern, a compromise between cost of simulation and diagnostic information is made by applying test vectors such that only the defective cell and its eight immediate neighbors are tested with all possible combinations of 0s and 1s.

In their experiments, Naik, et al. [120] have generated 67 unique bitmap patterns, each of which corresponds to a certain set of possible defects causing it. The smaller the cardinality of this set, the greater is the resolution of the pattern. Diagnostic resolution has been found to be improvable if the number of possible defects causing a certain pattern can be narrowed down using current (I_{DDQ}) measurement simulations.

2. **Experimental Results:** The method described can be used to associate bitmap patterns with defects that cause them. One example is as follows: one pattern obtained for single column failures has all cells in the faulty column stuck-at-zero. The vocabulary indicates five possible causes of such a pattern:

(i) Extra second metal: the bit and the $\overline{bit}$ could be shorted to each other.

(ii) Extra first metal: the $\overline{bit}$ line of the column could be shorted to V_{DD}.

(iii) Extra first metal: the bit line of the column could be shorted to GND.

(iv) Extra first metal: the bit and $\overline{bit}$ lines of the column could both be shorted to GND.

(v) Extra first metal: the bit and $\overline{bit}$ lines of the column could both be shorted to V_{DD}.

For tables showing their detailed experimental observations, see [120].

4.4.4 Testing of SRAMs by monitoring I_{DD}, the dynamic power supply current

While I_{DDQ} testing is fairly effective in detecting a general variety of CMOS defects not usually considered by functional testing, for example, gate oxide short, punch-through and leakage, it is found to have some shortcomings [157]. It is mostly limited to fully complementary logic (one that ideally draws no current from the supply during quiescent steady state conditions, and draws current only during switching transients), and is typically slower than voltage testing. Moreover, I_{DDQ} testing alone is often not adequate. It it is usually regarded as a good way of achieving quality control and of augmenting functional testing (stuck-at) to achieve greater defect coverage, rather than as a stand-alone test. To make I_{DDQ} testing faster, the soft-defect-detection ap-

Technology and Layout-Related Testing 179

proach [78], described earlier, uses the concept of built-in current monitors for fast current sensing.

The idea of I_{DDQ} testing has been recently (1992) extended to I_{DD} testing. In this technique, the *dynamic* power supply current is measured. As described a little later, there exists a relationship between the dynamic power supply current and a variety of SRAM defects such as pattern-sensitive faults and stuck open faults. The technique described by Su and Makki [157] uses a novel approach for distributing power to the cell array, resulting in low area overhead and a high fault coverage of the above faults.

Defect models for I_{DD} testing

Su and Makki [157] have studied the relationship between the dynamic power supply current I_{DD} and various RAM defects that induce pattern-sensitive faults. Some defects considered are cell open faults caused by opens in the wiring, gate, or along drain/source and cell short faults caused by breakdown in the insulating layer separating nodes belonging to different cells. Each class of defects results in a peculiar response of the dynamic power supply current I_{DD}. The various fault models for the defects considered are illustrated in Figure 4.8 [157].

The various pattern-sensitive defects modeled are enumerated below:

1. **Diode-Connected Transistor Short**: An electrical coupling is induced between two SRAM cells using the model in Figure 4.8 (*a*). This interaction is in one direction — logic changes in a cell c_1 affects the contents of another cell c_2, but not vice versa.

2. **Open Fault**: A PSF is induced as a result of an open drain fault in a cell (which prevents the pull-up transistor from reinforcing the bit-line voltage). This causes the cell to be easily disturbed by even a weak capacitive coupling to another cell. Such a cell is thereby very susceptible to pattern sensitivity. This fault model is shown in Figure 4.8 (*b*).

3. **Gate Oxide Short**: A gate oxide short is modeled by a resistor shorting the gate to the substrate. This causes current to flow into the substrate as leakage current, as shown in Figure 4.8 (*c*).

4. **Bridging Faults**: A resistive bridge between the inputs of two cross-coupled inverters in conjunction with a weak coupling capacitor between

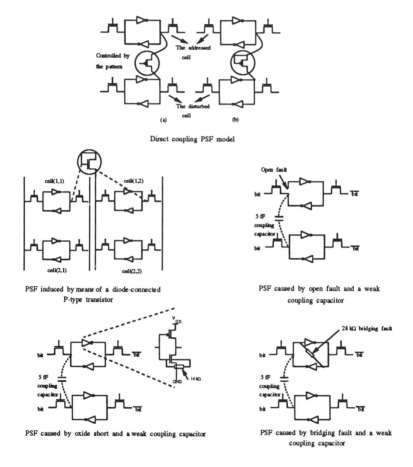

Figure 4.8 Defect models for I_{DD} testing; courtesy [157]

cells c_1 and c_2 having a value of about 20 fF, constitutes this fault model. Writing a 0 in cell c_2 causes cell c_1 to switch states.

I_{DD} response to the above defects

In a fully complemented CMOS circuit, a momentary current path, I_{DD}, between V_{DD} and ground is established only during the period of a state change. When a memory cell changes state, it is expected to produce a large I_{DD} pulse. However, because of PSFs, some other cell may also produce a large pulse. Thus a PSF can be detected under such conditions if the supply lines to each cell can be separately monitored.

The total I_{DD} response for each defect class has been studied in [157]. It is observed that each state transition (whether intended or erroneous) results in a significant I_{DD} pulse of finite duration (about 2 ns). These pulses are detectable at the V_{DD} and ground rails. The cell whose state is being changed is expected to have a significant pulse; however, some other cell which is coupled to the first cell might also produce a significant pulse.

Circuit techniques for I_{DD} testing

As described in [157], an I_{DD}-testable SRAM may be built using two current monitors called VTCM (V_{DD} Transient Current Monitor), and GTCM (Ground Transient Current Monitor). The system V_{DD} and ground lines are distributed to all parts of the array through Fault ISolation Control Circuitry (FISC). In this circuit, all cells on the same column share a common V_{DD} line, and all cells on the same row share a common ground line. The purpose of the FISC is to switch the supply lines to a given cell between the current monitors and the system power lines. The row and the column select lines are used to enable the FISCs. The circuit diagram is shown in Figure 4.9.

Circuit performance and overhead

Since a $\sqrt{n} \times \sqrt{n}$ SRAM array has a total of $2\sqrt{n}$ rows and columns, the number of FISCs required is $2\sqrt{n}$, using a total number of $8\sqrt{n}$ transistors, since each FISC requires 4 transistors. Since one VTCM and one GTCM are used, the cost of the FISCs becomes an overhead. The rise and fall time degradation due to the inclusion of the FISC is negligible. However, the individual cycle time for current monitoring may be about 25% longer than that for voltage

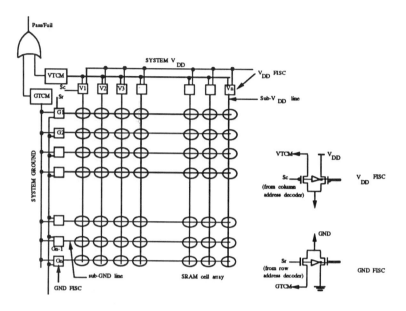

Figure 4.9 Circuit design for an I_{DD} testable SRAM; courtesy [157]

monitoring. Other circuit overhead like additional input/output is seen to be negligible from the circuit design.

4.5 GALLIUM ARSENIDE SRAMS: DESIGN, FAULT MODELING AND TESTING

Gallium Arsenide HEMT (high electron mobility transistor) devices, which are the fastest switching elements available today (with the exception of superconducting devices that operate at cryogenic temperatures), have now begun to make inroads into the world of commercial memory chip design. GaAs devices are preferred to superconducting devices because of their compatibility with ECL and other silicon logic families, and their ability to operate at room temperature. A good state-of-the-art survey of GaAs circuits is found in [2]; also Notomi, et al. [129] have described a 1 Kb SRAM using GaAs HEMTs (high electron mobility transistors), that has a 500 ps access time, clearly outperforming both ECL and CMOS in terms of speed. One major problem with GaAs memories is that the address access time over the whole address space varies as widely as 150 ps, that is, 30% of the nominal access time. This problem has been attributed to process variations and design problems which are not normally encountered with silicon SRAMs. Parasitic resistances are a much more serious problem with GaAs devices because of higher currents and lower operating voltages. Leakage currents are much larger relative to the operating currents and voltage margins are low. Hence GaAs devices have a number of distinct failure modes, which lead to faults unlike those commonly seen in silicon. Mohan and Mazumder [113] have explored several of these failure modes and identified fault models and parametric test procedures for the faults.

Since the basic structure and processing sequence of GaAs devices are different from those of silicon MOSFETs, the failure mechanisms are also different. Systematic variations in process parameters, such as threshold voltage across a wafer and across a chip, are observed more prominently in GaAs than in MOS. Moreover, new failure modes such as inter-electrode resistive path formation [37, 41] have been observed in GaAs. As a result, fault modeling and testing of GaAs memory devices requires the identification and characterization of process, design and layout-related faults.

Failure modes of GaAs SRAMs lead to different types of faulty operation. Several kinds of pattern-sensitive faults (PSFs) have been studied. Some of these faults cannot be detected even by efficient test algorithms; knowledge of the cause of such PSFs is required in order to design efficient test procedures. Variations in process parameters are seen to result in delay faults. Another peculiar fact observed in GaAs SRAMs is that data retention faults have an entirely new mechanism, apart from the stuck-open mechanism observed in silicon [34, 78]. Parametric test procedures for these faults are described shortly.

4.5.1 High electron mobility transistors

High electron mobility transistors (HEMTs), also known as MODFETs (Modulation doped FETs) or TEGFETs (Two dimensional Electron Gas FETs) are ultra-high-speed switching devices whose development came about as a result of advances in molecular beam epitaxy (MBE) that allowed heterostructures of GaAs and AlGaAs to be grown epitaxially [4, 5, 62]. HEMTs are field effect devices similar to MESFETs or JFETs and have the added advantage of being much faster. The increased speed results from the increased mobility of the electrons in the HEMT. In a regular MESFET or JFET channel, the mobility of the electrons is limited by impurity scattering and the number of impurity

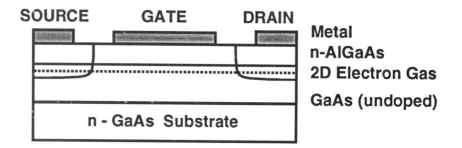

Figure 4.10 Structure of HEMT; courtesy [113]

atoms in the conducting layer is high because it is heavily doped for good conductivity. In HEMTs, the electrons in the conducting layer of GaAs (see Figure 4.10) come from the adjacent n-doped AlGaAs layer owing to the properties of the AlGaAs/GaAs heterojunction. Since the GaAs layer is undoped, impurity scattering is minimized and the electron mobility can potentially go up to $10^5 \mathrm{cm}^2/\mathrm{Vs}$ at 77 K and of the order of $10^4 \mathrm{cm}^2/\mathrm{Vs}$ at room temperature. This is much greater than the electron mobility of 500 $\mathrm{cm}^2/\mathrm{Vs}$ in silicon FETs.

The operating voltages and especially the output voltages of circuits cascaded to other HEMT circuits are limited by the fact that when the gate-source voltage increases above the Schottky barrier voltage of the gate, the gate begins to conduct and clamps the gate voltage at 0.8-0.9 V above the source voltage. Hence the typical HEMT direct-coupled logic circuit has a supply voltage range of 1 V with a logic high of 0.8 V and a logic low of 0.1 V. The drain-source leakage current flowing through a HEMT when the gate-source voltage is at logic 0 is about 1/200 of the current through the HEMT when the gate-source voltage is at logic 1; in a typical silicon NMOS logic circuit this ratio is about 1/10000. Hence leakage/threshold currents are much more likely to cause errors in HEMT RAMs.

4.5.2 Defects in GaAs SRAMs

A circuit diagram of a basic GaAs SRAM cell is given in Figure 4.11. Fault modeling for GaAs RAMs starts with the study of material and processing defects. This study can then be used to generate plausible circuit errors in the basic memory circuits; the observed erroneous behavior of the memory is a result of the errors in the component circuits. Reliability and aging studies

Technology and Layout-Related Testing

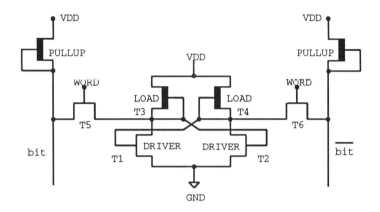

Figure 4.11 A Basic GaAs SRAM cell; courtesy [113]

[89] of HEMTs reveal the parametric and catastrophic failure modes of these transistors as a result of stress and aging. Fault effects of these failure modes may be studied to develop efficient tests for periodically testing these memories over their lifetime of operation. Study of the failure mechanisms also suggests ways to test for the presence of the resultant errors at the behavioral level.

The defects in GaAs wafers can be classified as primary and secondary defects. Primary defects are material-related ones affecting factors such as compositional purity, control of stoichiometry and crystalline perfection. These defects manifest themselves at the device level in the form of faults such as threshold voltage variation, mobility degradation and charge trapping. Deep donor levels associated with DX centers in the AlGaAs layer of the HEMT play a major role in electron capture and release [17], leading to various problems such as kinks in the V-I characteristics of the HEMT. Trap-related problems result in threshold voltage and transconductance shifts with temperature changes [68]. Surface defects known as *oval defects* are a by-product of the material growth process [3]. These defects range in size from submicron levels to a few microns. The effect of an oval defect in the gate region of a HEMT is to prevent the transistor from turning off.

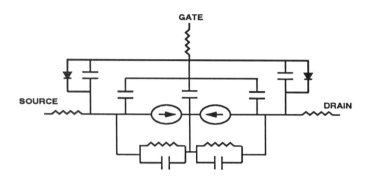

Figure 4.12 Equivalent circuit for HEMT; courtesy [113]

Threshold voltage variations over the wafer is a major problem in GaAs. With improved processing technology, this problem can be alleviated to some extent but never completely eliminated. Typical state-of-the-art parameters for threshold voltage variation over a wafer are as follows: $V_{TE} = 0.278$ V; $\sigma(V_{TE}) = 11.3$ mV; $V_{TD} = -0.602$ V; $\sigma(V_{TD}) = 14.2$ mV, the subscripts denoting enhancement and depletion mode devices respectively.

Secondary defects, on the other hand, are the ones introduced during wafer processing in the form of surface and sub-surface damage. These defects are responsible for most of the observed faults in the circuit, such as SAFs, bridging faults, and others.

A major secondary defect associated with GaAs devices is ohmic contact degradation and interdiffusion of gate metal with GaAs. This results in increased *on* resistance of the transistor, and decreased saturation current and pinch-off voltage; it may also bring about gate to channel short. Further, the depletion-mode transistors may develop interelectrode metallic paths due to electromigration and processing defects. Fracturing may also also some failure modes. For a discussion of these various failure modes, see [37, 41, 134, 139].

4.5.3 SRAM cell design and fault analysis

One of the principal issues in memory design is design centering. This is the process of choosing nominal design parameters so that they lie in the 'center' of their respective tolerance intervals so as to maximize yield. Statistical circuit design techniques [153, 156, 180] have been used in the past to solve the design

Technology and Layout-Related Testing

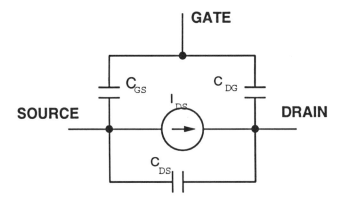

Figure 4.13 Simplified equivalent circuit for HEMT; courtesy [113]

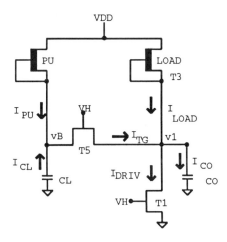

Figure 4.14 Devices involved in the read operation; courtesy [113]

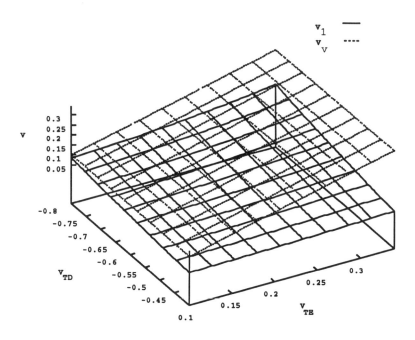

Figure 4.15 Variation of V_1 with threshold voltage; courtesy [113]

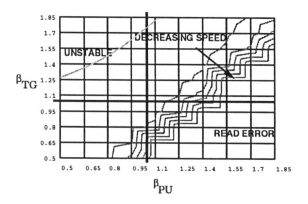

Figure 4.16 Read error as a function of pull-up(PU) and access transistor(TG) βs; courtesy [113]

Technology and Layout-Related Testing 189

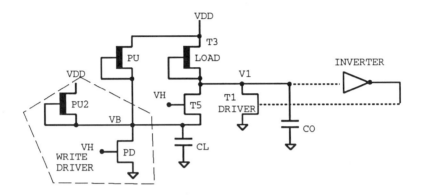

Figure 4.17 Simplified circuit for analyzing the write operation; courtesy [113]

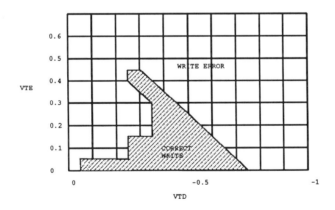

Figure 4.18 Write error as a function of threshold voltages; courtesy [113]

190   CHAPTER 4

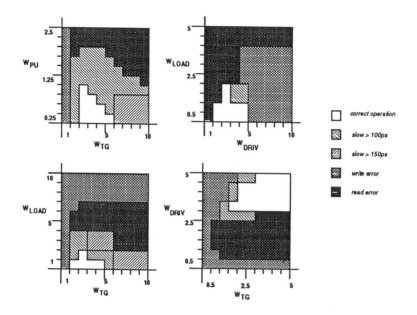

Figure 4.19 Effect of parameter variations: simulation results; courtesy [113]

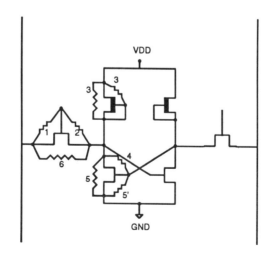

Figure 4.20 Canonical set of resistive paths; courtesy [113]

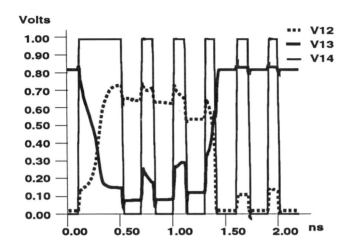

Figure 4.21 Simulation of cell with missing load: V12 and V13 are the cell storage nodes. The figure shows a write followed by 3 read operations the last one causing the cell to change state; courtesy [113]

centering problem. Parameters such as the threshold voltage and transistor length vary across the wafer, and across the chip. If the distribution of the parameters is known in terms of the type of distribution, its mean and variance, then statistical design attempts to assign nominal values m_{nom} to these parameters such that the design works satisfactorily for all actual values of the parameter m between m_{min} and m_{max}, where the values of m_{min} and m_{max} are chosen to maximize the probability that m lies within the range. Therefore, statistical design and yield optimization involve the computation of circuit parameters for various values of process parameters in order to compute better nominal values starting from some given value.

The present approach [113] is complementary to this design centering approach; starting from a nominal design value, the fault effect of parameter variations is studied. Various other defects like resistive bridging or missing devices are mapped on to equivalent parameter variations where possible, to present a simple, unified view of the problem in terms of parametric faults. *Parametric* yield and *catastrophic* yield are the two basic yield figures which measure the loss due to parameter variation and catastrophic defects (missing device, shorts, open circuits, and others) respectively [87]. These figures are usually obtained separately in yield estimation studies. The fault modeling approach presented here shows how the fault effect of catastrophic defects is mapped onto equivalent parameter variations and vice versa. Analysis of a single good circuit can

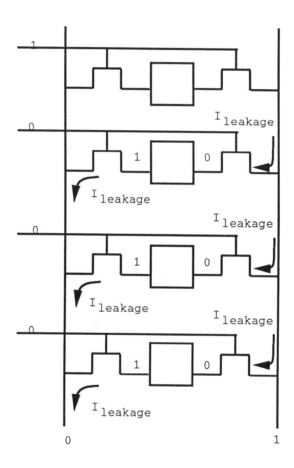

Figure 4.22 Parametric pattern-sensitive fault due to leakage current; courtesy [113]

Technology and Layout-Related Testing   193

provide an understanding of the fault effects of both parameter variations and catastrophic defects.

Let us refer again to Figure 4.11. The basic memory cell consists of two cross-coupled inverters connected to complementary *bit* and $\overline{bit}$ lines via pass gates. This circuit is analyzed with the help of a simple equivalent circuit model, to identify the fault effects of variations in element parameters. This simplified analysis is then verified with the help of SPICE simulations using a more complex numerical model [182]. The HEMT model is shown in Figures 4.12 and 4.13 and is a simplified version of the model presented in [182] and used in the simulator [181]. Typically, the gate-source capacitance C_{GS} is the dominant capacitance term; C_{GS} is about five times C_{GD} and ten times C_{DS} [151]. The dependent current source equations are as follows in the simplified model:

Cutoff $(V_{GS} < V_T)$: $I_{DS} = 0$

Saturation $(V_{GS} \geq V_T;\ V_{GS} - V_T < V_{DS})$: $I_{DS} = I_{DSAT} = \beta(V_{GS} - V_T)^2$

Linear $(V_{GS} \geq V_T;\ V_{GS} - V_T > V_{DS})$: $I_{DS} = I_{DSAT} \tanh(\alpha V_{DS})$

where $\beta = KW/L$, and W and L are the width and length respectively of the HEMT. Here K and α are constants and the capacitances are assumed to be constant over the range of operation, even though they are actually voltage dependent to a large extent. The validity of this assumption has been verified by simulating the circuits with the detailed HEMT model of [182]. The constant α has the dimensions of Volt^{-1}, and β and K have dimensions of Amperes/Volt^2.

4.5.4 Read and write operations

The memory cell stores a value 0 (1) with a high voltage V_H on the node adjacent to the $\overline{bit}$ line and a low voltage V_L on the node adjacent to the *bit* line. Prior to a read operation, the *bit* and $\overline{bit}$ are precharged to V_H. Then the *word* line goes high, turning on the pass transistor connecting the cell node which is at V_L to the adjacent $bit(\overline{bit})$ line. The memory cell is designed so that the cell does not change state during the read operation, the read/write operations are fast, and the cell area is minimized. The cell storage node voltages V_H and V_L are determined in the static case (when the access transistors are off) by the current ratios of the cell load and driver transistors and in the dynamic case by the sizes of the *bit* line pull-up transistors, the *bit* line capacitances and the access (pass) transistor currents.

Analysis of the noise margins and stability figures of silicon RAM cells consisting of four MOSFETs and two resistive loads has been performed in the past [6, 19, 145]. However, these analyses assume that the cell load resistor value is so high that the current through the resistor is insufficient to charge the cell nodes in a single access cycle or that the cell voltage V_L is greater than the threshold voltage of the FET so that the driver FETs of both inverters are conducting. While these are true for SRAM cells with a resistive load, they do not accurately model the behavior of HEMT RAMs with depletion mode loads and small operating voltages. The analysis by Mohan and Mazumder [113] is based on observations of the operating points of HEMTs in cells corresponding to published HEMT RAM designs [75]. In the static case, the driver HEMT whose drain is at V_H is turned off and the depletion loads provide enough current to recharge the cell nodes each cycle.

1. **Read Operation:** Analysis of the *read* operation starts with the assumption that the cell storage node connected to the *bit* line is at logic 0 and the node connected to the $\overline{bit}$ line is at logic 1. The *bit* and $\overline{bit}$ are precharged to V_H and the *word* line is at V_H. The corresponding operating points of various transistors are: the two depletion mode pull-up transistors on the *bit* and $\overline{bit}$ lines are in the linear region since $V_{GS} = 0$ V, $V_T = -0.6$ V (typical) and $V_{DS} = 0.1$ V. The access transistor $T5$ is in the linear region, the load transistor $T3$ is in saturation, the driver transistor $T1$ is in the linear region, the driver transistor $T2$ is in cutoff, the load transistor $T4$ is in the linear region and the access transistor $T6$ is in the linear region. During a correct *read* operation, the *bit* line is discharged by about 0.2 V and the sense amplifier (not shown) converts the voltage difference between *bit* and $\overline{bit}$ lines into a proper output voltage level. Since the $\overline{bit}$ line voltage V_H does not change significantly during the *read* operation on the cell storing a 0, it is sufficient to consider the devices connected to the *bit* line, $T1, T3, T5$ and the *bit* line pull-up. The transistor $T2$ is replaced by an equivalent load capacitor and the resultant circuit is shown in Figure 4.14.

   Static analysis of the memory cell provides information on the regions of correct operation. The aim of this analysis is to determine the range of V_{TE} and V_{TD}, the enhancement and depletion mode threshold voltages respectively, for which the cell storage node voltage at the end of the *read* operation does not increase above the threshold voltage of the transistor $T2$. The capacitors are ignored and the currents through the various transistors shown in Figure 4.14 are equated using Kirchhoff's current law at the two nodes corresponding to the *bit* line and the cell storage node. The resulting equation for the cell storage voltage is $V_1 =$

$(\beta'_{PU} + \beta_{LO})V_{TD}^2/\{\alpha\beta_{DR}(V_H - V_{TE})^2\}$, where $\beta'_{PU} = \beta_{PU} * \tanh(\alpha * v)$ and v, the drain-source voltage of the pullup, is about 0.1 V for the entire region of interest. A plot of the variation of V_1 with V_{TD} and V_{TE} is shown in Figure 4.15, along with a plot of $V_v = V_{TE}$. It can be seen that over a 3σ variation about the mean design values of the threshold voltages, the value of V_1 is less than $V_v(= V_{TE})$ for this particular design with a $3\mu m/0.5\mu m$ driver, a $0.5\mu m/3\mu m$ load, a $1.5\mu m/0.5\mu m$ access transistor, a $1.5\mu m/0.5\mu m$ pull-up and a 200 fF bit line capacitance.

Static analysis cannot catch the error that would follow if the time to discharge or precharge the bit line is too high, and the SRAM operates slowly as a result. More detailed analysis involving bit line and storage node capacitances is necessary. The simplified HEMT model can be regarded as a single current source except for the transistor T_2 which is replaced by its equivalent gate-source capacitance. Subsequent simulation results indicate that the capacitor currents in the ON transistors are much less than the corresponding drain-source currents. Using Kirchhoff's Current Law, the time t for the storage node voltage V_1 to fall below V_{TE} is found to be as follows [113]: $t = C_0(C' - A_1 \log(V_1 - u_1) - A_2 \log(V_1 - u_2))$, where C_0, C_1, C', A_1, A_2, u_1, and u_2 are constants that depend upon the values of V_H, V_{TE}, the various βs, and α.

If V_1 reaches the value V_{TE} before the *read* phase is complete, then there is a cell stability problem. In the worst case read situation, the cell is selected and the chip is enabled for an indefinite period of time, causing the static stability situation. More importantly, this analysis is used to predict the discharge time of the bit line. The bit-line is precharged to about 0.9 V and is discharged to between approximately 0.75 V and 0.7 V; at the same time, the cell storage node voltage varies slightly, as shown above, from 0.1 V. Assuming the node storage voltage V_1 is a constant, Mohan and Mazumder [113] solve the following KCL equation for the bit line:

$$\beta_{PU} V_{TD}^2 \alpha (V_H - V_B) - C_L dV_B/dt = \beta_{TG}(V_H - V_{TE} - V_1)^2,$$

with initial condition $V_B = V_H$ to obtain the discharge time of the bit line, which turned out to be as follows: $t = \log((V_B + K_1/K_0)/(V_H + K_1/K_0))/K_0$, where K_1, K_0 and V_B are constants depending on the βs of the access transistor and the pull up (see Figure 4.16), the threshold voltage of the depletion mode pull-up, and the threshold voltage of the enhancement mode n-transistors.

2. **Write Operation:** Suppose without loss of generality that the cell node connected to the $bit(\overline{bit})$ line stores a 1(0) and the write operation attempts to flip the cell state. In that case, transistor $T2$ in Figure 4.11 is turned

off while the $\overline{bit}$ line is driven high so that $T1$ is turned on. Since the load (pull-up) transistor of the cell is relatively weak compared to the cell driver, it is easier for a *bit*-line driver transistor to pull the cell node at logic 1 to a low voltage than for the *bit*-line pull-up to pull the cell node at logic 0 to a high voltage level V_H. Hence we look at circuits that cause the originally high cell node to be pulled low; the transistors on the opposite side ($\overline{bit}$ in this case) are replaced by a single capacitor loading the left half of the memory cell — the equivalent circuit is shown in Figure 4.17. The cell load transistor $T3$ is initially in the linear region and finally in saturation. The cell driver $T2$ goes from cut-off to linear while the access transistor $T5$ goes from saturation to linear. The *write* operation starts after the *bit* line is discharged and the *word* line goes high, turning on $T5$. The *bit*-line driver T_{PD} is linear. Once again, KCL equations can be used to compute the voltage V_1 at the storage node as a function of the βs, α, and threshold voltages V_{TD} and V_{TE}. Since we are trying to write a 0 in the storage node, V_1 should be less than V_{TE} for the cell to flip states. The dynamic analysis of the write operation in a manner analogous to read is quite complex. See Figures 4.18, 4.19.

4.5.5 Design-related errors

Simulation Studies: To observe these errors, the memory circuits are simulated with different values of process and design parameters such as transistor widths and threshold voltages. These variations are found to either cause the circuit to operate very slowly or to malfunction.

It is observed that as the thresholds for the depletion load and enhancement device are increased, the delay increases as predicted, finally resulting in erroneous operation. When the enhancement mode threshold voltage is reduced to nearly 0 V, erroneous operation due to the inability of the enhancement mode transistors to turn off is observed. The observed results are seen to be in agreement with the predicted ones.

Design-related errors are induced by varying the widths of the load resistor, transmission gate, the driver transistor and the bit-line pullup transistor. Since $\beta \propto W/L$, the β-ratio of two transistors is just the ratio of their widths if they have the same length. For studying design-related errors of GaAs SRAMs, therefore, Mohan and Mazumder [112, 113] consider the following ways of varying parameters of the circuit:

1. Variations of the width of the transmission gate and load transistors (with the lengths kept constant): The circuits are seen to produce faulty behavior if the width of the load transistor exceeds a certain value, correct operation being noticed only for a small range of values of W for the transmission gate and the load. If the width of the load transistor becomes much larger than that of the driver transistor, the circuit produces faulty behavior because the inverter operation is itself faulty (the output high and low levels fall outside allowable limits). When the load transistor becomes wide but not wide enough to cause circuit malfunction, its gate capacitance increases, causing slow operation. When the transmission gate becomes too wide, the speed decreases.

2. Variations of the width of the pullup transistor and the transmission gate (again with the lengths constant): The circuits are seen to operate correctly only for a small range of values. This can be explained as follows: when the transmission gate has very small width, the time taken to charge the cell node during a write operation and to discharge the *bit* line during a read operation become too large for correct operation. When the pullup transistor becomes wider, it prevents the *bit* line from discharging during a read operation, resulting in a read error. When the pullup transistor is much smaller than the transmission gate and the width of the transmission gate is increases, the read time is seen to increase, corresponding with predicted results using the simplified equivalent circuit. Keeping the pullup transistor width constant, an increase in the transmission gate width causes faster operation, in keeping with the mathematical prediction from the simplified equivalent circuit).

3. Variations of the width of the transmission gate and the driver: keeping the width of the driver constant, an increase in the width of the transmission gate produces faster operation, but beyond a certain point, the extra gate capacitance may cause the write operation to fail.

When matched devices are used as in the differential input stage of the sense amplifier or in the memory cell itself, variation of device parameters can cause erroneous operation due to mismatching. However, the intra-chip or intra-die variation of parameters is very small [2] and is not detrimental to proper circuit operation. Simulations have shown that a V_{TE} mismatch of 0.3 V is needed to cause wrong latching of the memory cell; the typical variation is of the order of a few millivolts. Similarly, the extent of the mismatch in transistor widths required to produce an error is equal to the nominal width of the transistor. These extreme conditions are more properly described as catastrophic failures rather than parameter variations, and will be described later.

As the operating temperature of the device increases, the electron mobility decreases and parameters such as threshold voltage are seen to shift [8]. While the temperature dependence of MESFETs is well defined in MESFET circuit simulators, there are no widely accepted temperature-dependent HEMT models and the HEMT simulator [182] assumes that the operating temperature is 300 K. Hence temperature effects must be approximated by variations of other parameters such as threshold voltage, β, and so on.

4.5.6 Catastrophic failure modes

The process and material-related defects may produce either slow operation (delay faults) or, beyond a certain point, read and/or write errors. Sometimes these defects may lead to catastrophic circuit modifications (faults), mostly manifested in the form of shorts and stuck-open faults. These are described in [113]. Figure 4.20 displays a canonical set of catastrophic resistive shorts. These may lead to the following circuit modifications:

- Resistive shorts between transistor electrodes /increased leakage current
- Bridging of metal lines — shorts between two adjacent signals
- Stuck-open transistors

Each transistor in the circuit could thus be stuck-open, or have a resistive path between any two of its electrodes. Any two lines in the circuit that lie within some arbitrary distance from each other in the layout, could be bridged. All these aberrations considered either individually or in groups, lead to a large number of modified, 'faulty' circuits. The coupling capacitance between lines, being small when air-bridge technology is used, is neglected, and the primary coupling mechanism is due to resistive bridging.

These faulty circuits have been analyzed and simulated to obtain the equivalent functional faults. For resistive shorts the basic memory cell together with the *bit* and *word* line circuits are simulated with a whole range of resistance values to simulate the 'short'.

4.5.7 Analysis of resistive/leakage current failure modes

Inter-electrode resistive paths and excessive leakage currents form the most commonly observed defects in processing as well as in aging and reliability studies. Since the memory cell is symmetric with respect to the *bit* and $\overline{bit}$ lines, it is sufficient to consider resistive/leakage paths between electrodes of transistors of one half of the memory cell. Inter-electrode paths could occur between the source and the gate, or between the drain and the gate or between the source and the drain of a transistor. The following is a list of all possible paths that can occur in a single half-cell (see Figure 4.20). Not shown in the figure are the gate-substrate paths that connect the gate to the substrate and may be modeled as increased gate-leakage current.

- Type 1: *bit* line to *word* line
- Type 2: *word* line to cell storage node
- Type 3: cell storage node to power supply
- Type 4: cell storage node to ground
- Type 5: *bit* line to cell storage node

The operation of the memory cell in the presence of a resistive path between two nodes may be analyzed using a simplified equivalent circuit similar to the circuit used to analyze normal circuit operation in the previous section. However, the addition of a resistor complicates the solution of the differential equations governing the cell operation and simple closed-form solutions cannot be obtained. However, when the resistive path occurs in parallel with an existing current path in an ON transistor, the conductance of the transistor in the linear region can be augmented with the conductance of the parallel resistive path to simplify the analysis. For example, a resistive path (with resistance R) between the power supply node and the cell storage node, across the drain-source nodes of the cell load transistor (resistor 3 in Figure 4.20) can be modeled as an increased β if the transistor is in the linear region. The transistor current is $I_{DS} = \beta_{LO} V_{TD}^2 \alpha V_{DS}$, approximating the tanh function by its argument, and the resistor current is V_{DS}/R. The total current between the drain and source nodes can thus be represented as $\beta' V_{TD}^2 \alpha V_{DS}$, where $\beta' = \beta_{LO} + 1/RV_{TD}^2\alpha$. However, this kind of simplification is often very difficult or impossible and simulation is the easiest solution.

Table 4.3 Read and write errors caused by different failure modes; courtesy [113]

Gate-source short on transmission gate - *bit* line to *word* line

Resistance (ohms)	Write	Read	Comments
100	Write 0 fails Write 1 delay < 100 ps	Weak 1	Write 0 fails Write/Read 1 slow
500	Write 1 delay > 100 ps Write 0 fails	Weak 1	
1000	Delay > 100 ps	Read 0 fails	
2000	Delay > 150 ps	Same as above	

Gate-source short on transmission gate: *word* line to cell

Resistance (ohms)	Write	Read	Comments
100	Cell follows *word* line		Cell follows *word* line
500	Weak 1 - decays after 22 ps	Read 1 error	
1000	Weak 1 - decays after 60 ps	Read 1 error	
2000	Weak 1 - decays very slowly	OK	Data retention problem
5000	OK	OK	OK

Coupling between cells of the same *bit* line due to a *bit* line to *word* line short in one of the cells

Resistance (ohms)	Write	Read	Comments
100	Write 1 fails	Read error	
500	Write 1 fails	Read error	
1000	OK	Read 1 error	
2000	OK	Read 1 error	Cell flips to 0 when read
5000	OK	OK	OK

The first table shows the result of a resistive path between the gate and source (*bit* line) nodes of the transmission gate connected to the *bit* line. The main effect of this error is to cause the *write*(0) operation to fail. The *read*(0) operation also fails because the *bit* line which is supposed to be at 0 is connected to the *word* line which goes high when enabled, causing the *bit* line voltage to increase. The second table shows the effect of the above fault on a different cell connected to the same *bit* line. The next one shows the effect of a short between the gate and source nodes of a transmission gate. This time the gate node (*word* line) is shorted to the cell directly. As a consequence, the cell follows the *word* line when the resistance is small. When the resistance is large there is a data retention problem. It may be noted that this problem occurs even though there is no missing pull-up which is the chief cause of data retention problems in silicon SRAMs, as we have seen before.

4.5.8 Stuck-open faults

A stuck-open fault on a transistor is characterized by the transistor being stuck in the cut-off region with an open circuit between the drain and the source nodes. Stuck-open faults are caused by catastrophic failures and usually result

Technology and Layout-Related Testing

in some major circuit malfunction. The stuck-open fault is equivalent to a missing transistor fault; it is also the limiting case of a device mismatch problem. Simulations of the memory cell with various stuck-open faults show that the behavior is equivalent to either a stuck-at fault or to a data retention problem.

4.5.9 Coupling and multiple faults

Defects due to the simultaneous failure of multiple transistors because of interelectrode metallic paths cause coupling and pattern-sensitive faults that cannot be detected by the standard NPSF algorithms found in [1] and [121]. It is possible that more than one device is affected by such defects. Another problem is that of increased leakage currents in normally off devices; this problem is accentuated in HEMT memories by the fact that the ratio of the ON to the OFF currents in HEMTs is much lower than in silicon transistors. This leads to row/column pattern-sensitive faults as described below.

1. **Increased Drain-Source Leakage Currents in Access Transistors**: Increase in drain-source leakage current is produced by a systematic variation in device parameters, by defects such as oval defects and by resistive paths between the drain and source of a transistor. Consider the situation where all the access transistors of the memory have increased leakage due to one or more of the above causes. Then consider the cells connected to one bit-line pair as shown in Figure 4.22. When all the cells store a 1 adjacent to the *bit* line and a 0 adjacent to the $\overline{bit}$ line and an attempt is made to write a 0 on one of the cells, two simultaneous effects combine to produce a write error. The leakage current from the 1 nodes to the *bit* line prevents the *bit* line from being pulled low enough and the leakage current from the $\overline{bit}$ to the 0 nodes prevents the $\overline{bit}$ line from being pulled high enough. This problem affects the read operation even more than the write operation. When the cell to be read stores a 0 and all the other cells (or a large majority of the other cells) store a 1, the cell 0 state causes the *bit* line voltage to be pulled down and the $\overline{bit}$ voltage to be constant at V_H but the leakage currents cause the *bit* and the $\overline{bit}$ line voltages to move in the opposite directions, causing a read error.

2. ***bit* line to *word* line short**: The effects of a single *bit* line to *word* line short and a single *word* line to cell storage node short have been documented in two of the three tables in Table 4.3. The final result is seen to be a function of the relative strengths of the *bit*-line and *word*-line drivers

and of the resistance of the path between the *bit* and *word* lines. The following behavior is observed:

- if the *bit* line driver is stronger than the *word* line driver, cell C (where the defect occurred) could get written even when not selected;
- if the *word*-line driver is stronger than the *bit*-line driver, all cells on the *bit*-line are stuck-at-zero;
- if the strengths of the two drivers are comparable, the read operation is slowed down;
- if more than one cells connected to the same *bit* line suffer a *bit*-line to *word*-line short, then depending on the resistance of each path from the *bit*-line to the *word*-lines, either all cells on the *bit*-line are stuck-at-zero, or the read operation is very slow; and
- if this defect occurs in more than one cell connected to the same *word* line, all bits on the *word*-line are stuck-at-one.

The dominant fault effect of this failure mode, considered in isolation, is the creation of multiple cell-stuck-at faults along the same *bit*-line, and the creation of delay faults for all cells on a single *bit*-line.

3. ***word* line to cell-storage-node short**: A single such defect could lead to data retention problems in a cell, and could also cause a stuck-at fault in the cell. However, it is the presence of multiple defects of this type that leads to interesting new pattern-sensitive faults.

The first case to be considered has multiple defects in cells connected to the same *word*-line. It is assumed that the resistance of the path from a cell storage node to the *word*-line is high enough to avoid data retention and stuck-at problems for the cell. Now, if all the defective cells store a 1, there are multiple resistive paths from the *word*-line to the nodes storing a 1. The effective resistance of the path from the *word*-line to a node at logic level 1 is thereby decreased. These so-called storage nodes are actually driven by an inverter — so the effect of multiple defects is to connect many drivers in parallel to the *word*-line. If the strength of these drivers is more than the strength of the *word*-line driver, the *word*-line is stuck-at-one. Another aspect of this problem is the fact that some cells might have a defect which connects the *bit*-line to the cell storage node, while some others might have the defect in the other half of the cell, connecting the complementary storage node with the *word*-line. In that case, the *word* line fault is sensitized by a pattern of 1s and 0s such that each cell with a path between the storage node and the *word*-line stores a 1, while each cell with a path between the complementary storage node and the *word*-line stores a 0, so that the *word* line gets a maximum strength 1 drive.

4.5.10 Test procedures for coupling and multiple faults

Well-known test procedures exist for virtually every kind of functional faults in memories. The objective of this section is to identify test procedures that test only those faults that have been described in earlier sections.

1. **Leakage current tests:** A sliding diagonal test is ideal to detect this type of fault. The fault is sensitized by a pattern of all-1 (all-0) on the cells in the *bit*-line and tested by writing and reading 0 to one cell.

2. **Tests for multiple *word* line to storage-node shorts:** If the cell defects were uniformly distributed between the two storage nodes, 2^n patterns would be necessary to guarantee that the fault is sensitized, where n is the number of cells on a single *word* line. An analysis of the causes of the gate-drain short reveals that this short is caused by a metallic path between the electrodes that develops as a result of stress, aging and the voltage difference between the two electrodes. Hence this defect can be made to manifest itself by continuously storing a pattern of all-1 and all-0. This would lead to the development of paths between the storage node and *word* line of every potentially defective cell. These failure modes can then be detected by simple stuck-at tests for each cell since the faults do not mask one another.

3. **Other multiple shorts and their tests:** There are three other types of shorts that can occur in a single memory cell. These are cell storage node to power supply, cell storage node to complementary cell storage node, and cell storage node to ground.

   A resistive path between the cell storage node and power supply obviously causes the cell to be stuck-at-one, and a resistive path between the cell storage node and ground causes the cell to be stuck-at-zero. A resistive path between a complementary pair of storage nodes would tend to equalize their voltages and thereby prevent them from being at opposite polarities. This would cause a write operation on a cell to fail; that is, the two node values cannot be modified by a write operation. Subsequent read operations would always give the same result regardless of what is written in the cell, assuming that the hysteresis of the sense amplifier can take care of fluctuations from its threshold voltage due to noise.

   Multiple faults of the same type for storage nodes shorted to power supply or to ground do not mask one another or create new classes of pattern-sensitive faults. Hence such faults can be detected by a march test or a classical RAM test pattern such as the $30n$ pattern of Nair [121].

Faults coupling different memory cells occur due to shorts between lines of adjacent cells. The effect of these faults is to cause coupling between a cell and one of its two neighbors on the same *word* line since the layout ensures that other cells are separated by power and ground lines. Hence a simple $8n$ test which reads and writes 1 and 0 on each cell in the presence of all combinations of 1 and 0 in the two adjacent cells is sufficient to cover all expected pattern-sensitive faults of this type.

4.6 FAULTS IN MOS DRAMS

The following faults noticed in MOS DRAMs are technology and layout-related. The tests for these faults are often based upon an evaluation of the conditions that trigger these faults. These tests and the fault models used are described below.

4.6.1 Retention fault, or sleeping sickness in DRAMs

This occurs when a DRAM cell is left unaccessed for some time, typically 100 μs to 100 ms and charge leaks away from the storage capacitances before the cell is refreshed. Upon read access for use or refreshing, such a cell may have too little charge to enable a correct read operation from taking place.

A condition under which maximum leakage will be encountered is when each cell is *physically* surrounded with cells that contain the opposite data value — the so-called *topological checkerboard pattern*, which looks like a checkerboard at the layout level of the memory cell array.

Franch, et al. [42] observed that there are two possible mechanisms that can cause a retention fault in a DRAM cell. Suppose the cell stores a physical 0, or 0 V, then the following two leakage paths (caused by breakdown of reverse biased pn junctions) may develop within the access transistor (an NMOS device) — (*a*) one path through the reverse-biased pn junction that exists between the p storage node (source) and the n well of the access transistor; (*b*) the other one from the drain of the transistor (connected to the bit-line BL) to the source (storing the physical 0) under subthreshold conditions, that is, when the gate voltage is below V_{TH}, and the bit-line BL is latched at V_{DD}.

To test for the first type of sleeping sickness, each cell of the memory should be left undisturbed (i.e., unaccessed) for at least one full refresh cycle. This encourages the maximum possible charge leakage to take place. A test algorithm for sleeping sickness consists of the following steps:

(1) Write a topological checkerboard pattern into the memory.

(2) Leave the memory unaccessed for a time period slightly larger than the time interval between two refresh cycles.

(3) Read the contents of the memory in the same order it is written and at the same operation rate and report any discrepancy between the value read and the value stored prior to the read operation. The same operation rate and order are maintained to ensure that no cells waits longer than the specified maximum.

(4) Complement the checkerboard pattern and repeat steps 1 through 3.

The two leakage mechanisms described for the second type of sleeping sickness are known as *isolation* leakage and *subthreshold* leakage [42] respectively. Subthreshold leakage, which is more predominant for megabit DRAMs, can be tested using the philosophy that a worst case subthreshold leakage (causing a data retention fault) can be created by latching the bit-line BL (connected to the drain of the access transistor) at V_{DD} and allowing the word-line (connected to the gate of the access transistor) to drop to a low value (below the threshold voltage). This would cause a large drain to source voltage (V_{DS}) under subthreshold conditions, provoking breakdown. The intent here is *not* to generate cell retention faults by exceeding the specified refresh interval, but to obtain such faults by maintaining a large V_{DS} across the access transistor for extended periods that are *well within* the required refresh interval. The algorithm is as follows:

(1) Initialize cells with physical 0s (i.e., 0 V stored in cell capacitors that are connected to the source of the access transistors).

(2) Repeat the following N times:

(*a*) Select one word-line in the quadrant being tested and hold it low. This would latch the sense amplifiers and hold the bit-lines at their latched levels (those that are latched at V_{DD} will be of interest because *only their* worst case subthreshold leakage will be provoked).

(b) Interrupt (a) periodically to perform burst refresh cycles, to allow other word-lines to be refreshed at the normal rate.

(3) Increment to next quadrant if one exists, and go to (2).

(4) Read back physical 0s from the memory.

A cell that passes its regular set of retention pattern tests, and fails the above test, can be specifically diagnosed as having unsuitable subthreshold characteristics for its access transistors. The 'large V_{DS}' pattern has the advantage of being able to specifically identify worst case access device subthreshold leakage, unlike other test patterns partly because there are very few active cycles during this test. The only active cycles are those for initialization and readback; when a word-line is selected, there are no write and read cycles.

The test pattern of the above algorithm has to be repeated with another choice of word-line so that cells previously connected to 0 V latched bit-lines BL are now connected to V_{DD} latched bit-lines. This completes testing each and every cell for subthreshold leakage. The time elapsed for performing the entire test is reported to be of the order of 1 s for a 1 Mb chip, making it feasible in a manufacturing environment.

A leakage phenomenon called *variable hold time* (VHT) has been reported by Yaney, et al. [179] (1987). They observed retention time (hold time) on many devices with planar cells and grounded field plates and noticed multi-valued and metastable leakage current for a small fraction of the bit population at room temperature. On such cells, stable periods are seen to last from seconds to hours and are punctuated by almost instantaneous transitions, the rates of which increase with temperature and substrate bias. The offending cells usually have silicon material defects, mostly process-induced. Based on the study of these microdefects, Yaney, et al. came up with a VHT model to detect and explain this phenomenon.

The VHT model proposed in [179] consists of two parts: one to account for the basic leakage causing short hold times, and the other to account for the variability of the hold times. Basic charge leakage can be explained by localized carrier multiplication and the variability is conjectured to have the same underlying mechanism as *burst noise* in large *pn* junctions. A single DRAM cell can be used to spatially resolve the point sources of burst noise integrated together by a large junction.

VHT is found to have a reversible and finite-state nature. This indicates that it is probably caused by a single or few defective spots instead of being the result of the quantum statistics of a large number of charge carriers. The model proposed in [179] regards VHT as being caused by the modulation of current through a major defect by the change in occupancy of a minor flaw located nearby. This qualitative model could not be described quantitatively because of experimental difficulties.

Noble, et al. [128] (1987) observed a parasitic voltage dependent junction leakage current inherent in the vertical gated diodes of DRAM trench storage cells. Their experimental observations demonstrate that in the normal range of operation, this current is limited by the diffusion of thermally generated carriers along the gated node surface.

4.6.2 Refresh line stuck-at fault

A refresh line SAF in DRAMs would adversely affect data retention. Also, the exact period of time after which the data will be lost is not known; it varies from chip to chip and is sensitive to temperature, however, it will take at least the maximum refresh time as specified by the manufacturer. The time for which one has to wait to test this fault is therefore impractical. So, such faults can be tested in quite the same fashion as sleeping sickness — that is, write a pattern, leave the memory untouched for a period of time (longer than the maximum refresh time), and then verify its contents.

4.6.3 Bit-line voltage imbalance fault

A typical memory array organization utilizes the differential amplifiers for sensing the signal partitioning of each array into two identical subarrays, as shown in Figure 4.23. Each bit-line in the array is split into two halves and they are sensed by a differential pair of sense amplifiers. One of the cells in each half acts as a reference cell and its voltage is compared with the selected cell on the same word-line, but on the other half. Thus in Figure 4.23 when a cell in the right-half bit-line is selected for reading, the reference cell on the left-half bit-line is utilized for comparison. The left-half bit-line is clamped to a reference voltage (which is close to the precharge voltage V_P) and is compared by the sense am-

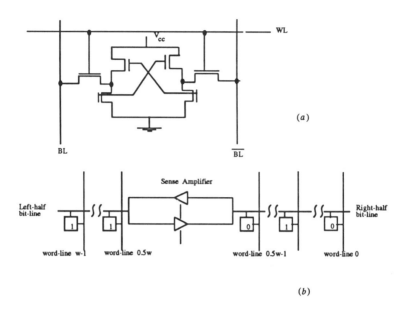

Figure 4.23 Bit-line structure for (a) an SRAM cell and, (b) a DRAM cell

plifier with the voltage of the right-half bit-line, which will be near to either V_{DD} or ground, according as the cell contains a 1 or 0. Thus if the difference of the voltages between the two bit-lines is larger than a threshold value, the sense amplifier can correctly distinguish the state of the selected cell during a read operation. When most of the cells in the left half of the memory subarray contain one type data (say, 1) and most of the cells in the right half contain the opposite type data (say, 0), the precharge voltage during a read cycle on the two halves of the bit-lines will be different. In Figure 4.23, all the cells connected to the left-half bit-line are at 1 and all but one of the cells connected to the right-half bit-line are at 0. If the cell containing 1 in the right-half bit-line line is read, both half bit-lines will at first be precharged to a voltage V_P. But the weak-inversion currents in the right-half cells would cause the precharge level of the right half to be degraded to $v_p = V_p - (1/C_B) \sum_{j=0}^{0.5w-1} I_{w_j}^\tau$, where C_B is the capacitance of the bit-line, I_{w_j} is the weak-inversion current in C_{ij}, and τ is the time interval for precharge during read operation. If the weak inversion currents are sufficiently large, the degradation in precharge voltage will be significant, that is, $v_p \ll V_p$. Consequently, when the word-line W_j is selected to read the contents of C_{ij}, which contains 1, a 0 will be read erroneously, because the difference between the precharge voltages in the right and left halves will be significant.

Technology and Layout-Related Testing

An algorithm to detect this imbalance fault should create the maximum voltage imbalance between the two precharged half-bit-lines (i.e., worst case conditions to stimulate voltage imbalance) and then verify the contents of the cell. A little thought tells us that the maximum imbalance will occur when cells on one half-bit-line contain a certain logic value (0 or 1) and cells on the other half-bit-line all contain the opposite value; the particular cell to be read should have a value opposite to that of the other cells on the same half-bit-line. The algorithm is described briefly below.

(1) Write 0 in all cells connected to the left half-bit-lines, except the cell to be read (c_1) which contains a 1.

(2) Write 1 in all cells connected to the right half-bit-lines, except the cell to be read (c_2) which contains a 0.

(3) Read c_1 and c_2.

(4) Repeat steps 1 through 3 with complementary data.

4.6.4 Coupling faults between bit-lines and word-lines

It is well-known that there is an overlap between the bit-line and the word-line, since they are orthogonal to each other. This overlapping forms a coupling between the bit-lines and the word-lines so that when the bit-line voltages change due to precharging and restoring operations during a read cycle, the unselected word-lines may be inadvertently turned on. This coupling is maximum if all the cells in the selected word line contain 1 and if some of the cells in the coupled unselected word line contain 0, which may be degraded due to the weak-inversion current of the access transistors. In the precharging phase, at first the bit-line voltage increases from 0 to V_P which is coupled as a noise voltage V_{n_j} to a word-line W_j, where $0 \leq j \neq i \leq w-1$. If all the cells in a word-line W_i contain 1, then if W_i is selected its voltage will increase, coupling a noise voltage V_{v_j} to W_j. The effect of these superimposed noise voltages may generate a sufficient weak-inversion current such that a stored 0 in a cell on the unselected word-line W_j may be degraded. It is seen quite easily that the algorithm we need is a walking-1 algorithm. The algorithm is sketched below.

(1) Write the all-1 pattern in one word-line, and 0 in all the other word-lines.

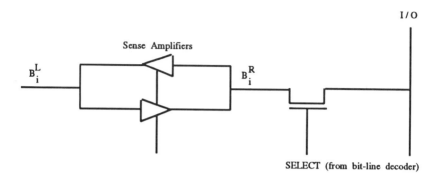

Figure 4.24 Single-ended write; courtesy [97]

(2) Read all word-lines.

(3) Repeat (1) and (2) for the next word-line. If all word-lines are processed in this manner, stop.

4.6.5 Single-ended write problems

In a DRAM employing single-ended write technique, a single I/O line is used to write into the bit-lines. From Figure 4.24, it can be seen that writing on the right bit-line is controlled by the I/O line, whereas writing on the left bit-line is controlled by the sense amplifiers. Therefore, the 0 level on the right bit-line is determined by the input driver, whereas the 0 level on the left one is determined by the sense amplifiers. Thus the level of 0 in the two halves may be different, and this asymmetry may result in pattern-sensitive faults.

4.6.6 Multiple selection faults in the decoder

This happens when there is an overlap between the precharge clock and the decoder enable clock. Multiple selection may occur if they overlap because of incorrect timing.

4.6.7 Faults related to the transmission-line effect

In the two-layer interconnect technology, the bit-lines are made of metal and the word-lines are made of polysilicon or diffusion, or vice versa. Usually, the delay through the poly and diffusion lines is proportional to the square of the length (because delay in these lines is due to the product of resistance and capacitance). Because of the high resistivity of these interconnects, the cells at the periphery of the chips, away from the sense amplifiers, are delivered a weak signal and may thereby fail. By inserting repeaters at suitable intervals, the signal strength and delay may be improved, but the layout may become more complex. In order to verify that all the cells in the array satisfy the limits of the stipulated memory cycle time, the transmission-line effect should be tested.

4.6.8 Testing of single-ended write, multiple selection and transmission line faults in DRAMs

In order to test all these faults, it is necessary to identify the circumstances in which each of them is likely to be maximum. It can be easily seen that the field-inversion current I_F which occurs between two adjoining storage cells is maximum if the four adjacent cells of a base cell contain opposite data to that of the base cell, that is, a checkerboard-type pattern can test the effect of a field inversion current. Similarly, the effects of a dark current and gate short can be tested by the checkerboard pattern, because the presence of these leakage currents manifests in the form of a cell stuck-at zero or one. The algorithm shown in Figure 4.25 tests the above three leakage currents in addition to the electrical faults due to the single-ended write and the transmission line effects. Testing for multiple selection faults in the decoder will be discussed a little later. Novel testable architecture for performing such tests efficiently will be discussed in the next chapter.

The effect of a weak-inversion current is maximum when all the cells in a bit-line, except one, contain 0. If the cell which contains 1 is addressed for a read operation, the weak-inversion currents in other cells will tend to degrade the precharge level of the bit-line and thereby the cell containing 1 will be sensed as 0 by the sense amplifier. Since the bit-line capacitance is typically 10-20 times the capacitance of an individual cell, the stored 1 may not be sufficient

1. Scanning each word-line, write a pattern of (01)* if the word line is even and a pattern of (10)* if the word-line is odd.
2. Freeze the clock for the entire refresh interval τ_R for testing *static refresh*.
3. Scan each word-line to verify that all even bit-lines have the same logic level, and that all odd bit-lines have the same level.
4. Read continuously any arbitrarily chosen word-line for the entire refresh interval τ_R and verify that its data is not lost. This test checks the effect of temperature rise and tests *dynamic refresh*.
5. Scan each word-line to verify that all even bit-lines have the same logic level, and all odd bit-lines have the same level.
6. Read continuously another arbitrarily chosen word-line for the entire refresh interval τ_R and verify that its data is not lost. This test checks the effect of temperature rise and tests *dynamic refresh*.
7. Scan each word-line to verify that all even bit-lines have the same logic level, and all odd bit-lines have the same level.
8. Repeat steps 1 through 7, with complementary data.

Figure 4.25 Algorithm 1: Checkerboard test to detect single-ended write and transmission line faults

1. Initialize the entire memory with 0.

2. Arbitrarily select two word-lines W_i and W_j and read them alternately for one refresh cycle. Check that the bit-lines have the same value.

3. For each word-line, compare all even bit-lines and all odd bit-lines to check that their values match.

4. Repeat steps 1 through 3 for the entire memory initialized to 1 in all locations.

5. Initialize the memory so that the left subarray contains 0 in all locations and the right subarray contains 1 in all locations.

6. For each word-line, do the following: (i) write a pattern of (01)* in the selected word-line; (ii) compare all even bits and all odd bits and check for error; (iii)initialize all the cells of the selected word-line to 0 if it is on the left half, otherwise to 1.

7. For each word-line, do the following: (i) write a pattern of (10)* in the selected word-line; (ii) compare all even bits and all odd bits and check for error; (iii)initialize all the cells of the selected word-line to 0 if it is on the left half, otherwise to 1.

8. Repeat steps 5 through 7, with complementary data.

Figure 4.26 Algorithm 2: Walking test to verify whether the weak-inversion current causes a bit-line voltage imbalance (serial version)

/* Bit-line Decoder Multiple Access Test */

1. Write 0 to all cells (bits) on an arbitrarily selected word-line W_j.

2. Read and compare all the cells on W_j.

3. Starting from the cell at the crosspoint of B_0 and W_j, for each cell on W_j, write a 1 and read the cell, and proceed in ascending order of bit-lines.

4. Starting from the cell at the crosspoint of $B_{\sqrt{n}-1}$ and W_j, for each cell on W_j, write a 1 and read the cell, and proceed in descending order of bit-lines.

/* Word-line Decoder Multiple Access Test */

5. Switch the roles of bit-lines and word-lines in steps (1) through (4) above.

Figure 4.27 Algorithm 3: Bit line and word-line decoder test (serial version)

to replenish the degraded precharge level. The testing strategy needs to test each memory cell so that when it is 1 all its bit-line neighbors will have 0.

In order to test the bit-line voltage imbalance, we have seen earlier that we need to write 0(1) on the cells at the bit-line on the left half of the subarray and 1(0) on the cells at the bit-line on the right half of the subarray. Thus the test to detect the weak-inversion current can be utilized to test the bit-line voltage imbalance by testing the left and right subarrays with opposite background data. It may be noted that the test also detects faults due to single-ended write. Algorithm 2 tests all the above faults. Algorithm 3 runs a marching pattern of 1 on a background of 0 and vice versa in each word-line. This algorithm detects the multiple-access faults in the word-line decoder by comparing the read data with the expected data. Algorithm 4 tests the memory for faults caused by voltage spikes due to noise. Noise spikes have high slew rates and they may occur during a read or write memory cycle causing the operation to fail. If the noise causes lowering of the supply voltage, the capacitive bias of the dynamic logic may be higher than the supply bias, and this may result in a failure of a read or write operation. Similarly, if the noise spike increases the supply voltage, the capacitive bias voltage may be sufficiently lower than the supply bias resulting in faulty read or write operation. The effect of this

1. Write 0 in the entire memory at maximum supply voltage.

2. For all word-lines starting from W_0, do the following: (i) write a pattern of $(01)^*$ in the selected word-line at maximum supply voltage; (ii) rapidly reduce the supply voltage to minimum; (iii) compare all even bits and all odd bits to check for error.

3. For all word-lines starting from $W_{\sqrt{n}-1}$, do the following: (i) write a pattern of $(10)^*$ in the selected word-line at maximum supply voltage; (ii) rapidly reduce the supply voltage to minimum; (iii) compare all even bits and all odd bits to check for error.

4. Write 0 in the entire memory at minimum supply voltage.

5. For all word-lines starting from W_0, do the following: (i) write a pattern of $(01)^*$ in the selected word-line at minimum supply voltage; (ii) rapidly increase the supply voltage to minimum; (iii) compare all even bits and all odd bits to check for error.

6. For all word-lines starting from $W_{\sqrt{n}-1}$, do the following: (i) write a pattern of $(10)^*$ in the selected word-line at minimum supply voltage; (ii) rapidly increase the supply voltage to maximum; (iii) compare all even bits and all odd bits to check for error.

Figure 4.28 Algorithm 4: Power supply voltage transition test – serial version

power supply voltage transition is evaluated for both the cases when the supply voltage rapidly increases and again when it rapidly decreases.

In the next chapter, we shall discuss BIST hardware for testing for the above faults using parallel versions of the algorithms described above.

4.7 CONCLUSION

This chapter describes the manner in which circuit design errors and layout defects for a given SRAM or DRAM technology can be used as a basis for SRAM testing. Conventional functional and electrical testing have a limited fault and defect coverage because they consider a simplified fault model based only on the input/output interface behavior of an SRAM cell. In the process, many layout defects that do not translate into observable faults are ignored. Technology- and layout-related tests can be performed in a variety of ways — by measuring the quiescent or steady-state power-supply current (I_{DDQ} or I_{DD}), by mapping hardware defects into faults, or by simulating the cell array to study the fault effect of parameter variations. There are various advantages of performing such tests — they are more circuit-specific and give a better idea of actual failure modes; besides, they achieve a higher fault coverage because the test techniques developed are targeted at the actual layout defects. Another important advantage of these techniques is that they can be used for *precise diagnosis* of defects, thereby being more useful for cell repair and reconfiguration purposes. Conventional testing, on the other hand, employs simple and generic fault models and is thereby quite cheap. It however fails to achieve precise defect diagnosis and usually has limited defect coverage.

Technology and Layout-Related Testing

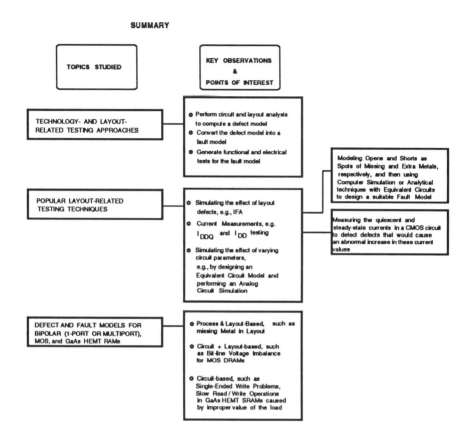

4.8 PROBLEMS

Worked-out Example

1. What is address skew for SRAMs and how can a test system study its effects? Describe, in particular, the effect of address skew on SRAMs with ATD (address transition detection).

   (**Sample Answer Sketch:**) In most cases, the address lines of an SRAM change logic levels at slightly different times, because of component and wire-length variations as determined by the layout characteristics. This phenomenon is known as address skew. If the magnitude of address skew is appreciable, an ATD (address-transition detecting) SRAM would interpret it as multiple transitions that usually result in multiple precharge pulses being generated. Hence, the length of the precharge cycle will be increased if address skew is present. However, it will not technically affect the access time because the access time is measured from the *last* address transition to the appearance of data.

   The effect of address skew may be studied if a test system is able to skew any address line relative to the remaining ones. One possible method of skewing an address line on a cheap tester is to use an alternate clock for the skewed addresses. This approach maintains timing accuracy and edge slew rates, but may require physical movement of wires on a performance card to examine skews on different address lines. A brute-force approach is to load the address input with a small lumped capacitance. This approach is not preferred because of its effect on edge slew rates and its relative lack of precision and control. Simulation results confirm the expectation that if one address input is skewed by a certain time delay, the data output will be delayed by the same amount as the address skew.

4.8.1 Exercises

1. The reason why data retention faults cannot be tested without introducing the two delays in Dekker's algorithm is because of the precharge phase accompanying every read operation. As a result, the $9N$ operations of Dekker's test, even though theoretically sufficient to test all data retention faults, will not detect them in practice, unless they are staggered at appropriate points by (long) delays. Explain, with the aid of circuit diagrams, how to simplify Dekker's test algorithm by modifying read operations of

the $9N$-test. Also show simulation results of circuit voltages at the storage nodes of an SRAM cell with a broken pullup.

2. Design SPICE models for the I_{DD} defect mechanisms shown in Figure 4.8. Simulate these defect mechanisms and tabulate the impact of these defect models on the dynamic power supply current.

3. What is the main advantage of using Inductive Fault Analysis (IFA)?

4. Perform a realistic IFA test using batches of actual wafers manufactured by a fabrication plant. Use a scanning electron microscope (SEM) to observe physical defects and perform statistical fault clustering. Then, on the basis of the most likely faults, suggest efficient test algorithms for functional and electrical testing.

5. Derive the equation of the cell storage voltage of a GaAs HEMT SRAM cell with the help of the simplified circuit for read operation, illustrated in Figure 4.14. Also derive the time taken for the storage node voltage V_1 to discharge below V_{TE}.

6. What are **chip select whiskers**? How can a tester study the effects of chip select whiskers?

5
BUILT-IN SELF-TESTING AND DESIGN FOR TESTABILITY

Chapter At a Glance

- Concurrent, Non-concurrent, and Transparent Testing
- BIST Architecture for RAMs
- Parallel Signature Analyzer
- BIST based on Functional and Electrical Tests
  -- TRAM, Serial BIST based on March Tests, BIST for Delay Faults
- DFT and BIST based on Parallel Testing
  -- Parametric Faults, Pattern-Sensitive Faults Address Maskable Parallel BIST
- Transparent BIST Algorithms and Architecture
- BIST based on Pseudo-random Pattern Testability
- Boundary-scan-controlled Pseudo-random BIST

5.1 INTRODUCTION

A *built-in self-test* (BIST) approach allows all the functions comprising a test to be performed autonomously within a chip, with no external control or test data. A *design for testability* (DFT) approach, on the other hand, allows part of the test algorithms to be executed off-chip. Usually, the inner loops of a test algorithm (the ones that produce the largest reduction in test time) are implemented on-chip in a DFT approach.

BIST provides a number of advantages over conventional off-chip memory testing techniques — first, the chip does not need to spend much time on a tester; secondly, this technique can be used for embedded RAMs for which neither are the address, read/write and data input lines controllable nor are the data output lines observable.

However, there are several design considerations that need to be examined to prevent the reduction in test cost from being neutralized. They are: silicon area overhead, speed of the overall hardware, extra pins or time-multiplexed pins, fault coverage, and test time. They are discussed below:

(1) The silicon area required for the BIST circuitry would reduce the area available for memory cells, or would cause an increase in the die size, thereby decreasing yield.

(2) The addition of BIST circuitry should be done so as to cause minimal impact on the memory access time — in other words, the BIST circuitry should be placed away from the critical path within the chip.

(3) Extra pins or multiplexed pins are often necessary to make the device switch from *operating mode* to *test mode*, because extra control signals are required to achieve this. The use of extra pins should be minimized because pins are an important cost item and have packaging consequences. There is a performance penalty for multiplexed pins due to the multiplexer delay.

(4) BIST should try to achieve a high fault coverage, at least for the most likely faults.

In addition, some designers prefer that the BIST circuitry should be self-checking; in other words, if the memory is fault-free but the BIST circuitry is itself faulty, an error signal should be produced. However, this requirement is often relaxed because of higher area overheads that may arise as a result, and most designers try to ensure that the BIST circuitry can be checked using some (external) means and not necessarily through self-test.

An important advantage of using BIST is that it eliminates the need for expensive test equipment. Optimizing the silicon area might cause an increase in the test time, but this can be somewhat offset by exploiting the parallelism within the chip.

5.2 CONCURRENT, NON-CONCURRENT AND TRANSPARENT TESTING

A test is said to be *concurrent* if it is performed during normal use of the chip — that is, faults occurring during normal use can be detected and even corrected, if the test mechanism is powerful enough. Concurrent testing usually relies on some form of information redundancy used in the form of a parity checking or an error-correcting code (ECC). Concurrent testing allows all faults in the memory, whether permanent or non-permanent, to be detected and/or corrected, providing a certain level of fault tolerance and obviating a special test mode of operation. The disadvantage is the large hardware overhead due to the logic circuitry required to generate the redundant information upon a write, for checking the normal and redundant information upon a read, and for storing the redundant information. Moreover, every memory access has an associated performance penalty due to the time required for performing the generation of a codeword or for checking the read data. Implementation problems force the use of a relatively simple fault model for such testing.

A *non-concurrent* test is performed only in test mode and not during normal use. Most BIST implementations are non-concurrent. The advantage of this test mode over the concurrent mode is that the test does not have to preserve the data stored in the chip, thus allowing a maximum freedom in the test data and testing for complex fault models. Thus, hardware for non-concurrent testing is cheaper. The disadvantage of non-concurrent BIST is that faults not covered by the fault model and intermittent ones that occur between BIST periods, will not be detected.

A non-concurrent test is said to be *transparent* if at the end of the test session, the contents of the RAM are equal to its initial contents. Transparent tests are very suitable for periodic testing because they ensure that normal operation on the memory data can be continued between periodic test sessions. We shall

Figure 5.1 A RAM test configuration

soon discuss some transparent BIST techniques that accomplish the same fault coverage as their non-transparent counterparts, with only a slight increase in the silicon area overhead. This chapter will deal mostly with non-concurrent and transparent testing. Concurrent testing or the use of error-correcting codes, has applications primarily in ensuring reliability and fault-tolerance of RAMs and is beyond the scope of this book.

5.3 AN OVERVIEW OF BIST APPROACHES FOR RAMS

The most high-level description of a test configuration is a cascade of three stages, the *test data generator* (TDG), the *RAM under test* (RUT), and the *response data evaluator* (RDE). This cascade of stages is shown in Figure 5.1 [165].

5.3.1 Classification

From a test acceleration point of view, non-concurrent BIST architectures are classified according to the number of bits and the number of memory cell arrays accessed. Accordingly we can have SA or MA (single array or multiple array), and SB or MB (single bit or multiple bit access) and four combinations of the above pairs. These four combinations are described below [43].

- **Single array, single bit architecture (SASB):** A single bit is accessed at a time with no parallelism being used. This architecture is needed for certain fault models such as the coupling fault model, which allows any cell to be coupled with any other cell.

- **Single array, multiple bit architecture (SAMB):** A single array is tested while accessing multiple bits — this is done by writing the same data into multiple cells in a single access, or employing multiple reads and

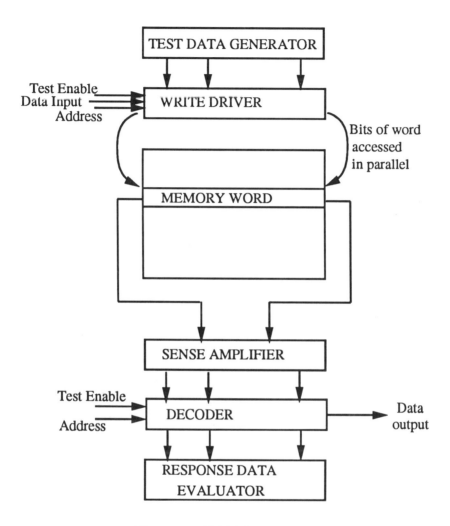

Figure 5.2 Multiple bit architecture

properly managing the read data (for example, by the use of compression techniques). The mechanism used for accessing multiple bits is shown in Figure 5.2.

If the memory array has n cells, consisting of $\sqrt{n}$ rows and $\sqrt{n}$ columns, the SAMB architecture can reduce the test time by a factor of $\sqrt{n}$, if the parallelism in the row access can be exploited completely. This form of testing is known as *line-mode testing* [63, 88]. Some tests, for example,

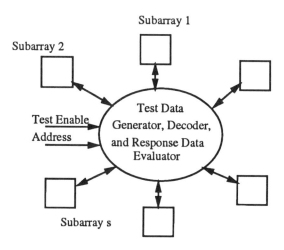

Figure 5.3 Multiple array architecture

those for coupling faults between cells within a row, are hard to implement using this approach.

- **Multiple array, single bit architecture (MASB):** If the memory chip is composed of several subarrays, they can all be tested independently and in parallel. If a single bit is accessed from each array, the BIST architecture is an MASB architecture. The block diagram of an MASB BIST is shown in Figure 5.3 [165]. If there are s subarrays, the MASB BIST architecture reduces the test time by a factor s.

- **Multiple array, multiple bit architecture (MAMB):** The MAMB architecture combines line-mode testing with parallel test of s subarrays. As with the SAMB architecture, tests for certain fault models, such as the coupling fault model, are difficult to implement for faults coupled within a row.

5.3.2 Test data generation

BIST hardware need to be controlled to enter test mode, to communicate test commands, to enter test data and/or to retrieve response data, and to go back to normal mode on completion. Various ways of doing this are: by using extra pins, by using extra signal levels on pins, and through the use of an illegal control sequence which would fail to cause a normal row/column access.

BIST and DFT

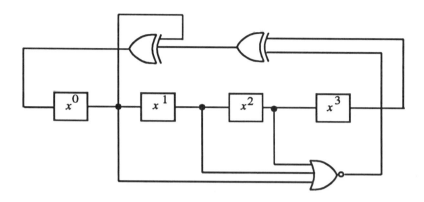

Figure 5.4 A self-testable LFSR

Test data is generated using an external or an internal test data generator (TDG). Usually, the generation of addresses is the major task of the TDG, and the test data is derived from the addresses or generated using a small finite state machine. There are three common methods for address generation — by using a counter, an LFSR, or a microprocessor.

A counter generates physical addresses sequentially and is thereby useful for BIST applications such as tests for NPSFs which require sequential memory access. The two main disadvantages of the use of counters are: (a) more hardware requirement than LFSRs because of their carry propagation logic; (b) they cannot readily be made self-testable.

Because of their smaller area, LFSRs are preferred to counters for address generation when march tests are used. The theory of LFSR pattern generation can be found in [165]. An LFSR has the property that the probability of a particular address bit change is equal for all address bits; hence, use of an LFSR permits the detection of some address-dependent delays or write recovery faults. LFSRs require less silicon area than counters and can easily be made self-testable. An LFSR is a serial-in-serial/parallel-out *pseudorandom pattern generating* (PRPG) shift register with feedback paths through XOR gates, the location of these paths determining the *characteristic polynomial*. Often some additional hardware, such as a parity checker, or a NOR gate, is added to the design. An example of an LFSR is shown in Figure 5.4. The characteristic polynomial determines the patterns that will be generated by the PRPG.

For the LFSR shown in Figure 5.4, the characteristic polynomial is $x^4 + x + 1$. This characteristic polynomial is obtained as follows: we notice the following relationship between the contents of the D flip-flops (making the LFSR) at time t and the input at time $t + 1$:

$$[x^3]_t + [x^0]_t = [x^0]_{t+1},$$

for all $x^0 x^1 x^2 x^3$ combinations except the combination 0001, + denoting modulo-2 addition, or XOR.

It can be noticed that $[x^i]_t = [x^{i+1}]_{t+1}$, and hence, manipulating the above equation to obtain the same subscript $t + 1$ throughout, we obtain the polynomial $x^4 + x + 1$, which takes the value 0 for every possible bit-string that can be generated by the LFSR from an initial bit-string. It is therefore easily seen that the LFSR cannot generate the all-0 string.

The characteristic polynomial, therefore, determines the sequence of patterns that will be generated if the initial pattern is provided. To force the LFSR to generate the all-0 string, the NOR gate is provided. It is easily seen that the all-0 string (0000) will be produced immediately after the string 0001 is generated by the LFSR.

5.3.3 Response data evaluation

If the response data evaluator (RDE) is external to the chip, it is impossible to exploit the inherent parallelism within the chip because of a constraint on the number of pins. However, if the RDE is internal to the chip, the parallelism within the chip can be exploited. There are several methods for parallel on-chip response data evaluation — deterministic comparison, mutual comparison, the use of a multiple input shift register, and a microprocessor analyzer.

In *deterministic comparison*, the response data from a group of cells accessed simultaneously from the same row is compared against the expected data, using two-input XOR gates. The outputs of the XOR gates are OR-ed together to produce an error output, which goes high if at least one of the cells stores a faulty value. This method is called 'deterministic comparison' because the expected data is known beforehand, and all cells in the row are expected to contain the same data value. This method requires little additional hardware and can be used to test for NPSFs by using the multiple bit (MB) architecture, where the memory array is filled with *a few* different patterns per row of cells. This method will not be useful for tests in which there are *a large number of* different patterns per row of cells, for example, march tests, GALPAT, and others.

The comparator consisting of XOR gates can be easily tested using the theory of combinational logic testing. The OR gate can also be simultaneously tested by choosing suitable test patterns. An external tester writes the required test patterns in normal mode and then the vector is read out in test mode from the memory cell array.

In *mutual comparison*, a multiple array (MA) architecture is used, the outputs of the arrays being compared mutually. As before, this is done using a mutual comparator consisting of a set of two input XOR gates whose outputs are fed into an OR gate. One disadvantage of this method is that if all the subarrays are equally faulty (which is rather improbable but not impossible), the fault will go undetected, for obvious reasons. The advantages of this method are: low hardware costs, no reference values to be known and provided *a priori*, and no delays in the normal data path of the chip. The mutual comparator is itself tested in the same way as a deterministic comparator.

A *multiple-input shift register* (MISR) can be used to reduce the amount of response data using the technique of *response compression*; this method, described in [154, 155] consists of of the following steps: computing a compressed version, called the *signature* of the response data by passing it through the MISR in *compress mode*, this response signature being the remainder obtained when the polynomial represented by the response data is divided by the characteristic polynomial of the MISR; and then comparing the response signature with a *reference signature* in *scan mode*. When an MISR is used for response compression, it is known as a parallel signature analyzer (PSA), described in the next section.

Another technique for response data evaluation is the use of a *microprocessor* to analyze the response. This technique, proposed by Ritter [138], takes into account spare rows and columns. It has a large area overhead and low performance.

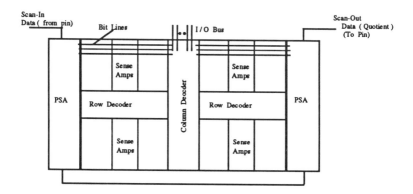

Figure 5.5 Block diagram of a DRAM with an integrated PSA; courtesy [155]

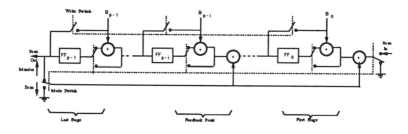

Figure 5.6 Functional diagram of a parallel signature analyzer; courtesy [155]

5.4 PARALLEL SIGNATURE ANALYZER (PSA): INTEGRATING TEST APPLICATION WITH SIGNATURE ANALYSIS

In a high-density, large-capacity DRAM, the memory cells are arranged into two or more two-dimensional subarrays. In Figure 5.5, a testable memory with four subarrays is shown with a parallel signature analyzer (PSA) integrated at the periphery of the chip. The column decoder selects a bit line and the row decoder selects a word line. The sense amplifiers drive the bit lines which run across each subarray from its outside quadrant to the column decoder. The PSA is embedded in the RAM chip to access all the bit lines in parallel. Two more pins are added in the chip — an input pin to scan data into the PSA and

BIST and DFT

an output pin to scan the signature out. The concept of fast data compaction by using PSA was originally proposed by Benowitz, et al.[11].

The functional diagram of the PSA adapted for the testable memory is shown in Figure 5.6. The PSA has a pair of controls — a $MODE$ control and a $R/\overline{W}$ (read/write) control. It principally works in three modes, as described below:

- a **shift** mode ($MODE = 0$ and $R/\overline{W} = 1$), during which the PSA is configured as an m-bit shift register, with bit-data serially loaded into the PSA through the scan-in input, and simultaneously the stored data serially shifted out through the scan-out pin;

- a **parallel load** or **write** mode ($MODE = 1$ and $R/\overline{W} = 0$), during which an m-bit pattern is written in parallel into the PSA in one memory cycle;

- a **signature** mode ($MODE = 1$ and $R/\overline{W} = 1$), during which the content of m-cells are simultaneously read into the PSA to generate a signature whose value depends on the characteristic (feedback) polynomial of the PSA. Every time a signature is generated, the quotient bit is scanned out and compared to check any discrepancy. In order to achieve high parallelism in testing, a signature (or its cyclic shifted pattern) is usually used as an input pattern for writing into the m cells of the subsequent word lines. After several write and read operations, the whole signature is scanned out. Thus, by continuous monitoring of the quotient bit and the final signature, the occurrence of any fault in the memory can be detected.

In the technique proposed by Sridhar in [155], the same PSA can be used for test application as well as for response evaluation. In 1985, Sridhar proposed a scheme [155], by which a conventional algorithm, such as a march test, can be modified so that patterns are applied and responses are captured via the PSA. He considers the memory array of $N \times d$ cells to consist of N/F words of size p each, where $p = Fd$. Each of these p-bit words can be accessed via the k-bit PSA. In other words, all Nd memory cells can be read in Nd/p cycles, and all Nd/p p-bit words can be written with the same p-bit pattern in Nd/p cycles. This causes the read access time and the number of write cycles to be reduced by a factor of p. As mentioned above, the PSA has a "scan" mode of operation, by virtue of which it functions as a k-bit shift register whose contents can be scanned out at the end of a test experiment using k cycles through the scan-out pin. The scan-in pin is used to load the PSA with any desired k-bit word. This will be useful to write known data into the p cells during the write mode. If separate pins are used for input and output, both scan-in and scan-out of data could be achieved concurrently in k cycles only. A PSA performs a self-test during normal use, a hardware error in it producing an incorrect signature. It can be used with any test, because no assumptions are made about the nature of response data.

The fundamental idea behind the use of PSAs to test large memories is to reduce the total test time by reducing the number of read and write cycles. Any functional algorithm can be modified into a parallel algorithm (an algorithm that reads and writes from/into multiple bits in parallel) that can be used in conjunction with a PSA. Hence, each write operation of the test algorithm is modified into a *parallel write* operation via the PSA, reducing the write cycle time by a factor of F. Each read-operation of the test algorithm is modified into a signature analysis operation via the PSA. Sridhar illustrates the modification of a $6n$ march test and a Walking 1s and 0s algorithm using this approach. In any such modification, before a memory write operation, the PSA should be initialized with the data to be written. This can be achieved in two ways: by *scanning in* the required data from outside in the *scan mode*, which requires k-scan cycles for the k-bit PSA, or to derive the write data from the current signature by inverting selected bits. This is often possible because the signature sequence is periodic in nature in the fault-free case. The inversion of selected bits causes the PSA initialization to be significantly quicker than with the scan-in approach which takes k cycles.

Sridhar [155] examines the manner in which signatures and quotient bit streams are generated. In the case of memory test algorithms, test data patterns are often repetitive. Therefore, the data compressed by the PSA is also repetitive in nature. However, it is found [155] that when the parallel data compressed by the PSA is *periodic*, the signature sequence is not necessarily so. As described earlier, a periodic signature sequence is useful in generating test data to be written into the memory from PSA contents, instead of scanning in test data bits into the PSA. It is conjectured that for a given k-bit PSA with maximal-length feedback polynomial, exactly one initial PSA state will produce a periodic signature sequence with the same periodicity as that of the parallel data input being compressed. For this initial state of the PSA, the signature sequence and hence the quotient bit stream are periodic. An interesting application for having a periodic signature sequence is the testing for faults which destroy the periodicity, and are thus detectable.

Sridhar [155] proves that the test times due to $O(n)$ and $O(n^2)$ algorithms can be reduced by a factor of F and F^2 respectively, where $F = p/d$, even if the *scan-in* PSA initialization is used. For RAMs of capacity such as 256 Kb or 1 Mb, therefore, this approach is advantageous in reducing the test time.

A disadvantage of a PSA or an MISR is its larger area overhead (caused by a higher transistor count) than a comparator. This produces an additional area overhead of 0.4% for a 256 Kb DRAM [97]. Another disadvantage is that the layout does not fit well within the inter-cell pitch of the memory cell array. A third disadvantage is that of *aliasing*. Aliasing is said to occur when, because of loss of information that results during response compression, identical signatures are produced in the PSA for a faulty memory array and a fault-free array. Fortunately, for certain simple kinds of faults, such as bit lines stuck at 1, there is no reduction in fault coverage due to aliasing. Another shortcoming of Sridhar's approach is that for a memory array with hardware redundancy, redundant columns (bit lines) may cause unpredictable or unknown data from a bad column in a repaired memory chip to be received by the PSA. This will invalidate the parallel test, because the correct expected signature will be unknown. A workaround for this problem is to place the PSA *after* the column decoding stage that selects the redundant bit lines; this way, the PSA will get no bad data from the bad columns that are replaced. A possible drawback of this scheme is that the redundant bit line decoders are typically towards the end of the hierarchical column-decoding system. For example, suppose the redundant columns are multiplexed at the final stage of the column decoding, when the final 1-out-of-4 selection is made. Then the ratio $F = p/d$ will be only four. However, more parallelism can be obtained by moving the redundant decoders to an earlier column decoding stage, which, in turn, may cause other problems that are layout-related, such as wiring and extra silicon area.

A second solution [155] to the problem of capturing the data stored in faulty bit lines is to integrate the PSA at an early stage of the column decoder and select the redundant column (or bit line) at a later stage. Let the PSA be placed after a 1-out-of-n decoding of bit lines, that is, every PSA bit receives data from one out of n bit lines. The parallel test will test all the bit lines *before* any redundant column is selected. However, after the selection of the redundant bit line, some j-th bit position of the PSA can still receive bad data from the bad column every time that column is addressed. At the same time, potentially good data from the redundant bit line will be multiplexed after the PSA. Therefore, it will not be possible to apply the parallel test to a repaired memory, because the correct expected signature is unknown. One solution to this problem [155] is to force the expected signature to a known value by writing a constant value (0 or 1) into the affected bit position of the PSA. This constant value may be written using a fuse-blown circuit or multiplexing logic. This will necessitate testing of all the good bit lines and the selected redundant one outside the parallel test mode, because the j-th bit position of the PSA will be unresponsive to data from the $(n-1)$ good bit lines.

For a repaired memory chip therefore, Sridhar's parallel signature analysis technique has problems associated with it. Sridhar recommends the use of conventional serial algorithms if these problems cause the parallel signature analysis to be infeasible. His proposed PSA can still be useful in reducing the overall cost of testing of a mixture of repaired and unrepaired chips.

5.5 DFT AND BIST ARCHITECTURES BASED ON DETERMINISTIC TESTING

A number of DFT and BIST approaches for efficient implementation of various deterministic tests are popular. Both functional and electrical fault models are used in such tests. For example, some BIST techniques use various march tests on the memory, while others perform testing for delay faults [144]. Various BIST and DFT approaches are described below.

5.5.1 Tree RAM (TRAM) DFT

In this approach, the memory array is partitioned into subarrays that are tested in parallel [66], with decoders placed between the subarrays. Decoder delays are thereby reduced drastically because of shorter bit and word lines. Conceptually, a TRAM is as shown in Figure 5.7 [66].

An external tester is used for testing the memory array of a TRAM chip — the approach is thereby a DFT one. The TRAM operates in one of two modes: a normal mode, and a test mode. In normal mode, the decoder selects a memory node and a single cell inside that node. In test mode, it selects all nodes simultaneously, and accesses a single cell within each node. The following pseudocode describes the operation of the DFT algorithm in the test mode:

(1) Write the same data via the data input pin on all s selected cells of s nodes. Data is thus written in a bit-serial fashion, and the architecture is thereby MASB.

(2) Cells from pairs of adjacent nodes are compared, as shown in Figure 5.8. The pairs are: nodes 0 and 1, nodes 1 and 2, nodes 2 and 3, and so on, upto nodes $n-1$ and n (where $n+1$ is a power of 2, for example, $n = 15$). When

BIST and DFT   235

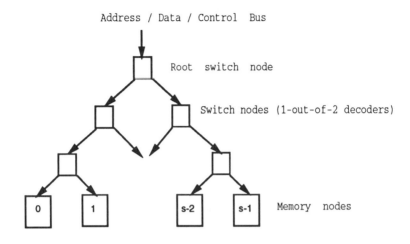

Figure 5.7 A tree RAM organization; courtesy [66]

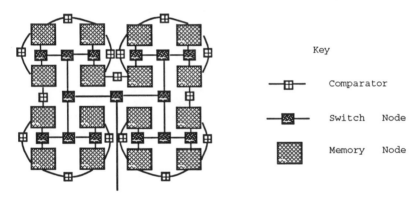

Figure 5.8 DFT features of a TRAM chip; courtesy [66]

the output values don't match, the $\overline{FAIL}$ becomes active. This completes the testing of the memory cell array.

(3) Testing the tree decoder is done by verifying that it is possible to uniquely access one cell in each memory node.

(4) Testing the DFT structure consists of verifying the outputs of the comparators consisting of XOR gates and a multi-input NOR gate. Simple combinational logic testing techniques are employed for this.

5.5.2 BIST based on march and other functional tests

There is a basic difference between external testing and BIST. In external testing, the first priority is to minimize the length of the test algorithm, whereas in BIST, the first priority is to make the tests as simple and regular as possible, so that the self-test circuitry can be made to require a minimum of silicon area.

As mentioned earlier, march tests are particularly amenable to BIST hardware, because they are regular and symmetric and based upon sequential address generation and application of a fixed set of operations at each address. Besides, march tests designed intelligently can give a high fault coverage. The hardware components to a BIST design for applying march tests on a memory consist of the following:

- A finite state machine M, that can be used to generate read and write control signals for the selected march test, the number of states of this machine being the total number of reads and writes in all the march elements taken together. This finite-state machine must step through the states of the selected march test.

- An address generator which receives a control signal from M and generates the next address — this control signal determines whether the current address will be incremented or decremented.

- Two copies of the memory may be used, and whether or not the correct data has been read is determined by comparing the two copies using a comparator. An error signal is sent to M if they don't match. The obvious bug in this arrangement is that both copies can have an identical error at the same location, which is very unlikely but possible nevertheless.

Several hardware implementations of march-test based BISTs and the fault models on which they are based, have been described in [119]. Nadeau-Dostie, et al. [119] have presented an approach based on serial memory access. The fault model they use consists of the basic stuck-at-fault model for the memory array, the address decoder, and the read/write circuitry. In addition, they consider two types of coupling faults in memory cells — state coupling faults (static) and dynamic coupling faults (CFids and CFins). The algorithms they consider for BIST implementation are March C- (similar to that described by Marinescu), GALPAT and Walk tests. Also, they consider a modified March

Table 5.1 Serialized march elements

Operation	SDI	Memory Word	Data Latches	SDO
read	-	$abcd$	$abcd$	d
write e	e	$eabc$	$abcd$	d
read	-	$eabc$	$eabc$	c
write f	f	$feab$	$eabc$	c
read	-	$feab$	$feab$	b
write g	g	$gfea$	$feab$	b
read	-	$gfea$	$gfea$	a
write h	h	$hgfe$	$gfea$	a
read	-	$hgfe$	$hgfe$	e

test to incorporate stuck open faults observed in static address decoders. Their method and algorithms are described below.

The serial shifting operation

The method described in [119] is based on applying march elements with *shifting* — this concept is useful in the context of embedded memories, where read and write operations cannot directly be performed on memory cells. The shifting is performed by adding multiplexers to the inputs of the write drivers, with each multiplexer selecting between the normal data input originating outside the memory or the value stored in the transparent latch (originating from the sense amplifier) from its left-neighboring bit. During normal mode of operation, the normal data input is selected, and during test mode, the memory is configured as a shift register and memory writes produce shifting. The operation is described in Table 5.1 [164].

In Table 5.1 SDI and SDO denote Serial Data In and Out respectively. From this table, we can make the following four observations:

- During a read operation, the contents of the accessed memory word is identical to that of the data output latches.

- During a write operation, only the contents of the accessed memory word are changed — the output data latches still contain the old value, and will only change their contents during the next read operation. A write operation causes each cell to obtain data from its left neighbor instead of from outside; this has the effect of 'shifting' the data one step to the right.

- The SDO content is the same as the rightmost bit in the output data latches, and will therefore change only during read operations, and can never change during write operations.

- The SDI is meaningful during write operations (obviously, SDI is identical to the contents of the leftmost bit in the memory word), and is thus irrelevant during a read operation.

Algorithms used

Modifications of simple functional algorithms are used with this serial access BIST approach. These algorithms are SMarch, SMarchdec, SGalpat, and SWalk. A special notation may be used for describing these algorithms, as follows: each word has c bits, and there are a total of w words in the memory; there are three kinds of read operations, $r0$, $r1$, and rx, depending on the expected values, and three kinds of write operations $w0$, $w1$, and wx, depending on the value actually shifted in from the SDI, x denoting a don't-care. A notation such as $(r0w1)^c$ means that the operation $r0$ followed by $w1$ is repeated c times. A write operation causes only the leftmost bit to be directly controlled, and a read operation causes only the rightmost bit to be directly observed. SMarch, SMarchdec, and SGalpat are *word-oriented*, which means that the entire word is shifted out at each address. SWalk, in contrast, is *bit-oriented*, which means that a single shift operation (a read-write pair) is performed at each address.

The pseudocodes for these algorithms are shown in Figures 5.9, 5.10, 5.11 and 5.12[119].

SMarch is similar to Marinescu's Algorithm C-, covering all stuck faults and some coupling faults. Its complexity is $24cw$. SMarchdec is an extension of SMarch to cover stuck-open faults observed in static address decoders. Its complexity is $24cw+2cw(1+2\log w)$. Since the SMarchdec test adds about 30% area overhead, it is used sparingly. SGalpat is a serial implementation of the classic GALPAT test, and the pseudocode is self-explanatory. Its complexity is $O(cw^2)$.

SWalk is somewhat different from the other three algorithms. To run SWalk, the memory is configured as a huge shift register, as follows: the w-word c-bit-per-word memory is converted into a cw bit shift register, by adding a D

(1) *for* address $= 1$ *to* w *do*
   {count address forward}
   $(rxw0)^c(r0w0)^c$
   {initialize memory with 0s}
(2) *for* address $= 1$ *to* w *do*
   $(r0w1)^c(r1w1)^c$
   {read 0s and replace with 1s}
(3) *for* address $= 1$ *to* w *do*
   $(r1w0)^c(r0w0)^c$
   {read 1s and replace with 0s}
(4) *for* address $= w$ *to* 1 *do*
   {use reverse address sequence}
   $(r0w1)^c(r1w1)^c$
   {read 0s and replace with 1s}
(5) *for* address $= w$ *to* 1 *do*
   $(r1w0)^c(r0w0)^c$
   {read 1s and replace with 0s}
(6) *for* address $= w$ *to* 1 *do*
   $(r0w0)^c(r0w0)^c$
   {only first read operation is important}
   {final RAM state can be selected}

Figure 5.9 Pseudocode for the SMarch algorithm

(1) Execute SMarch.
(2) *for* refadd $= 1$ *to* w *do*
    {RAM assumed to be clear initially}
    $(r0w1)^c$
    {write 1s at reference address, denoted by *refadd*}
    for $j = 1$ *to* $\log w$ *do*
    {for each bit of the address}
        newadd $=$ refadd $\oplus 2^{j-1}$
        {flip one bit of the reference address}
        {to produce an address *newadd*}
        $(r0w0)^c$
        {write 0s at this new address}
        $(r1w1)^c_{refadd}$
        {check contents of reference address}
    $(r1w0)^c$
    {reset contents of reference address to 0s}

Figure 5.10   Pseudocode for the SMarchdec algorithm

(1) Initialize memory with 0s by executing the march element $(rxw0)^c$ for each address.
(2) *for* refadd $= 1$ *to* w *do*
    $(r0w1)^c$
    {write 1s at reference address}
    for newadd $= 1$ *to* w *do*
        if newadd $\neq$ refadd
            $(r0w0)^c$
            {check if 0s are still there}
        $(r1w1)^c$
        {check if 1s are still at reference}
        {write 0s at this new address}
    $(r1w0)^c$
    {reset contents of reference address to 0s}
(3) Repeat steps 1 and 2 with complementary data.

Figure 5.11   Pseudocode for the SGalpat algorithm

flip-flop from the SDO to the SDI. The concept is similar to the one used by You and Hayes [183], but achieves the same result with much less hardware. The last bit of one word is fed into the first bit of the word at the following address.

SWalk covers all the faults in the fault model described except the decoder stuck-open faults. The test length is $O(cw^2)$, as in SGalpat, but covers fewer multiple-cell coupling faults.

The novelty of the SWalk test is the simplicity of its BIST controller and the ease with which this approach can be extended so as to allow a single, compact BIST circuit to be shared among multiple embedded memories. A boundary-scan controlled testing strategy for multiple embedded memories will be described shortly. This is a pseudorandom test strategy whose implementation resembles this approach very closely.

(1) Initialize all words in the memory to 0s.
(2) Set first bit of first word to 1, by applying the
    march element $(r0w1)$ at the first address.
    {Let i denote the number of bit shifts.}
(3) *for* $i = 1$ *to* wc
    for address $= 1$ *to* $w - 1$
        (rxw)
        {shift each word by 1 bit}
        {the value read at SDO is used as the serial input SDI
        for the next word}
    for address $= w$
        if $i < wc$ *then*
            $(r0w)$
            {expect 0 from the last bit of the last address w}
        else
            $(r1w)$
            {expect the 1 introduced at step (2) to come out}
(4) *for* address $= 1$ to w
        $(r0w1)^c$
        {initialize all words to 1s and verify that the}
        {0s are still there in case of backward coupling}
(5) Repeat steps (2) to (4) with complementary data.

Figure 5.12 Pseudocode for the SWalk algorithm

Advantages of a serial BIST approach

The serial BIST approach just described is implemented using (i) data-path multiplexers to set up the serial shifting mode; (ii) multiplexers on the address and control lines to switch from/to normal mode to/from test mode; (iii) a set of counters whose length depends on the RAM's dimensions and the algorithms being used; we need at least two such counters, one which steps through memory addresses (address counter) and another which steps through the bits of a single word (position counter); (iv) a finite-state automaton, that executes the actual test algorithm, controls the counters and generates the serial data stream and expected data for comparison with the RAM output; and (v) a timing generator for the memory controller. As seen, the BIST circuitry is fairly simple and cheap.

One advantage of the serial BIST circuit is the ease with which it can be shared by several RAM blocks embedded within the same ASIC or microprocessor. The sharing can be done in any of four different ways — namely *daisychaining*, *test multiplexing*, *address windowing*, and *word decomposition*. In *daisychaining*, the embedded RAMs are chained together by connecting the SDO of one RAM to the SDI of the next RAM. A small circuit associated with each memory tests if an address reference belongs to the memory block; if not, the memory is not accessed and the serial input is passed on to the next memory block along the chain. In *test multiplexing*, all the SDIs are connected together, and so are all the SDOs. The address, serial data and control are supplied to all the RAMs simultaneously, and a small additional circuitry chooses which RAM has to be tested. The first method typically is less complex to implement, but is slower and uses more chip area than the second one. In *address windowing*, a group of RAMs that have an identical number of bits per word are treated as one big block and tested. *Word decomposition* is useful for memories with very wide words. With this scheme, a word is decomposed and tested using more than one serial bit stream, allowing for a degree of built-in parallel processing during testing.

5.5.3 BIST for stuck-at and coupling faults in SRAMs

One of the earliest BIST approaches to detect stuck-at and coupling faults for embedded memories is described by Jain and Stroud in [65]. They present BIST implementation of two simple algorithms for functional testing of SRAMs. The hardware requirement for their methods requires a hardware test pattern generator, which produces address, data, and read/write inputs. They have used a parallel input signature analyzer for compressing the memory responses during test mode read. The salient feature of their BIST hardware, to be described below, is that it considers the underlying memory layout and the address scrambling from the logical address space to the physical address space, and generates patterns accordingly.

Fault models considered in Jain and Stroud's approach

The RAM array is divided into three functional blocks — the memory cell array, the decoder logic including row decoders, read and write column decoders, and multiplexers for the input and output data, and the read/write logic including the clock generator, sense amplifiers, and write drivers. The failures modes include: memory cells stuck at 0 or 1, capacitive coupling between adjacent cells, breaks and shorts in buses, faulty decoders accessing multiple cells or failing to access the intended cells; and sense amplifiers and write drivers stuck at 0 or 1. The fault model comprises: one or more memory cells stuck at 1 or 0; one or more memory cells having transition faults; one or more pair of *physically* adjacent memory cells having a coupling fault or a bridging fault between them, in the latter case, the fault causing an AND/OR bridge; one or more memory cells have data retention faults; and destructive read operation.

When a subset of the most likely faults is considered, the physical layout of the memory cell array must be explored to obtain appropriate test strategies. Therefore, any useful test must achieve a compromise between two considerations: the logical addressing scheme of reading from and writing into the memory, and the physical (or layout-related) problem of setting up conditions to trigger different faults. The physical layout considered in this approach is characterized by physically adjacent bits belonging to the same bit position in different memory words. Thus the bitmapping from logical word to physical word is as shown in Figure 5.13. In this approach, it is also assumed that each bit (flip-flop) is a master-slave flip-flop, with possible faulty interaction (coupling, bridging, and others) between the master and the slave flip-flops within each bit.

BIST techniques and implementation

The BIST sequence should satisfy some conditions to detect all faults in the fault model described above, namely, each cell must be made to undergo a $0 \rightarrow 1$ transition followed by a $1 \rightarrow 0$ transition, or vice versa; every pair

BIST and DFT

Addr. 4 D1	Addr. 3 D1	Addr. 2 D1	Addr. 1 D1	Addr. 4 D2	Addr. 3 D2	Addr. 2 D2	Addr. 1 D2	Addr. 4 D3	Addr. 3 D3	Addr. 2 D3	Addr. 1 D3	Addr. 4 D4	Addr. 3 D4	Addr. 2 D4	Addr. 1 D4
Addr. 8 D1	Addr. 7 D1	Addr. 6 D1	Addr. 5 D1	Addr. 8 D2	Addr. 7 D2	Addr. 6 D2	Addr. 5 D2	Addr. 8 D3	•	•	•	Addr. 8 D4	•	•	•
•	•	•	•										•	•	•
•															•
•	•	•	•												•
Addr. 28 D1	Addr. 27 D1	Addr. 26 D1	Addr. 25 D1	Addr. 28 D2	•	•	•	Addr. 28 D3	•	•	•	Addr. 28 D4	•	•	•
Addr. 32 D1	Addr. 31 D1	Addr. 30 D1	Addr. 29 D1	Addr. 32 D2	•	•	•	Addr. 32 D3	•	•	•	Addr. 32 D4	•	•	•

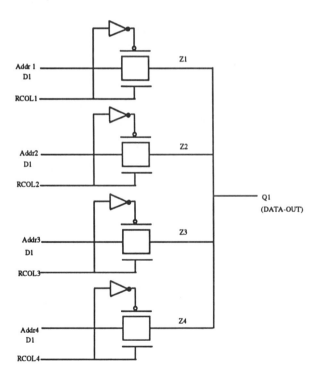

Figure 5.13 Translation from logical space to physical space; courtesy [65]

of adjacent cells must be exercised by writing a value into one cell and the complementary value into the other one, and then reading the first cell; each memory cell must be written with 0 and 1 and read after h units of time, to test for retention faults; unique data should be stored in each memory word and the contents subsequently read, to test decoder faults; a special test sequence of data patterns should be written to detect stuck-at faults in the read column decoder logic — these stuck-at faults might cause multiple lines (columns to be selected) and the data wire-AND'ed or wire-OR'ed; and every pair of adjacent input data (bit) lines on a memory word should be written and read with a pair of complementary values, like 01 or 10, to test bridging faults between adjacent bits.

The address, data and control signals used in test mode are provided through a scan register. The initialization patterns for the algorithms used (described below) are scanned into the memory using the scan register, and the memory addresses are generated by an address counter. The response from the memory obtained after a read operation is compressed either by a parallel entry signature register (in both the test schemes described below), or by a comparator (only in Test Scheme 2).

1. **Test Scheme 1:** The basic idea in this scheme is to write address i with the value i or $\bar{i}$; that is, each memory location stores the value of its address or the bitwise complement of its address. Therefore, each location has a unique value stored in it. The procedure is a very easy one, consisting of making up and down marches through the memory and verifying the content of each address and also verifying both transitions. This test covers stuck-at faults and transition faults.

2. **Test Scheme 2:** The basic idea in this scheme is to write either the all-0 word or the all-1 word into each location. This would cause a checkerboard pattern to be set up in the physical memory cell array. This can be easily verified if we look again at Figure 5.13. This test covers coupling, bridging and retention faults.

5.5.4 BIST for delay faults

Scholz, et al. [144] (1993) describe BIST circuitry to test for access delay and other kinds of electrical faults in RAMs, described in Chapter 2. They have proposed an extension of standard BIST to incorporate the testing for delay faults.

BIST and DFT 247

For the purpose of testing, delay faults can be categorized according to their effect on timing specifications [144]. Accordingly, the types of such faults are: Input to Input Faults (IIF), Input to Output Faults (IOF), and Output to Output Faults (OOF). For example, an IIF may be caused by a delayed address input to an asynchronous SRAM that stretches the set-up time required for the address input to be stable with respect to the write-enable signal. An example of an IOF is an unexpected output delay caused by abnormally high clock skew. Finally, an OOF example includes a fault that adds a delay to one output of a six-transistor SRAM cell which, in turn, affects its complemented output.

In a stand-alone mode, delay fault testing is straightforward. For example, the address to output delay of an asynchronous RAM can be tested on a digital tester by looking at the output at the maximum specified delay t_D. If the outputs have the correct value, no delay faults are present. However, if some or all of the outputs have not yet settled to the correct values, a delay fault is detected. For this test to be valid, the tester's speed must match that of the circuit being tested. For example, if the tester is too slow and is unable to examine the output within time t_D, the delay fault will go unnoticed.

The disadvantage of external testing is that circuits become faster with improving technologies, and the test apparatus needed to test these circuits *at-speed* becomes increasingly expensive. Scholz, et al. [144] propose delay fault self-testing by extending the test pattern generator and the output data compactor (or signature analyzer) of a standard BIST, adding some extra latches and multiplexers in the design. The BIST circuitry with and without delay fault detection as proposed by Scholz, et al. is illustrated in Figure 5.14.

The test pattern generator is extended by adding delay latches and often multiplexers. The BIST control logic block is made more powerful to deal with new testing requirements. The extended test pattern generator produces the timing needed to detect IIFs. For example, the setup time between address and write enable for an asynchronous RAM is tested by delaying the write enable input by a pre-calculated amount. This is achieved with a delay cell designed for the purpose. Similarly, a hold time (for address with respect to write enable) test can be performed if all address signals are delayed by precise amounts.

IOFs and OOFs can be detected by the extended output data compactor. Two different modifications of the compactor are proposed, one or both of which may be incorporated within the same compactor. The first technique is independent of the clock frequency, and requires the addition of latch(es) and a delay cell. A delay cell provides a clock to the latch at the appropriate time to detect a specific delay fault. Sometimes, these latches can be inserted in the compactor

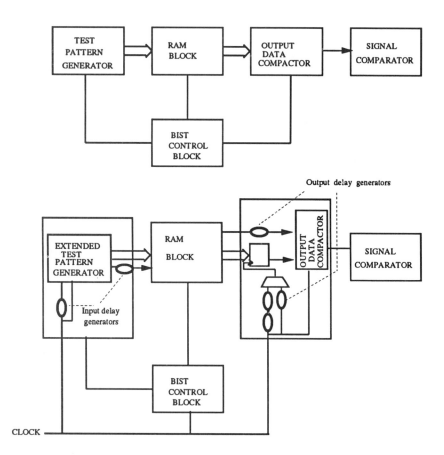

Figure 5.14 BIST without and with delay fault detection; courtesy [144]

BIST and DFT

logic to reduce the number needed. Otherwise, one latch per output may be required. For example, any delay fault in the access time of an asynchronous RAM can be detected if the output is captured at the maximum specified address to output delay after addresses are applied. In some cases, there may be different delay faults to the same output, in which case they are multiplexed before being passed on to the latch. The BIST control block (which is a finite-state machine) ensures that the multiplexer uses the correct delay input during the test.

The second method of detecting IOFs and OOFs is to add a delay cell to delay the output until the next active clock edge is received by the compactor. This has the obvious disadvantage that a delay fault may be aliased if its result is not presented to the compactor prior to the active clock edge. This method also has the disadvantage of needing a delay cell for every output, and also a constant clock frequency.

Therefore, delay faults can be tested by augmenting standard BIST hardware for memories with some extra circuits (latches and multiplexers) and redesigning the BIST control block. The addition of delay fault detection to BIST will improve the quality of the blocks being tested. However, improper addition may result in otherwise functional blocks to fail BIST, reducing yield and increasing cost. It is assumed that the delay cells themselves have the same or similar scaling of delays as the RAM block being tested with variations in process, temperature, and voltage.

The delay fault testing circuitry has been added to the BIST of a 1024 × 8 asynchronous SRAM, to detect the delay between addresses and the write enable signal (setup and hold times). The additional area overhead for delay testing is just 10% of the standard BIST area. For a 1024 × 8 SRAM, this translates to an additional area of less than 1% of the RAM block + BIST. For smaller memories, this percentage area overhead will be more, but the absolute area remains the same (that is, it is independent of the memory size).

5.6 DFT AND BIST FOR PARALLEL TESTING OF DRAMS

Mazumder and Patel [97, 98, 102] have proposed a testable design of DRAM architecture which allows the simultaneous access and test of multiple cells in a word line. The technique efficiently utilizes the 2D structure of the DRAM and achieves considerable speedup. They have also investigated failure mechanisms of 3D DRAM with trench-type capacitors. The scheme has a time complexity of $O(\sqrt{n/p})$, where p is the number of subarrays within the DRAM chip, whereas conventional sliding diagonal type approaches are $O(n^{1.5})$.

The approach to parallel testing of parametric faults [97] can be described as *intra-subarray multiple cell comparison*; that is, the memory array has multiple subarrays and cells *within* each subarray are tested in parallel using a parallel comparator.

The problem of parallel testing has also been addressed by other researchers like McAdams, et al. [105] who fabricated a 1 Mb CMOS three-dimensional DRAM with DFT functions. They partitioned the memory into eight subarrays and tested them concurrently. In their scheme, the same data is simultaneously written on eight cells that are identically located inside the different subarrays. During read operation, access to any cell would result in all eight cells being simultaneously accessed and compared using a two-mode 8-bit parallel comparator. This scheme is described as *inter-subarray single cell comparison*. McAdams, et al. observed a test time reduction by a factor of 5.2 with this scheme. Shah, et al. [147] used a similar 16-bit parallel comparator in their 4 Mb DRAM with trench-transistor cells and reduced the test time complexity to that of a 256 Kb DRAM.

You and Hayes [183] introduced the concept of testing of multiple bits in parallel within the subarrays, and their approach can be called *inter-subarray multiple-cell comparison*. Their basic idea is to reconfigure the memory subarray of size s bits into an s-bit cyclic shift register in which the data recirculate whenever a read operation is done. This reconfiguration is achieved by incorporating pass transistors on the bit lines as a result of which (*a*) the sensitivity of the sense amplifiers is deteriorated by V_T (threshold voltage of the MOS devices); and (*b*) the access time of the DRAM is degraded because of extra logic delays and routing overheads. Therefore, to reduce the routing complexity, only a pair of adjacent cells could be compared in practice. This has the obvious disadvantage of failing to detect the fault when a pair of adjacent cells are *identically* faulty. Moreover, this scheme fails to accommodate a large class of commonly encountered faults, like coupling and pattern-sensitive faults. A similar model of BIST by serial shifting, proposed by Nadeau-Dostie, et al. in 1990 [119] is described earlier in this chapter.

Table 5.2 compares the three schemes mentioned above for parallel memory testing.

BIST and DFT   251

Table 5.2 Comparison of three schemes for parallel memory testing

Criterion	Scheme 1	Scheme 2	Scheme 3
Performance — Parallelism	p	$O(\sqrt{pn})$	$O(\sqrt{pn})$
Degradation in Access Time	None	Large	Negligible
in Sense Amp Sensitivity	None	Large	None
Architecture Modification — Decoder	Not Modified	Modified	Modified
Memory Plane	Not Modified	Modified	Not Modified
Comparator Size	One p-bit	$\sqrt{n/p} \times p$-bit	One $\sqrt{n/p}$-bit 0/1-Detector/Subarray
Routing Complexity	None	High	Very Low
Reliability	Moderate	Low, if only two cells are mutually compared; Moderate, if all p cells are mutually compared	Very High
Fault coverage	Functional and Parametric	Only Parametric	Functional and Parametric

The hardware proposed by Mazumder and Patel [91, 96, 97, 98, 102] for parallel testing of parametric faults in 3D DRAMs is briefly described in the following sections.

5.6.1 Testable DRAM design

The DRAM design is shown in Figure 5.15. The memory is organized as a $b \times w = n$ matrix, b being the number of bit lines and w the number of word lines. The decoder design is modified from a 1-out-of-b decoder to a 1-out-of-b/g decoder during test mode. The bit lines in each word are divided into g groups; during testing, all bit lines i that belong to the same residue class modulo g are selected for simultaneous read or write (i.e., the bit lines numbered i_1 and i_2 will be selected simultaneously during testing if and only if $i_1 \pmod{g}$ = $i_2 \pmod{g}$). During simultaneous read operation, if the multiple-accessed cells contain a 0 (1), a 0 (1) is entered into the data-out buffer. If there is a disagreement in the contents of the multiple-accessed cells, then it so happens that the value stored in the majority of the cells is entered into the data-out buffer and an error latch gets triggered.

The word-line decoder has the conventional design, because accessing multiple words for read with the same sense amplifier would make little sense (no pun intended!), because it would result in a wired-OR or a wired-AND operation. Unlike parallel bit line operations, simultaneous write operation through multiple word lines would result in high write-cycle time delay because of the greater drive required. Increasing physical size of the sense-amplifier driver would re-

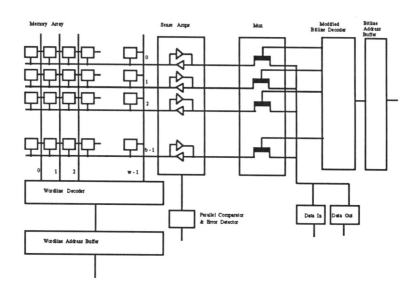

Figure 5.15 A testable DRAM system organization; courtesy [97]

sult in less delay but increased power consumption and lower sense-amplifier slew rate because of large gate capacitance.

5.6.2 Modified DRAM circuit

Figures 5.16 and 5.17 illustrates the modified decoder. As seen quite easily, it is an ordinary decoder with two additional transistors Q_8 and Q_9 which can bypass the address lines $a_0 \ldots a_{k-1}$ during test mode — this can be ensured by setting $\overline{\text{SELECT}}$ to 1.

Figure 5.18 illustrates the parallel comparator. It is basically a multi-bit 0/1 detector which monitors the output of sense amplifiers connected to bit lines that are selected in parallel and detects the concurrent occurrence of either 0s or 1s. A discrepancy would cause the error latch to be set. The design shown is for $g = 2$, that is, the pass transistors used compare simultaneously only the odd or even bit lines; signals L_1 and L_2 select these bit lines. The rest of the circuit is quite self-explanatory. For a detailed explanation, see [97]. The comparator is preferred to a parallel signature analyzer (PSA) because it does not have the aliasing problem.

BIST and DFT

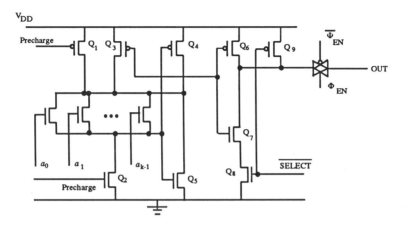

Figure 5.16 Modified bit line decoder; courtesy [97]

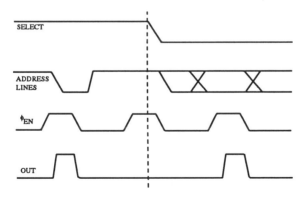

Figure 5.17 Operation of modified decoder; courtesy [97]

254   CHAPTER 5

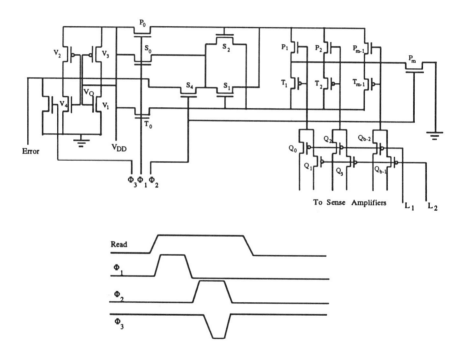

Figure 5.18 Parallel comparator and error detector; courtesy [97]

5.6.3 Parametric faults in a 3D DRAM with trench-type cell

A typical configuration of the 3D trench-type memory cell [82, 147] with $p+$ sidewall doping is shown in Figure 5.19. The access transistor is a PMOS transistor located within an n-well diffused over the $p+$ substrate. A deep trench capacitor extends from the planarized surface through the n-well into the $p+$ substrate. A conducting strap connects the $p+$ doped polysilicon storage electrode inside the trench to the $p+$ source region of the access transistor. With a thin composite insulator separating the polysilicon from the bulk silicon surrounding the trench, the storage capacitance comes primarily from the portions of the four trench sidewalls in the $p+$ substrate region and the trench bottom. Some additional capacitance results from the four trench walls intersecting the n-well. The grounded $p+$ substrate provides a very solid reference potential to the capacitor plate. The leakage currents due to process parameter variation have also been shown in the figure. These currents are divided into four components: *weak-inversion current* I_W from the storage area to the bit line; *field-inversion current* I_F between the two adjacent cells; *gate leakage current* I_G due to pinhole defects in the gate oxide; and the *dark current* I_B between the storage area and the p-type substrate. The weak-inversion current can degrade a stored 0 by flow of minority carriers from the trench capacitor to the positively biased bit lines. Dark current which flows from the trench capacitor to the $p+$ substrate can degrade a stored 1. It may be observed that the cell forms a vertical parasitic FET device which occurs between the storage node and the substrate along the trench wall, gated by the node polysilicon as shown in Figure 5.19.

The effects of the leakage currents result in parametric faults such as the bit line voltage imbalance and the bit line to word line crosstalk. The other types of parametric faults are caused by a wide variation of timing signals in the decoders, address buffer, and peripheral circuits, such as the sense amplifiers. Incorrect timing between decoder enable, precharge clock, and decoder address signals may cause multiple-address selection. These parametric faults are described below.

1. **Bit-line voltage imbalance**: This manifests itself as a functional fault; when one cell of a column (defined by a pair of complementary bit lines) contains a certain logic value and most of the other cells on the same column contain the opposite value, the sense amplifier may produce the opposite value when a read operation is performed. This precharge voltage imbalance is caused by leakage currents which are state-dependent. Figure

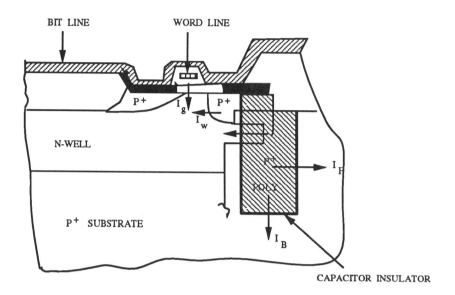

Figure 5.19 3D trench-type memory cell; courtesy [97]

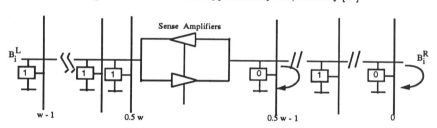

Figure 5.20 Bit-line voltage imbalance; courtesy [97]

5.20 shows how bit lines are logically arranged in a typical memory. Each bit line consists of two half-bit lines.

A memory cell is connected to one of the two half-bit lines, while a dummy cell is connected to each half-bit line. A read operation first precharges both half-bit lines to a reference voltage V_p which has a value midway between the high and low voltages; next, the cell to be read is logically connected to its half of the bit lines such that the voltage on it may increase or decrease slightly depending on the stored information. The differential sense amplifier will be able to sense a small difference in voltage between the two half-bit lines. Because of the existence of imperfect transistors,

BIST and DFT

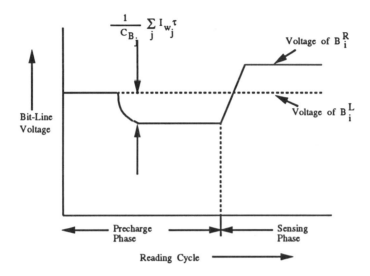

Figure 5.21 Precharge voltage level degradation due to leakage current; courtesy [97]

small leakage currents flow from the cell capacitors to the bit lines. In the worst case, all cells connected to the same half-bit line contain the same logic value and the leakage current is the sum of the individual leakage currents. This total leakage current may be sufficiently large so as to prevent the two half-bit lines from being precharged to the same voltage V_C. For instance, for a given word, if all the left half-bit lines are say, at logic level 1, and all but one of the right half-bit lines are at level 0, then reading the cell with logic level 1 on the right half may cause erroneous operation. This is because the precharge level of the right half-cells will be degraded to

$$v_p = V_p - (1/C_B) \sum_{j=0}^{0.5w-1} I_{W_j}\tau,$$

because of the weak inversion currents, and thereby the precharge levels of the two halves may become considerably different, and the the sense-amplifier may erroneously conclude that the right half cell stores a 0. In the above formulation, C_B denotes the bit line capacitance, I_{W_j} is the weak-inversion current in C_{ij}, and τ is the time interval for precharge during read operation. This is illustrated in Figure 5.21.

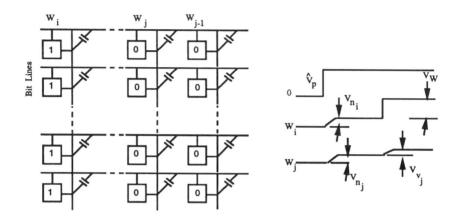

Figure 5.22 Bit-line to word-line crosstalk; courtesy [97]

2. **Bit-line to word-line crosstalk**: Since bit lines and word lines overlap perpendicularly to one another, there exists the possibility of coupling between them (Figure 5.22).

   Consider a pair of coupled word lines, one of which stores the all-1 pattern in its bits, and the other, the all-0 pattern. The 0s in the first word may be degraded due to the weak inversion current of the access transistors connected to the bit lines which straddle both word lines — the selected one and the unselected one.

3. **Single-Ended Write Fault**: For DRAMs that employ single-ended write technique (Figure 5.23), a single I/O line writes into the two half-bit lines instead of symmetrically placed I/O lines on two sides. For such DRAMs, writing on the right half-bit line is controlled by the I/O line driver, whereas writing on the left half-bit line is controlled by the sense-amplifier. Thus the 0 level in the two halves may be different, and this asymmetry may result in pattern-sensitive faults.

4. **Multiple Selection**: This happens when the precharge clock Φ_1 and the decoder enable clock Φ_{EN} overlap due to incorrect timings and delays, causing faulty outcomes.

5. **Transmission-line effects**: Signal propagation delays in poly and diffusion lines (because of the RC effect, causing delay to be proportional to l^2, where l is the length) and high resistivity in such interconnects lead to weak signals being delivered to cells along the periphery of the chip. These weak signals may cause incorrect read/write operations.

BIST and DFT

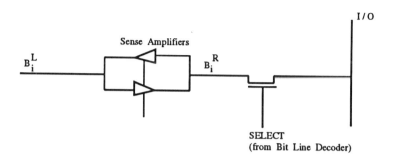

Figure 5.23 Single-ended write; courtesy [97]

5.6.4 Testing algorithms for parametric faults

As explained in Chapter 4, testing the above faults can be facilitated if conditions are set up for maximizing the probability of occurrence of these faults. It can be easily noticed that the field-inversion current I_F which occurs between two adjoining storage cells is maximum if the four adjacent cells of a base cell contain opposite data to that of the base cell; that is, a checkerboard-type pattern can test the effect of a field-inversion current. Similarly, the effects of a dark current and gate short can be tested by the checkerboard pattern, because the presence of these leakage currents manifests in the form of a cell stuck-at zero or one. Algorithm 1, shown in Figure 5.24 is a parallel version of the checkerboard test, which tests the above three leakage currents in addition to parametric faults caused by single-ended write and the transmission line effects. Algorithm 1 tests each cell in a word line for logic-0 and logic-1 data, and thereby tests for all stuck-at faults. Moreover, if a bit line or a word line is faulty (namely, broken, stuck to ground, or cannot be precharged), Algorithm 1 will detect the fault. It may be noted that a faulty bit line will be incorrectly compared by the parallel 0/1 detector to set ERROR = 1. A faulty word line where all the bits are identical may not be detected by the parallel 0/1 detector. However, such a fault can be easily detected by monitoring the output of the data-out buffer, if the entire memory consists of a single array , otherwise, by mutually comparing the data-out values of all the subarrays. It may be added that if the entire word line is faulty, then the expected value at the data-out buffer will be different from the obtained one if and only if the word line is tested for both logic-0 and logic-1 values. If a word line is completely fault-free, then the expected value may be different from the obtained one, but it will certainly be detected by the 0/1 detector.

1) Use complementing address sequence from word line W_0 until all word lines are scanned; write in two steps a pattern of $(01)^*$ if the word line is even and in two steps a pattern of $(10)^*$ if the word line is odd.

2) Freeze the clock for the entire refresh interval τ_R for testing *static* refresh.

3) Use a complementing address sequence from word line W_0 until all word lines are scanned; compare in parallel all even and odd bit lines to check ERROR = 0.

4) Read continuously any arbitrary word line for the entire refresh interval τ_R to check ERROR = 0. This test checks the effect of temperature rise and tests the *dynamic* refresh.

5) Use the complementing address sequence from word line W_0 until all word lines are scanned; compare in parallel all even and odd bit lines to check ERROR = 0.

6) Read continuously another distinct word line for the entire refresh interval τ_R to check ERROR = 0. This test checks the effect of temperature rise and tests the dynamic refresh.

7) Use the complementing address sequence from word line W_0 until all word lines are scanned; compare in parallel all even and odd bit lines to check ERROR = 0.

8) Repeat steps 1-7, with opposite data.

Figure 5.24 Pseudocode for the parametric checkerboard test (Algorithm 1)

Algorithm 1 is described in Figure 5.24.

The effect of a weak-inversion current is maximum when all the cells in a bit line, except one, contain 0. If the cell which contains 1 is addressed for a read operation, the weak-inversion currents in other cells will tend to degrade the precharge level of the bit line and thereby the cell containing 1 will be sensed as 0 by the sense amplifier. Since the bit line capacitance is typically 10-20 times the capacitance of an individual cell, the stored 1 may not be sufficient to replenish the degraded precharge level. The testing strategy needs to test each memory cell so that when it is 1 all its bit line neighbors will contain 0.

BIST and DFT

Table 5.3 Algorithms and their coverage of parametric faults

Fault type	Algorithm 1	Algorithm 2	Algorithm 3
Weak-inversion current	No	Yes	Yes
Field-inversion current	Yes	No	No
Dark current	Yes	No	No
Gate short	Yes	Yes	Yes
Multiple selection	No	No	Yes
Single-ended write	Yes	Yes	Yes
Bit line voltage imbalance	No	Yes	No
Bit-line to word-line crosstalk	No	Yes	No
Transmission line effect	Yes	No	No

In order to test the bit line voltage imbalance, it is necessary to write 0(1) on the cells at the bit line on the left half of the subarray and write 1(0) on the cells at the bit line on the right half of the subarray. Thus the test to detect the weak-inversion current can be utilized to test the bit line voltage imbalance by testing the left and right subarrays with opposite background data. It may be noted that the test also detects faults due to single-ended write. Algorithm 2, shown in Figure 5.25 tests all the above faults.

Algorithm 3, which runs a marching pattern of 1 on a background of 0s and a marching pattern of 0 on a background of 1s in each word line, will detect multiple-access faults in the word line decoder by comparing the read data with the expected data. Since Algorithm 2 employs parallel writing by accessing all the even bit lines and odd bit lines in a single memory cycle, multiple access in the bit line decoder will not be tested by Algorithm 2. A separate algorithm is needed to test bit line decoders. This algorithm is described in Figure 5.26.

Algorithm 1 is seen to take a total of $10w\tau_A + 6\tau_R$ time to complete all the steps, where τ_A is the average memory cycle time and τ_R is the refresh interval. Algorithm 2 takes $20w\tau_A + 2\tau_R$ time to test the entire DRAM. Finally, Algorithm 3 takes $(4w + 2)\tau_A$ time to test for multiple-access faults in the decoder logic. Hence, altogether $(34w + 2)\tau_A + 8\tau_R$ time is needed to test all the parametric faults in the DRAM. Table 5.3 depicts the different types of parametric faults and how they are covered by these algorithms.

1) Initialize the entire memory, writing 0 in all locations.

2) Arbitrarily select two word lines W_i and W_j and read them alternately for one refresh interval. Check if the ERROR = 0 during the entire refresh interval.

3) For all word lines starting from W_0, compare in parallel all even and odd bit lines to check ERROR = 0.

4) Initialize the entire memory, writing 1 in all locations.

5) Arbitrarily select two word lines W_p and W_q and read them alternately for one refresh interval. Check if the ERROR = 0 during the entire refresh interval.

6) For all word lines starting from W_0, compare in parallel all even and odd bit lines to check ERROR = 0.

7) Initialize the memory such that the left subarray contains 0 in all locations and the right subarray contains 1 in all locations.

8) For all word lines starting from W_0, do the following: i) write a pattern of $(01)^*$ in the selected word line; ii) compare cells in parallel and check if ERROR = 0; iii) initialize every cell in the selected word line to 0 if it is in the left half, otherwise initialize it to 1.

9) For all word lines starting from W_0, do the following: i) write a pattern of $(10)^*$ in the selected word line; ii) parallel compare and check if ERROR = 0; iii) initialize every cell in the selected word line to 0 if it is in the left half, otherwise initialize it to 1.

10) Repeat steps 7-9 with complementary bit patterns.

Figure 5.25 Pseudocode for the parallel parametric walking test

- /* Bit Line Decoder Multiple Access Test */

1) Write in parallel 0 in all cells on the arbitrarily selected word line W_j.

2) Read and compare in parallel all the cells on W_j.

3) Starting from the cell at the cross-point of B_0 and W_j, for each cell on W_j, at first write 1 and read the cell (one cell at a time in ascending order of the bit line).

4) Starting from the cell at the cross-point of $B_{\sqrt{n}-1}$ and W_j, for each cell on W_j, at first write 1 and read the cell (one cell at a time in descending order of the bit line).

- /* Word Line Decoder Multiple Access Test */

5) Write in parallel 0 in all cells on the bit line B_i.

6) Read and compare in parallel all the cells on B_i.

7) Starting from the cell at the cross-point of W_0 and B_i, for each cell on B_i, at first write 1 and read the cell (one cell at a time in ascending order of the word line).

8) Starting from the cell at the cross-point of $W_{\sqrt{n}-1}$ and B_i, for each cell on B_i, at first write 1 and read the cell (one cell at a time in descending order of the word line).

Figure 5.26 Bit-line and word-line decoder test

1) Write in parallel 0 to all cells of the memory at maximum supply voltage.

2) For all word lines starting from W_0, do the following: i) write a pattern of $(01)^*$ in the selected word line at maximum supply voltage; ii) rapidly reduce the supply voltage to minimum; iii) compare in parallel and check that ERROR $= 0$.

3) For all word lines starting from $W_{\sqrt{n}-1}$, do the following: i) write a pattern of $(10)^*$ in the selected word line at maximum supply voltage; ii) rapidly reduce the supply voltage to minimum; iii) compare in parallel and check that ERROR $= 0$.

4) Write in parallel 0 in the entire memory at minimum supply voltage.

5) For all word lines starting from W_0, do the following: i) write a pattern of $(01)^*$ in the selected word line at minimum supply voltage; ii) rapidly increase the supply voltage to maximum; iii) parallel compare and check if ERROR $= 0$.

6) For all word lines starting from $W_{\sqrt{n}-1}$, do the following: i) write a pattern of $(10)^*$ in the selected word line at minimum supply voltage; ii) rapidly increase the supply voltage to maximum; iii) compare in parallel and check if ERROR $= 0$.

Figure 5.27 Power-supply voltage transition test

The above algorithms are tested at the rated maximum and minimum power-supply voltages. But, in addition to the worst-case measurements, it is necessary to test the memory when the supply voltage changes rapidly due to an impressed noise voltage. Noise spikes have high slew rates and they may occur during a read or write memory cycle causing the operation to fail. If the effect of the noise spike is to lower the supply voltage, the capacitive bias of the dynamic logic may be higher than the supply bias, and this may result in a failure of a read or write operation. Similarly, if the noise spike increases the supply voltage to a high value, the capacitive bias voltage may be sufficiently lower than the supply bias resulting in a faulty read or write operation. In order to test the effect of this power supply voltage transition, the entire memory should be tested for both the cases when the supply voltage rapidly increases and again when it rapidly decreases. The typical slew rate is about a few microseconds. Algorithm 4 (Figure 5.27) tests the memory for the above faults.

The above BIST architecture [97] has several advantages. It enhances the speed of testing the occurrence of parametric faults in a 3D DRAM employing trench-type storage capacitor. This novel testable architecture tests multiple cells in a word line simultaneously, resulting in a speedup by a factor of $O(\sqrt{n/p})$. The testable design has been proposed to fit within the inter-cell pitch width of 3λ in the 3D DRAM. It employs only an additional $2\sqrt{pn} + p \lg n - p \lg p + 12p$ transistors and has very low overhead.

5.6.5 Parallel testing of pattern-sensitive faults

Mazumder and Patel [101] have also presented a design strategy for testing in parallel a broad class of NPSFs in RAMs. By using this strategy, we can test an n-bit RAM organized as a $\sqrt{n} \times \sqrt{n}$ array in $97\sqrt{n}$ memory cycles. This DFT approach does not modify the RAM architecture to a great extent. It can be easily implemented as a BIST and the overhead is low. It speeds up conventional fault testing algorithms by a factor of $O(\sqrt{n})$.

Many efforts have been made in the past to characterize DFT requirements for pattern-sensitive faults. Recall our discussion of PSFs in Chapter 3. We have seen in Chapter 3 that the complexity of the algorithms for NPSFs is linear with an $O(k2^k)$ constant multiplier [54, 158]. Fuentes, et al. [47] and Mazumder and Patel [104] have independently proposed random testing as an alternative to deterministic algorithms for PSFs, and have analyzed Markov chain models of PSFs (to be discussed later) to obtain a fault coverage of 99.9%. However, test lengths are very large for random testing.

The testable DRAM design described in the previous section is also useful for testing NPSFs. The self-testing algorithms running on such hardware are different for NPSFs and parametric faults. Mazumder and Patel have used deterministic algorithms for testing NPSFs and discuss ways of implementing these with their special architecture (i.e., a modified bit line decoder and a parallel comparator). These algorithms are briefly described below.

Algorithms for testing NPSFs in parallel

In Chapter 3, the testing of NPSFs using Eulerian tours in Type-1 and Type-2 neighborhoods is discussed. All these algorithms are based upon creation and look-up of tables that give the sequence of transition writes in cells of

1) Write 0 into every cell of the RAM.

2) Set $m = 1$.

3) Look up the table and find the cell number p ($p \in \{0, 1, 2, 3, 4\}$) in operation #m on which the transition write is to be made.

4) Set $j = 0$.

5) Do in parallel the operation #m on all cells C_{ij}, i being the bit line number such that $(2j + i)(\bmod 5) = p$. (A little analysis would tell us that this would indeed write the same data on identically located cells in different neighborhoods.)

6) Increment j.

7) If $j < w$, go to (5), else go to (8).

8) Set $j = 0$.

9) Read in parallel all cells C_{ij} and check the ERROR signal from the parallel comparator (Figure 5.18) to see if ERROR = 0, (if ERROR = 1, a fault is detected).

10) Increment j.

11) If $j < w$, go to (9), else go to (12).

12) Increment m (to go to the next table entry).

13) If $m \leq$ (the number of table entries), go to (3), else stop.

Figure 5.28 Parallel NPSF test in RAM

neighborhoods to complete an Eulerian tour (or tours). The key idea behind parallel testing for NPSFs is that cells that are identically located in disjoint neighborhoods can be written in parallel. Based on these look-up algorithms, a parallel NPSF test on a 5-cell Type-1 neighborhood in RAMs is shown in Figure 5.28. We call this procedure "NPSF-Algorithm 1."

The complexity of this algorithm is $O(w)$ where w is the number of words in the memory, since the sequential loop runs twice over all words, performing at most a constant amount of operation at each pass. Now $w = \sqrt{n/(pe)}$, where $n = pbw$ is the total number of memory cells, p is the number of identical square submatrices comprising the memory, and e denotes the ratio of the number of bit lines b to the number of word lines w, that is, the ratio $b : w$, called the eccentricity of the memory device.

In parallel testing of NPSFs, the computation of the effective neighborhood size is an interesting task. The cell number assignment allows one to describe a number of memory plane *tessellations* as shown in Figure 5.29. Neighborhoods in these tessellations are the corresponding tile shape.

It can be seen that by linear translations of the tile geometries, the memory plane can be covered without overlapping, (i.e., *tessellated*). Since there are eight such tile geometries, a parallel test such as NPSF-Algorithm 1 will effectively affect eight distinct neighborhoods of a base cell simultaneously at any read/write step. Therefore, the effective physical size of the neighborhood can be estimated by superimposing these tile geometries. But it is necessary to invoke a constraint that all cells numbered i ($0 \leq i \leq 4$) are mutually non-interactive and consistent in the sense that a transition write on a cell numbered i does not affect the content of another cell numbered i in the same neighborhood, nor does it mask the coupling effect between any other coupling cell i and a coupled cell $j \neq i$, both in the same neighborhood.

Parallel test procedures for symmetric pattern-sensitive faults

In order to detect all the static NPSFs over a 9-cell Type-2 neighborhood, every cell in the memory should undergo both an ↑ and a ↓ transition in the presence of all $2^8 = 256$ patterns in the memory. Thus, a total of 512 transition writes are necessary for each cell in the memory to detect the static PSFs. In order to detect all the active NPSFs, each base cell should be tested by a read operation whenever a transition write is made over a cell in its neighborhood. Since each cell can make two types of transition writes or all the possible binary patterns in the other eight cells, there are a total of $2 * 2^8 * 8 = 4096$ transition writes in the neighborhood of a cell to test all the ANPSFs. This requires a large amount of time.

In [101], Mazumder and Patel have adopted a new approach in which cells in the 9-neighborhood are classified into four logical groups, namely, the base cell, bit line neighbors (N_b), word line neighbors (N_w), and diagonal neighbors (N_2).

268   CHAPTER 5

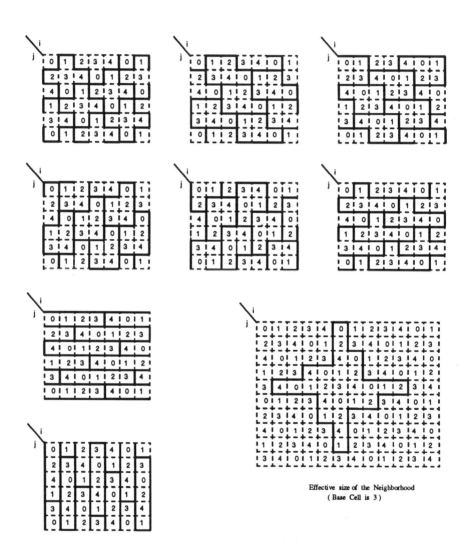

Figure 5.29 Effective neighborhood size; courtesy [101]

It may be recalled that a pattern-sensitive fault models the adjacency effect between a base-cell and its physically neighboring cell. It is typically due to leakage effects between a memory cell and its adjoining cells in the presence of a particular data pattern in the neighboring cells [20, 80, 97]. It has been found that the leakage is maximum when the symmetrically-located cells contain the same bit patterns [80]. By the above classification, many unnecessary binary combinations are avoided. For example, suppose there is a situation when a read operation is made to verify the transition write operation $\uparrow$ on a cell C_{ij} when its bit-line neighbors $C_{i,j+1}$ and $C_{i,j-1}$ contain 0. Clearly, at first, the bit line i will be precharged to some high potential. Now, if any of the access transistors of the bit-line neighbors is weak, then, in the presence of 0s in the bit-line neighborhood, the precharge level in the bit line will be degraded, and the sense amplifier on the ith bit line will fail to detect a 1 in the base cell. Similarly, if there is a weak transistor in cell $C_{i+1,j}$, which does not allow the base cell to make a $\uparrow$ transition because $C_{i+1,j}$ is at state 0, then it suffices to test the fault when the symmetrically located cell $C_{i-1,j}$ is also at state 0, since then the leakage effect will be maximized. Thus, if a fault does not occur when both the cells $C_{i+1,j}$ and $C_{i-1,j}$ are at state 0, then it will not occur when they have opposite states.

A class of NPSFs called symmetric PSFs are of special interest. Such a fault is said to have occurred if either (a) a read or write operation, denoted by $\psi(C_{ij})$ in a cell C_{ij}, located at the cross-point of bit line i with word line j is faulty in the presence of a fixed *symmetric* pattern in its deleted Type-2 neighborhood — this is called a *symmetric static PSF (SSPSF)*; or (b) the state of a cell C_{ij} undergoes a spurious modification because of an operation on one or more cells in the deleted Type-2 neighborhood storing a fixed symmetric pattern — this is called a *symmetric dynamic PSF (SDPSF)*. A SSPSF is mathematically described by the notation: $\langle \psi(C_{ij})/s(N_b)s(N_w)s(N_2) \rangle$, where $s(N_b)$, $s(N_w)$, and $s(N_2)$ represent the states of all cells in N_b, N_w, and N_2, respectively. A SDPSF is denoted by a notation such as: $s_{ij}/s(N_b)s(N_w)\psi(N_2)$, which corresponds to the situation that the state of the cell C_{ij} changes because of a symmetric operation $\psi(N_2)$ on its diagonal neighbors, in presence of fixed patterns written symmetrically into the bit line and word line neighbors.

SSPSFs and SDPSFs are detected in a manner analogous to that of detecting SNPSFs and ANPSFs, as described in Chapter 3. The cells in a 9-cell neighborhood are labeled in a different manner, however. All neighbors of the same kind are assigned the same label in this scheme. Thus, for example, if the base cell is labeled 3, we may choose to label all four diagonal neighbors (N_d) 0, both word line neighbors 1, and both bit line neighbors 2. In this manner, the rectangular cell array is filled with a pattern of labels. If any transition write operation is performed in this new scheme, all cells having the same label will be written (in parallel) with the same data, and so the pattern will still remain symmetric. Thus, this labeling scheme provides a convenient way to ensure that all (and only) symmetric PSFs are stimulated during the test procedure.

It is seen quite clearly that $s(N_x) = 1$ if all the cells in the neighborhood $N_x \in \{N_b, N_w, N_d\}$ are 1 and $s(N_x) = 0$ if all the cells in N_x are 0. It will be assumed that during testing of a pattern-sensitive fault, the contents of all the cells in N_x are always identical. Therefore, a read operation to test a fault is made only after all the cells in N_x are treated identically. In order to detect SDPSFs, it will be assumed that the fault will cause at most one transition in the base cell, whether the operations on the cells of N_x are applied one cell at a time or two cells at a time.

If the base cell i is in state $s_i \in \{0, 1\}$ and a transition write ψ is made on one cell or simultaneously over two cells belonging to N_x, then an SDPSF is said to be *consistent* if the state of cell i does not change back from $\overline{s_i}$ to s_i after the successive ψ operation on the remaining cells in N_x.

Utilizing this notation, the SSPSFs and SDPSFs can be represented by Table 5.4. The four-tuples are as described before. Since there are 16 SSPSFs and 48 SDPSFs associated with each cell in the memory, it can be easily seen that at least $4 * 2^4 = 64$ transition writes in the neighborhood will be necessary to sensitize all the SSPSFs and SDPSFs of a DRAM cell. A straightforward extension of this idea to an n-cell memory will require $64n$ transition writes. However, in this section, it will be shown that by cleverly combining the transition write sequence in overlapping neighborhoods, the number of transition writes can be reduced to $16n$.

In order to accomplish the minimal number of transition writes per memory cell, at first, each cell C_{ij} in the memory is assigned a positive number $k \in \{0, 1, 2, 3\}$ such that $k = 2i' + j'$, where $i' = i \pmod 2$ and $j' = j \pmod 2$. Thus, the memory cells are divided into four types of cells 0, 1, 2, and 3, as shown in Figure 5.30. It will be shown later that a transition write on a cell followed by a suitable sequence of read operations on the adjoining cells will simultaneously sensitize four pattern-sensitive faults and thereby the number of transition writes on each cell can be reduced by a factor of 4.

In order to obtain a test procedure which needs only 16 transition write sequences per cell, a graph-theoretic approach similar to that used in "NPSF-Algorithm 1" can be used. The 4-tuple, $\langle s(C_{ij})s(N_b)s(N_w)s(N_2)\rangle$ describes a state space of 16 nodes (states) where each node represents the symmetric

BIST and DFT

Table 5.4 All possible SSPSFs and SDPSFs

Fault Type	Fault Notation			
SSPSF	↑/000,	↑/100,	↑/010,	↑/110,
	↑/001,	↑/101,	↑/011,	↑/111,
	↓/000,	↓/100,	↓/010,	↓/110,
	↓/001,	↓/101,	↓/011,	↓/111,
SDPSF	0/↑00,	0/↑10,	0/↑01,	0/↑11,
	1/↑00,	1/↑10,	1/↑01,	1/↑11,
	0/↓00,	0/↓10,	0/↓01,	0/↓11,
	1/↓00,	1/↓10,	1/↓01,	1/↓11,
	0/0↑0,	0/1↑0,	0/0↑1,	0/1↑1,
	1/0↑0,	1/1↑0,	1/0↑1,	1/1↑1,
	0/0↓0,	0/1↓0,	0/0↓1,	0/1↓1,
	1/0↓0,	1/1↓0,	1/0↓1,	1/1↓1,
	0/00↑,	0/10↑,	0/01↑,	0/11↑,
	1/00↑,	1/10↑,	1/01↑,	1/11↑,
	0/00↓,	0/10↓,	0/01↓,	0/11↓,
	1/00↓,	1/10↓,	1/01↓,	1/11↓,

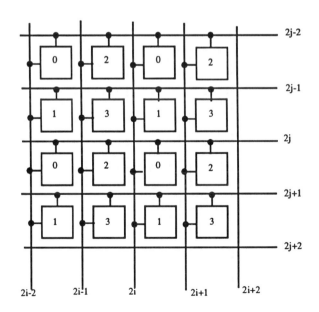

Figure 5.30 Cell type assignment; courtesy [101]

binary pattern in the 9-neighborhood. These nodes are numbered 0-15, depending on the binary pattern in the 4-tuple. The base cell state is the MSB in the 4-tuple. By complementing the pth bit in the 4-tuple, the memory state of the neighborhood will change from k to $k - 2^p$, if the pth bit changes from 0 to 1, and from k to $k + 2^p$, if the pth bit changes from 1 to 0. All these transition write operations from one state to another represent edges in the state-space graph, and the resulting graph describes a symmetric four-dimensional cube as shown in Figure 5.31. The set of thick edges corresponds to changing the state of the base cell and will pertain to sensitizing the SSPSFs. Other edges pertain to sensitizing the SDPSFs. Similar to Algorithm 1, an algorithm with optimal transition writes can be obtained by deriving the write sequences from an Eulerian tour over the symmetric 4-cube. The resulting algorithm will have a complexity of $97w$, involving $32w$ transition writes, $64w$ read, and w write operations corresponding to memory initialization, where $w = n/4$. It may be noted that even though 64 transition writes over the symmetric 4-cube are needed to sensitize all the SSPSFs and SDPSFs, only $32w$ transition writes are needed over the entire memory.

Even though the test algorithm derived from an Eulerian tour has an optimal test length, for many applications like in an embedded environment, it is desirable to made a tradeoff between the test length and the BIST hardware. It is shown in [95] that the test generator circuit can be simplified considerably if the sequence of transition writes is derived by decomposing an Eulerian tour over the symmetric 4-cube into the following eight disjoint Hamiltonian cyclic tours described over a subgraph of the symmetric 4-cube:

- **H1**: $\langle 0, 2, 6, 14, 15, 13, 9, 1, 0 \rangle$
- **H2**: $\langle 0, 1, 9, 13, 15, 14, 6, 2, 0 \rangle$
- **H3**: $\langle 0, 4, 6, 7, 15, 11, 9, 8, 0 \rangle$
- **H4**: $\langle 0, 8, 9, 11, 15, 7, 6, 4, 0 \rangle$
- **H5**: $\langle 12, 4, 5, 7, 3, 11, 10, 8, 12 \rangle$
- **H6**: $\langle 12, 8, 10, 11, 3, 7, 5, 4, 12 \rangle$
- **H7**: $\langle 12, 13, 5, 1, 3, 2, 10, 14, 12 \rangle$, and
- **H8**: $\langle 12, 14, 10, 2, 3, 1, 5, 13, 12 \rangle$.

BIST and DFT

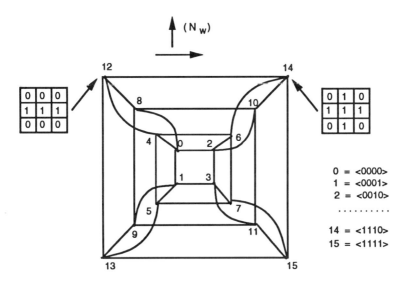

Figure 5.31 State space graph for neighborhood with $k = 4$; courtesy [101]

These tours are shown in Figure 5.32, where each set of dark edges constitutes two Hamiltonian cycles, one clockwise and the other counter-clockwise. Thus all the 64 edges of the symmetric 4-cube are traversed, sensitizing all 16 SSPSFs and 48 SDPSFs. Initially, the memory is initialized to zero and the transition writes corresponding to the first four Hamiltonian cycles are applied as indicated above. Then the memory is re-initialized such that it contains a column bar pattern of 0s and 1s, so that the content of the Type-2 neighborhoods in the memory is represented by node 12. This needs an additional w write operations which for a 256 Kb square memory sub-array having a memory cycle time of 50 ns will take an extra 25 μs. After memory re-initialization, the remaining Hamiltonian tours are performed. Thus, the overall tour can be represented as $\langle H1, H2, H3, H4, (0, 12), H5, H6, H7, H8 \rangle$, and it consists of 65 transition-write operations. Initially, all the cells contain 0 and successively the transition-write operations are made to change the state of the neighborhood. Thus, after the first operation, the memory state changes to 2, after the second transition write, it changes to 6, and so on. In general, the m-th transition write $(m > 0)$ will change the state S_{m-1} to S_m, where $S_m \in \{0, 15\}$ denotes the state of the neighborhood after the m-th operation is applied. After each operation, the whole memory is read to find out whether any SSPSF or SDPSF has occurred. The above algorithm has been described as "NPSF-Algorithm

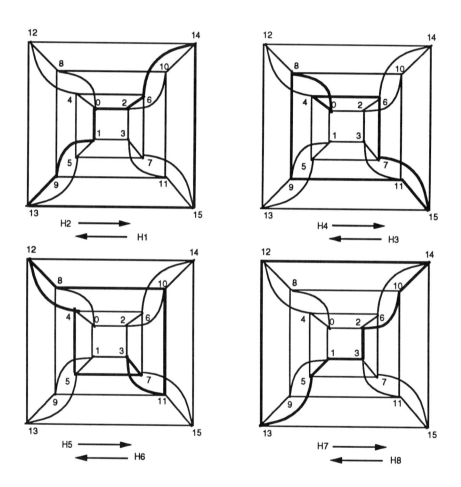

Figure 5.32 Hamiltonian cycles on subgraphs of symmetric 4-cubes; courtesy [101]

BIST and DFT

2" which uses these write sequences to test each memory cell for SSPSFs and SDPSFs over its 9-neighborhood.

In this algorithm, all cells numbered $k \in \{0, 1, 2, 3\}$ make an upward ($\uparrow$) and a downward ($\downarrow$) transition write in the presence of all binary patterns in all other cells whose numbers are different from k. There are altogether eight upward transitions in cells numbered k for eight distinct binary patterns. Also, there are eight downward transitions in cells k for all eight distinct binary patterns. Hence, all 16 operations to sensitize SSPSFs are performed. Since all the cells are read after each transition write, any SSPSF will be detected by the parallel comparator and error latch. Also, because of the neighborhood relationship in Figure 5.30, every transition write on cells numbered k will also sensitize the SDPSFs for the other three cells for which the number is not k. For example, in the third operation in **H1**, the state of the Type-2 neighborhood changes from 6 to 14 by writing $\uparrow$ in the cell numbered 0 while the contents of the cells numbered 1,2, and 3 remain unchanged. The following read operation in step 4 of the algorithm detects an SSPSF in all neighborhoods for which the base cell is 0, and SDPSFs in all other neighborhoods for which the base cell in not 0. Thus, this algorithm (Figure 5.32), which makes $65w$ transition writes over the memory, will sensitize both SSPSFs and SDPSFs for each cell in the memory.

5.6.6 Procedures for testing other fault types

Parallel test procedures that can employ the modified decoder and the parallel comparator described can also be designed for other types of faults such as faulty non-transition writes, destructive read operations, and others. These faults are briefly described below, and procedures for testing them are explained.

1. **Non-transition write operations**: We have assumed in most of the discussion thus far that only transition writes can be faulty. If we relax this assumption for a moment and allow the possibility of faulty non-transition writes, then a very simple modification of the algorithms described above will work — namely, to follow each transition write with a non-transition write, after which the entire neighborhood is tested using the parallel comparator and sense amplifiers to detect the occurrence of any pattern-sensitive faults.

1) Initialize all cells to 0.

2) Set $m = 1$.

3) If the transition write in the mth operation is on cells numbered $p \in \{0, 2\}$, write on all cells that are labeled p and are on even word lines.

4) Read the memory in parallel and monitor the ERROR signal of the comparator.

5) If the transition write in the mth operation is on cells numbered $p \in \{1, 3\}$, write on all cells that are labeled p and are on odd word lines.

6) Read the memory in parallel and monitor the ERROR signal of the comparator.

7) Increment m.

8) If $m \leq$ (the number of table entries), go to (3), else stop.

Figure 5.33 Parallel test of PSFs over type-2 neighborhood

BIST and DFT

(a) Write in parallel 0 into all cells at an arbitrarily selected word line W_j.

(b) Using the parallel comparator, read and compare in parallel all cells on W_j.

(c) Starting from the cell at the cross-point of B_0 and W_j, for each cell on W_j, a 1 is first written and the cell is then read, one cell at a time in ascending order of bit line.

(d) Starting from the cell at the cross-point of $B_{\sqrt{n}-1}$ and W_j, for each cell on W_j, a 0 is first written and the cell is then read, as in step (3).

(e) Do steps (1) through (4) for an arbitrarily selected fixed bit line and all word lines that cross it.

Figure 5.34 Pseudocode for parallel testing for faults in the modified bit-line and the word-line decoders

2. **Destructive read operation**: This can be tested if every read operation is followed by reading the entire neighborhood to detect (possible) faulty transitions in other cells.

3. **Modified bit-line decoder faults**: As described in Chapter 4, a faulty decoder may produce a *no-access*, a *multiple-cells access*, or a *multiple-address access* fault. The BIST algorithm shown in Figure 5.34 can be used to self-test the modified bit-line decoder and test the word-line decoder.

4. **Parallel comparator faults**: The algorithm for detecting SSPSFs and SDPSFs would also partially detect faults in the parallel comparator, because when all the even or the odd lines contain 0 and are compared, the n-type transistors $P_0, \ldots, P_{m-1}$ are tested for stuck-at-one faults, and when all of them contain 1, all the p-type transistors $T_0, \ldots, T_{m-1}$ are tested for stuck-at-zero faults. The rest of the stuck-at faults can be detected by the algorithm, whose pseudocode is shown in Figure 5.35.

5.6.7 Address-maskable parallel BIST for DRAMs

Morooka, et al. [114] describe a memory array architecture and its related test method named column address-maskable parallel-test (CMT) architecture, suitable for ultra-high density DRAMs. The basic objective for their work is to

(a) Initialize all odd cells on the first word line to 0, and all even cells to 1.

(b) Use the parallel comparator to compare in parallel odd bit lines and verify if ERROR = 0.

(c) Use the parallel comparator for even bit lines to verify if ERROR = 0.

(d) For each odd bit line in the selected word line, write a 1 and see if ERROR = 1, and then write a 0 back in and see if ERROR = 0.

(e) For each even bit line in the selected word line, write a 0 and see if ERROR = 1, and then write a 1 back in and see if ERROR = 0.

Figure 5.35 Pseudocode for parallel testing for faults in the parallel comparator

reduce the area penalty associated with other parallel BIST approaches, such as those that involve the use of a parallel comparator or a modified decoder.

The architecture of this parallel BIST approach consists of separate read and write data-lines. The read data-line is also used as the match line during test mode operation, and no additional circuit and routing lines are required to perform the parallel test. No additional area is required for the test control circuit — the read/write control block and column address controller contains all the lines needed during the test. As a result, the area overhead of this test is not too high, and is also independent of the memory density.

Figure 5.36 [114] shows the architecture of the CMT. Figure 5.37 [114] describes the error detection circuit. There are separate read and write data lines in this architecture. The read data lines are used for read operation in normal mode and as match-lines in test mode. The switches SW0 and SW1 determine the mode of operation.

In normal mode of operation, only one column decoder output is selected. As usual, the write data goes into the selected bit line pair in the write cycle, and the selected bit line data is sensed during the read cycle.

In test mode, multiple column decoder outputs are triggered. The same data is written into multiple bit line pairs simultaneously during test mode. The read operation during test mode has the effect of matching the data with the expected values. These match lines are precharged to a high level initially. If there is no error, all selected bit-line pairs should hold the same data, and one

BIST and DFT

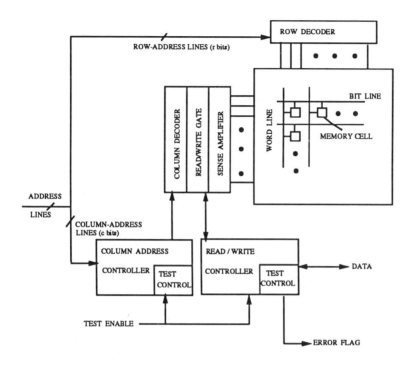

Figure 5.36 The column address maskable parallel test architecture (CMT); courtesy [114]

line of the match-line pair gets discharged, while the other one does not. The error detection circuit, which is a simple AND gate, does not generate the error flag in this case. However, when a single error occurs, the faulty bit line pair holds a value opposite to that stored in the other bit line pairs, so that both lines of the pair are discharged, causing the error flag to be generated.

As shown in the figure, if there is an error, both lines of the match-line pair will be discharged, otherwise exactly one line will be discharged.

Column address masking

By adding some combinational circuitry to the column address controller, the normal mode decoding is overriden by a special test mode decoding that activates multiple column decoder outputs instead of only one. This masking operation is carried out in the column address controller and the masking sig-

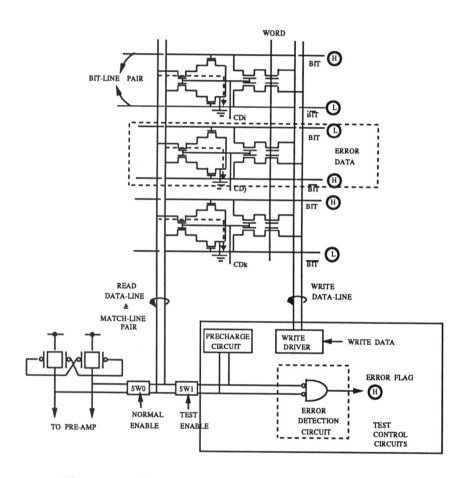

Figure 5.37 The error detection circuit used in CMT; courtesy [114]

nals produced are gated to the decoder gates. No modification is done on the column decoder.

Detection of pattern-sensitive faults

The CMT architecture provides various kinds of test patterns, including disturb-data patterns resulting from masking the column addresses during the test mode operation. An algorithm for NPSF detection in a Type-1 neighborhood using this architecture is as follows:

(1) Write a checkerboard pattern into all memory cells. A base cell stores 0 and surrounding cells store 1. In CMT mode, this requires a total of $2^R * 2$ operations (R being the address input size of the row decoder) because for each row, one operation writes the base cell, and one operation writes the entire neighborhood of 4 cells, whose column addresses are activated simultaneously by column address masking.

(2) Consider each row address in turn. In test mode, write a 'disturb-0' in all the surrounding cells (this requires one step for each row because of column address masking). Then write a 'disturb-1' in all the surrounding cells. The complexity for this step is $2^R * 2$.

(3) Read data from all cells. This requires a total of $2^R * 2$ operations, because in test mode, one read operation reads the base cell and one read operation reads the entire neighborhood of 4 cells for each row.

The complexity of the above test is thus seen to be $6 * 2^R$.

Address searching capability

A failed column address can be easily tested by this scheme using the following algorithm (where the column address decoder has an address input size of K):

(1) Set i to 0.

(2) Mask column address bits j where $j \neq i$. Then check column address input CA_i for 0 and 1 respectively. An error at this stage denotes failure of the column address check and causes the algorithm to terminate. This column address masking of $K-1$ bits causes half of all memory cells along a word line to be tested simultaneously.

(3) If $i > K - 1$ the RAM passes the column address check, else go to (2).

The above algorithm requires only $2K$ cycles as opposed to 2^K for normal mode column addressing which generates each of 2^K possible addresses and verifies that they are correct.

Application to 64 Mb DRAM

The CMT architecture has been applied to the design of a 64 Mb DRAM. The memory array is divided into sixteen 4 Mb blocks, each block being further divided into sixteen 256 Kb subarrays. Half of the sixteen 256 Kb subarrays, in two 4 Mb arrays, are activated simultaneously during the same cycle in CMT mode. The chip size is $12.5 \times 18.7 mm^2$, and the CMT overhead is less than 0.1%.

5.7 DFT FOR EMBEDDED DYNAMIC CONTENT-ADDRESSABLE MEMORIES (CAMS)

A content-addressable memory (CAM) can be tested in a far simpler manner than a RAM [94, 99]. The reading operations in a CAM can be replaced by one content-addressable search operation, and instead of using a signature analyzer, as is typically proposed for RAM BIST [65, 124], simple hardware can be used to monitor the match lines in a CAM. In contrast to the signature analyzer approach, simple testable designs have been proposed by Mazumder, et al. [94, 99] which require at most $2w + 25$ transistors in a w-word CAM of any arbitrary word size. Three methodologies are proposed for testing stuck-at faults, adjacent cell coupling faults, and pattern-sensitive faults using simple, testable DFT and BIST hardware. Some advantages of this approach, to be described shortly, include: a simple testable design for CAMs, high fault coverage of three kinds of functional faults, and an efficient approach to testing CAMs utilizing a single associative search operation as opposed to the conventional w read or associative search operations [49, 50].

We shall now introduce the design of the dynamic CAM (DCAM), and then discuss the design for testability (DFT) and built-in self-testing (BIST) approaches used.

BIST and DFT

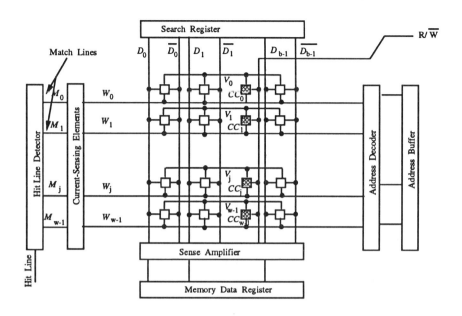

Figure 5.38 Basic organization of a dynamic CAM; courtesy [99]

5.7.1 Overview of dynamic CAM design

A typical organization for a dynamic content-addressable memory is shown in Figure 5.38. A b-bit, w-word CAM is organized as an $n = b \times w$ two-dimensional array where all word lines (rows) and bit lines (columns) can be accessed randomly and in parallel. Each memory cell is connected to a pair of horizontal lines, forming a row, and a pair of vertical lines, forming a column. A memory cell is denoted by C_{ij} if it is located at the cross-point of the i-th bit-line and the j-th word line, and is connected to the horizontal lines W_j, V_j, and to the vertical lines D_i, $\overline{D_i}$. Lines D_i and $\overline{D_i}$ are the i-th bit and $\overline{bit}$ lines driven by the sense amplifiers. Line W_j is the j-th word line and is selected by the word-line decoder, which decodes the contents of the lg w-bit memory address register (MAR). Line V_j is derived by the control cell (CC_j) and its value depends on the logical value of W_j and the $R/\overline{W}$, called the read/write line.

The basic memory cell is shown in Figure 5.39. It consists of four transistors. Transistors Q_1 and Q_2 form the left half-cell, which is somewhat similar to a switched capacitor dynamic RAM cell. As in DRAM, the logic state is stored by the parasitic gate capacitance of Q_1. The right half-cell consists of transistors

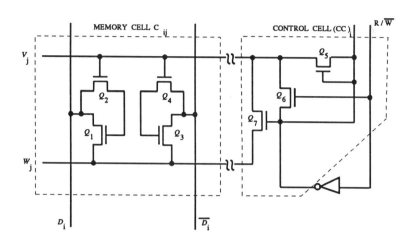

Figure 5.39 Dynamic CAM cell; courtesy [99]

Q_3 and Q_4 and is identical to the left half-cell. Unlike DRAMs, both halves are necessary to constitute a single CAM cell. Otherwise, as will be evident subsequently, the search operation cannot be done with a half-cell.

The left and right half-cells contain complementary data. The cell is said to store a logic 0(1) if the parasitic capacitance of Q_1 (Q_3) is discharged and that of Q_3 (Q_1) is charged. In order to replenish the charges due to leakage, periodic refreshing is necessary. Line V_j in C_{ij} is supplied by the control cell CC_j, which consists of three transistors, Q_5, Q_6 and Q_7. These three transistors realize the Boolean function $V_j = \overline{R}.(W_j + R)$, so that transistors Q_2 and Q_4 can be turned on only in write mode when $R = 0$. Depending on the values of R, D_i, $\overline{D_i}$, and W_j, the memory cell operates in the modes shown in Table 5.5. These modes are explained below.

1. **Write Mode:** In the write mode, $W_j = 1$ and $R = 0$. Thus $V_j = 1$ and transistors Q_2 and Q_4 are turned on so the values of D_i and $\overline{D_i}$ will be stored in the parasitic gate capacitances.

2. **Masked Write Modes:** If $W_j = 0$ and $R = 0$, then $V_j = 0$ and transistors Q_2 and Q_4 do not turn on. Hence no write operation can be made on the cell. Also if $W_j = 1$ and $R = 0$, with both D_i and $\overline{D_i}$ at high impedance, the content of the memory will remain unchanged.

3. **Read Mode:** In the read mode, $W_j = 0$ and $R = 1$. The bit-lines D_i and $\overline{D_i}$ are precharged to 1. Depending on the content of the cell, either Q_1 or

Table 5.5 Truth Table for CAM; courtesy [99]

Function	Input				Output
	D_i	$\overline{D_i}$	W_j	R	
Write 0	0	1	1	0	
Write 1	1	0	1	0	
Masked Write	x	x	0	0	
Masked Write	z	z	1	0	
Read 0	1	1	0	1	Current in $\overline{D_i}$
Read 1	1	1	0	1	Current in D_i
Masked Read	1	1	1	1	
Match 0	0	1	1	1	Current in W_j
Match 1	1	0	1	1	if Mismatch occurs
Masked Match	1	1	1	1	

Q_3 (but not both) will turn on and either D_i or $\overline{D_i}$ will be discharged by the word line. Thus the sense amplifier can read a 0 or a 1.

4. **Content-Addressable Search Mode:** In this mode, $W_j = 1$ and $R = 1$. The data in the search register drive the D_i and $\overline{D_i}$ lines, and these values are compared with the stored content of the cell. In case of a mismatch, current flows from the W_j through Q_1 or Q_3 depending on whether D_i or $\overline{D_i}$ stores 0. This current is detected by the current-sensing elements, which pull up the corresponding MATCH line M_j to a high value in case of a match, and pulls it down to a low value for a mismatch. If any mismatch occurs, the HIT line goes high (i.e., inactive). Hence, the HIT line is a logical NAND of all the MATCH lines.

5. **Masked Search Mode:** A memory cell can be masked in search mode if both the D_i and $\overline{D_i}$ lines are set to 1. In this case, even if Q_1 or Q_3 turns on due to a mismatch between the state of the cell and D_i or $\overline{D_i}$, respectively, no current will flow through the word line W_j since both the bit lines and word line are at value 1.

In addition, an *internal masking* of cells occurs if both the cells contain a 0 during a read operation [117]. In such cases, no current will flow through D_i and $\overline{D_i}$, and the reading will be masked.

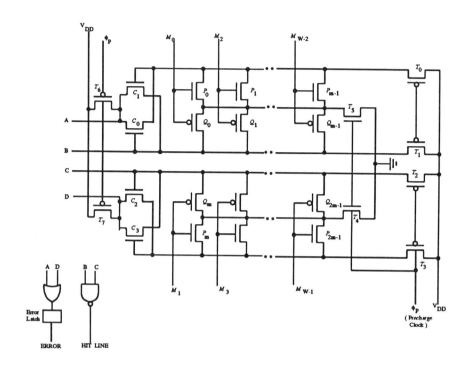

Figure 5.40 Testable CAM circuit; courtesy [99]

5.7.2 DFT approaches for Dynamic CAMs

The proposed testable design [94, 99] exploits the fact that test algorithms often generate test patterns which repeat after each p-word line by doing mutual comparison of all p word lines. Thus, the proposed technique replaces the w read/associative search operations by a single associative search. Giles and Hunter [49] have proposed the modification of the HIT line detector to augment the testability of stuck-at faults in a CAM. In [99], a similar approach is adopted to test a large class of functional faults, and the odd and even word lines have been grouped into two different classes. The modified HIT line detector is shown in Figure 5.38 and have the proposed testability hardware is shown in Figure 5.40.

In Figure 5.40, the transistors Q_0 through Q_{m-1} ($m = b/2$) detect the simultaneous mismatch of all even word lines, while transistors Q_m through Q_{2m-1} detect the simultaneous mismatch of all odd word lines. Transistors $P_0, \ldots P_{m-1}$ detect the simultaneous match of all even lines, while transistors P_m

BIST and DFT

through P_{2m-1} detect the simultaneous match of all odd lines. Transistors C_0 through C_3 are comparator transistors and $A = 0$ if all the even-numbered word lines either match or mismatch simultaneously with the data in the search register. Likewise, $D = 0$ if all the odd-numbered word lines either match or mismatch simultaneously with the search register data. $A = 1$ if some of the even lines match and others do not. Similarly, $D = 1$ if some of the odd lines match and others do not. If either A or D is 1, then the error latch triggers and ERROR = 1. If both A and D are 0, then ERROR = 0. Therefore, ERROR = A OR D. The HIT line is seen quite easily to be the Boolean NAND of B and C, because $B = 0$ if at least one even line matches and $C = 0$ if at least one odd line matches. Transistors T_0 through T_7 are used for precharging with the clock ϕ_P.

5.7.3 Parallel test for pattern-sensitive faults in Dynamic CAMs

The technology and layout-related mechanisms responsible for NPSFs in DCAMs are somewhat similar to those responsible for NPSFs in DRAMs. Hence, no new fault modeling approach is necessary. However, the test algorithm, known as Algorithm 1, for NPSFs in dynamic CAMs will have some peculiar features, and is given in Figure 5.41. As with DRAMs, the cells in the memory are each numbered 0 through 3, such that if 0 is a base cell of a square 9-cell neighborhood, then its north-south neighbors are numbered 1, east-west neighbors are numbered 2, and diagonal neighbors are numbered 3. As before, we consider both SSPSFs and SDPSFs in the memory array, that is, all cells having the same number belong to one group and are modified/accessed simultaneously and identically. We also construct a table consisting of a sequence of 64 successive states that result when an Eulerian sequence is traversed; that is, cells belonging to exactly one group undergo transition writes from one state to the next in the sequence. This table, Table 5.6 is looked up in steps 7 and 9.

Algorithm 1 has a complexity of $33w + 2b + 64$, and if the technique of simultaneous access of even and odd word lines [49] is employed in addition to the bit lines, the test speed can be further improved to $97 + 2b$ for any arbitrary-sized CAM. For moderately small CAMs, this will not deteriorate the write cycle time, however, for relatively large-sized CAMs, this would cause an increased write cycle time and greater power consumption.

It should be noted that Algorithm 1 does not test the testable hardware in Figure 5.40 exhaustively. When all the even and odd lines match, the N-type

1. Initialize all the cells to 0.
2. Set $k = 0$.
3. Set the k-th bit in SR (search register) to 1 and rest of the bits to 0. Associatively read the memory. Check if ERROR = 0, else an error is detected.
4. Increment k.
5. If $k < b$, go to step 3.
6. Set $i = 0$.
7. If the transition write in i-th operation, as given by the Table 5.6 is on cells numbered $p \in \{0, 2\}$, write on all cells having number p and are on even word lines.
8. Set SR = $(s_0 s_2)*$, where s_0 and s_2 are the contents of cells numbered 0 and 2, respectively. Associatively read the memory. Check if ERROR = 0, else an error is detected.
9. If the transition write in i-th operation is on cells numbered $p \in \{1, 3\}$, write on all cells having number p and are on odd word lines.
10. Set SR = $(s_1 s_3)*$, where s_1 and s_3 are the contents of cells numbered 0 and 2, respectively. Associatively read the memory. Check if ERROR = 0, else an error is detected.
11. Increment i.
12. If $i = 10$, then go to step 13, else go to step 17.
13. Set $k = 0$ (to begin testing the CAM with a marching pattern of 0s in a background of 1s).
14. Set the k-th bit in SR (search register) to 0 and rest of the bits to 1. Associatively read the memory. Check if ERROR = 0, else an error is detected. Complement the k-th bit in SR.
15. Increment k.
16. If $k < b$, go to step 14.
17. If $i < 65$, go to step 7, else exit successfully.

Figure 5.41 Algorithm 1: test of NPSFs in embedded CAMs

BIST and DFT

Table 5.6 Content of Memory Cells; courtesy [99]

Op#	$s_3s_1s_2s_0$	Op#	$s_3s_1s_2s_0$	Op#	$s_3s_1s_2s_0$	Op#	$s_3s_1s_2s_0$
1	0001	17	0001	33	0001	49	0001
2	0011	18	0000	34	1001	50	0101
3	0010	19	1000	35	1101	51	0111
4	0110	20	1010	36	1100	52	1111
5	0111	21	0010	37	1000	53	1101
6	0101	22	0011	38	0000	54	1001
7	0100	23	1011	39	0100	55	1011
8	1100	24	1111	40	0110	56	1010
9	1101	25	0111	41	0010	57	1000
10	1111	26	0110	42	1010	58	1100
11	1110	27	1110	43	1110	59	1110
12	1010	28	1100	44	1111	60	0110
13	1011	29	0100	45	1011	61	0100
14	1001	30	0101	46	0011	62	0000
15	1000	31	1101	47	0111	63	0010
16	1001	32	0101	48	0011	64	0000

transistors Q_0 through Q_{2m} will be tested for stuck-at 1 faults, and when they show a mismatch, the P-type transistors P_0 through P_{2m} will be tested for stuck-at 0 faults. The remaining stuck-at faults in the testable hardware can be tested by the procedure illustrated in Figure 5.42.

In [93], an alternative algorithm to test *static* CAMs is described. The NPSFs in which are different from those described here. NPSFs found in static CAMs are tested by verifying whether a 0 and a 1 can be stored and held in the base cell in the presence of all valid patterns in the neighborhood, and the corresponding algorithm has a complexity of $33w + 2b + 160$. In the present approach, however, we are interested in verifying transition writes at all the cells in the memory.

5.7.4 BIST implementation for dynamic CAMs

The operations in Table 5.6 can be generated within the memory chip by using a sequential circuit consisting of some simple flip-flops. In [99] one such imple-

1. Initialize the memory with all-0.

2. Set $(SR) = (00)*$ and check if ERROR = 0.

3. For each word line, do
   Write 1s in all its bits;
   Set $(SR) = (00)*$ and check if ERROR = 1.
   Write 0s in all its bits;
   Set $(SR) = (00)*$ and check if ERROR = 0.

4. Initialize the memory with all-1.

5. For each word line, do
   Write 0s in all its bits;
   Set $(SR) = (11)*$ and check if ERROR = 1.
   Write 0s in all its bits;
   Set $(SR) = (11)*$ and check if ERROR = 0.

Figure 5.42 Parallel testing algorithm for stuck-at faults

BIST and DFT

mentation, shown in Figure 5.43, is described. The basic idea is to generate the reflected Gray code corresponding to operations 1 through 15, followed by 64, in Table 5.6. This is done using flip-flops (such as JK), with suitable feedback logic. The remaining operations in the table can be obtained by reordering the outputs (complemented and uncomplemented) of the flip-flops appropriately, using multiplexers. Reads and non-transition writes are disabled by providing a disable control signal to the sense amplifiers. The $w/2$ word addresses (for the even or odd word lines) are generated using a $\lg w - 1$-bit synchronous counter connected to the address input lines of the word-line decoder. The LSB of the address lines of the word-line decoder is set to 0 if the even word lines are to be accessed; it is set to 1 if the odd word lines are to be accessed.

The BIST generator circuit can be further simplified by decomposing the Eulerian tour into eight disjoint Hamiltonian cycles on subgraphs of the symmetric 4-cube. In Algorithm 1, the sequence of transition writes is generated by constructing an Eulerian tour over the symmetric 4-cube. The hardware overhead in Figure 5.43 can be reduced by a factor of 2 by decomposing the Eulerian tour into eight disjoint Hamiltonian cyclic tours over subgraphs of the symmetric 4-cube:

- $\langle 0, 1, 9, 13, 15, 14, 6, 2, 0 \rangle$,
- $\langle 0, 2, 6, 14, 15, 13, 9, 1, 0 \rangle$,
- $\langle 0, 8, 9, 11, 15, 7, 6, 4, 0 \rangle$,
- $\langle 0, 4, 6, 7, 15, 11, 9, 8, 0 \rangle$,
- $\langle 12, 4, 5, 7, 3, 11, 10, 8, 12 \rangle$,
- $\langle 12, 8, 10, 11, 3, 7, 5, 4, 12 \rangle$,
- $\langle 12, 13, 5, 1, 3, 2, 10, 14, 12 \rangle$,
- $\langle 12, 14, 10, 2, 3, 1, 5, 13, 12 \rangle$.

Each of the above Hamiltonian cycles over a subgraph of the symmetric 4-cube has a length of 8, and all these cycles are disjoint. Thus all the 64 edges of the symmetric 4-cube is traversed, and in this way, all the 16 static NPSFs and 48 dynamic (active) NPSFs are sensitized. Initially, the memory is initialized to zero and the first four Hamiltonian cycles are performed as indicated above. Then the memory is reinitialized such that it contains a column bar pattern of 0s and 1s. This needs an additional w write operations, and for a 1 KB memory

292 CHAPTER 5

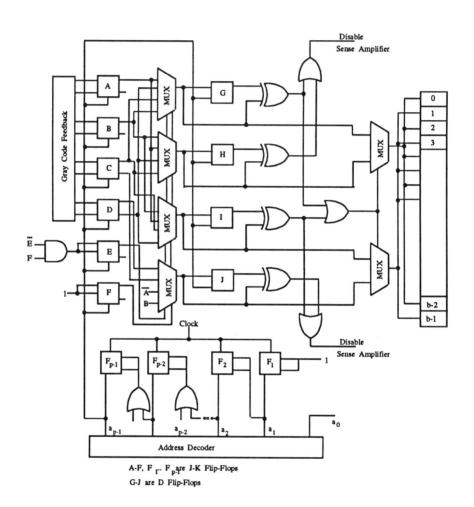

Figure 5.43 BIST implementation of the circuit; courtesy [99]

Table 5.7 Test Size Optimality versus BIST Hardware

Component	Hamiltonian Tour	Eulerian Tour	Extra Component
	$34w + 2b + 64$	$33w + 2b + 64$	
Flip-flop	$7 + p$	$10 + p$	3
MUX (4 to 1)	0	4	4
MUX (2 to 1)	1	2	1
XOR Gate	3	4	1
OR Gate	1	8	7
AND Gate	3	15	12
Inverter	1	1	0

having a memory cycle time of 50 ns will take less than an extra 50 μs. But the economy in hardware is considerable [95] and is shown in Table 5.7.

5.7.5 Parallel test for stuck-at faults and checkerboard tests in dynamic CAMs

The above testable hardware can also be used to test for stuck-at faults in the cell array of the embedded CAM. Moreover, it can be used to perform parallel checkerboard test. It may be recalled here that a major advantage of these tests is that all the memory cells are associatively compared and tested by a single search operation, instead of by w read operations, as is commonly the practice. This causes a significant reduction in the time complexity of these tests. For example, stuck-at faults may be easily tested for by writing at each address, the unique binary value of that address, and then searching the memory associatively (note: search, not read) for that value. This is done by simply writing the value to be searched in the search register and monitoring the HIT and ERROR lines. After this, we may write the complement of the binary value of the address and then search the memory associatively for that value once again. A checkerboard test can be easily performed in an analogous fashion.

5.8 TRANSPARENT BIST ALGORITHMS AND ARCHITECTURE

In 1986, B. Koeneman introduced transparent BIST for RAMs at the Design for Testability (DFT) workshop. The technique he proposed can be used only with linear compactors for the memory signature. Nicolaidis and Kebichi [126] proposed a BIST technique which preserves the initial contents of the RAM and works with any RAM test algorithm. They also propose an architecture that allows the efficient implementation of BIST-ed and transparent BIST-ed RAMs. The motivation for their technique, and its implementation, are described in the sections below.

5.8.1 The need for transparent BIST

In several applications, the contents of a RAM must not be destroyed when the RAM is tested. For example, the RAM may have to be tested periodically using various test algorithms, with normal operation continuing between tests. One naive approach to solve this problem is to save the chip contents to some spare chip and later restore these contents. This approach would become prohibitively difficult if the RAM is embedded within a more complex circuit. In this case, addressing the RAM and manipulating the RAM data from outside of the circuit can turn out to be very difficult. Besides, this approach requires the use of a spare chip and additional circuitry to access both the chips from outside. This would increase the hardware cost considerably. An efficient solution to this problem is the use of a transparent BIST scheme which preserves the RAM contents.

5.8.2 Transparent BIST algorithm for linear signature analyzers

In 1986, B. Koeneman presented a transparent BIST algorithm at the DFT workshop in Vail, Colorado. This technique works only for linear signature analyzers. Suppose TP denotes a test pattern. The algorithm is described below.

1. Reset the signature MISR (multiple input signature register), which is assumed to be linear.

2. Read each memory word ($WORD$) and save its signature $s(WORD)$ in a temporary array called $TEMP1$. For each word ($WORD$), write ($NEW\_WORD = WORD$ XOR TP) back into the memory and reset the signature LFSR.

3. Read each memory word (NEW_WORD), save its signature $s(NEW\_WORD)$ in a temporary array called $TEMP2$, and write NEW_WORD XOR TP back into the memory.

4. Output the result of XOR-ing each element of $TEMP1$ with the corresponding element of $TEMP2$.

In the above algorithm, the final contents of each RAM cell are: NEW_WORD XOR $TP = (WORD$ XOR $TP)$ XOR $TP = WORD$, hence the RAM contents are not destroyed after the algorithm ends. Also, due to the linearity of the MISR, we have $TEMP2 = s(NEW\_WORD) = s(WORD$ XOR $TP) = s(WORD)$ XOR $s(TP) = TEMP1$ XOR $s(TP)$. Therefore, the final signature is $TEMP2$ XOR $TEMP1 = s(TP)$, and is thereby predictable. (Note: by $TEMP1$ and $TEMP2$ here, we mean their corresponding elements, not the whole array).

There are four drawbacks of this algorithm. First, it works only if the output response is verified by a linear compactor. Secondly, this scheme is not free from the possibility of fault aliasing. For example, suppose an error string E is present during the read operation in steps 2 and 3. In such a case, the resulting signatures will be $TEMP1 = s(WORD$ XOR $E)$ and $TEMP2 = s(NEW\_WORD$ XOR $E)$, and the final signature will be $s(TP)$ which is the same as the correct signature. The third drawback is that the proposed algorithm is composed of only one read and one write operation at each cell in each of the steps 2 and 3. However, the majority of existing RAM test algorithms include more complex read and write sequences, hence, this technique cannot be adapted to any arbitrarily chosen RAM testing scheme. The fourth drawback is that the fault coverage is not very high, because of aliasing, and moreover, it is difficult to calculate the exact fault coverage.

Nicolaidis and Kebichi [126] have recently described a technique in which most of these drawbacks are removed. Their approach can be used with any RAM test algorithm and any signature compactor. Moreover, their technique, described in the following sections, allows a very easy computation of the fault coverage with or without transparent BIST.

Conversion of a non-transparent test into a transparent one

The basic idea behind the conversion of a non-transparent test into a transparent one is to use the stored data in the RAM for testing in such a way that the stored data is complemented an even number of times. So instead of initializing the memory with a fixed pattern (as done by a typical non-transparent test algorithm) and thereby losing the original data, we perform an initial *read* operation, and write back the data or its complement subsequently. So the actual data written into each memory cell over a period of time is a function of the initial data read from that cell, as described below. In the following approach, a bit-oriented memory is assumed; however, the approach can be easily extended to a word-oriented memory.

1. Add an initial read operation to each test pattern of the (non-transparent) algorithm. Suppose that for a given memory cell c, the initial read produces the value a.

2. Substitute every write-0 operation on cell c with write-a, and every write-1 operation on cell c with write-$\bar{a}$, or vice versa, for each test pattern. One convenient way of re-stating this is to write the exclusive-OR of the test pattern bits and the initial contents for each cell. Each read operation on a cell c in the original non-transparent test is converted into the operation "read the data written by the most recent write operation into cell c."

3. If for any test pattern, the data stored in a cell c during the last write operation happens to be the complement of the initial data read from c, then add a read operation and a transition write operation at the end of the pattern so that the initial data is restored. In other words, each cell must be complemented an even number of times when the test terminates.

4. Signature prediction for the fault-free memory is performed next. This is done by examining only the read operations of the test obtained after performing the above steps. The fault-free signature is predicted by concatenating a sequence of expected values to be read from each memory cell during the test and compacting this sequence with a signature analyzer. This sequence is a function of the initial contents of the memory cells.

5. The transparent test computed in steps 1 to 3 is then applied to the memory, and the actual signature is computed by compacting the sequence of values produced by the read operations within the test. The two signatures are compared. Barring aliasing in the compactor, if the original,

BIST and DFT

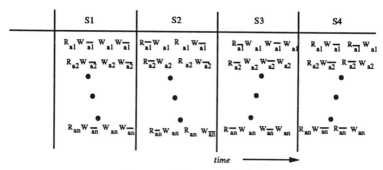

Figure 5.44 Transparent test algorithm derived from Marinescu's algorithm; courtesy [126]

non-transparent test could detect a certain fault in the memory, the new test would produce a signature which is different from the fault-free signature, and would thereby also detect the same fault.

The correctness and fault coverage of the above approach have been examined by Nicolaidis and Kebichi [126]. An example of converting a non-transparent test (Marinescu's algorithm B [83]) to a transparent test is shown in Figure 5.44.

It should be noted that the above approach can be easily extended to a word-oriented memory by replacing each bit-read operation with a word-read operation, that is, reading all the bits of a word in parallel; and replacing each bit write-0 operation with writing the all-0s vector and each bit write-1 operation with writing the all-1s vector.

5.8.3 BIST implementation for a transparent RAM test

Standard BIST hardware include the BIST sequence controller, read and write instance generator and controller, address and test generator, and response verification hardware such as an MISR. A transparent BIST architecture needs some modifications. First, a register must be added in order to store the contents of the addressed cell. The contents of this register will be re-used for the test data generation.

To control the sequence of operations, a **sequence control block** is used. This may be implemented as a counter with enough bits to generate not only the test sequence but also the signature-producing sequence (consisting of read operations) before the test.

The type of operation (read/write), the address generation clock signal, the clock of the sequence control block, and a control signal for the test data generator are all produced by combinational control logic. A modification is needed for the read and write **instance generator** of the standard BIST. The new instance generator is designed so as to switch the system between the test application mode and the signature prediction mode. The task of this unit is to avoid the activation of write cycles during the signature prediction mode. This may be achieved by using a sequential control logic implemented with D flip-flops. The exact manner of implementing this is dependent on the transparent test being performed.

Except for the extra register used for test generation, the above technique uses a fixed amount of extra hardware which is independent of the memory size, in order to implement the transparent BIST. The area overhead due to the BIST and to the transparent BIST implementation has been found to be 6% and 8.8%, respectively, for a 16 Kb RAM. The area overhead decreases to as low as 1% and 1.2% respectively for the two kinds of BIST implementation, as the RAM size increases to 64 Kb. Besides, this technique does not assume linearity of the response compaction scheme and does not involve error-masking above that which is inherent in the compaction mechanism itself.

5.8.4 Transparent BIST scheme for detecting single k-coupling faults

Cockburn and Sat [25] (1994) describe a transparent BIST technique for detecting k-coupling faults in RAMs. Their test algorithm is described in Chapter 3. It may be recalled here that their algorithm is based on the result that all single k-coupling faults in an $n \times 1$ RAM can be detected by a test if a certain bit-matrix M corresponding to the test is $(n, k-1)$-exhaustive. Each row of this matrix denotes a succession of bits written into a memory cell over a period of time; each column, therefore, denotes a test pattern applied to the RAM cells. Based on this idea, the matrix M can be constructed either deterministically or probabilistically, as mentioned in [25]. Here we shall briefly describe the BIST structure proposed by Cockburn and Sat.

The test pattern generator is a combinational circuit (the detailed design for which is discussed in [25]) that generates successive bits of an (n, k)-exhaustive code. The address generator for the BIST is a counter of width $(\lg n + \lg k)$, with $\lg n$ bits forming the word address, and $\lg k$ bits forming the bit address. A counter known as the *background* counter counts the number of the *data background* (or test pattern) currently being used, and the BIST controller increments this count after each test pattern application. For a RAM cycle time of 100 ns and $k = 2$ (i.e., the test is for single 3-coupling faults), this self-test is found to run for 56 ms, 1.14 s, 21.4 s and 393 s for $n \times 1$ memories of size 4 Kb, 64 Kb, 1 Mb and 16 Mb, respectively. For $k = 3$ and a RAM cycle time of 100 ns, the self-test is found to run for 486 ms, 13.5 s, 332 s and 126.3 min, respectively, for memories of the above capacities. For 64 Kb RAMs, the area overhead is found to be above 10%, but it tapers down exponentially to almost 0.01% for 256 Mb RAMs.

5.9 BIST BASED ON PSEUDORANDOM PATTERN TESTABILITY OF EMBEDDED MEMORIES

Memory devices embedded within larger chips, such as microprocessors, special-purpose controllers, high-density ASICs and others, are difficult to test, because of the fact that the address, data and read/write control lines are not directly controllable and/or observable. For such devices, built-in self-testing is a must because physical test access is limited. Various BIST techniques and standards are becoming increasingly popular for testing embedded memories these days. Some of these approaches use pseudorandom pattern testability because pseudorandom patterns are often easier to generate than deterministic ones. Response verification is frequently done by performing mutual comparison between subblocks on which identical test patterns have been applied. The scheme is illustrated in Figure 5.45.

Bardell, et al. [9] examined the problem of pseudorandom pattern testing of embedded RAMs for stuck-at faults. They also studied the manner in which stuck-at faults in embedded RAMs affect the detection probability of faults in the combinational logic feeding the data lines or the address lines of these RAMs. Their testability analysis is based on simple Markov chain models for these faults. The simplified block diagram used for this modeling is shown in Figure 5.46.

5.9.1 Modeling stuck-at faults in embedded RAMs

The model used is a single stuck-at fault model, because single stuck-at faults are the hardest to detect using random patterns [106]. The other two assumptions are that reading from memory is non-destructive and there are no sequential circuits in the logic preceding the memory. Since embedded RAMs are mostly SRAMs, these assumptions are realistic.

Let p_d be the signal probability (the probability of having the value 1) for each data-in line at the memory boundary — it is assumed that the signals on these lines can be independently asserted. Suppose each address line has an independent signal probability of p_a, and the read/write control line has a probability p_c of being in the write mode and $1-p_c$ of being in the read mode, at any given clock.

BIST and DFT

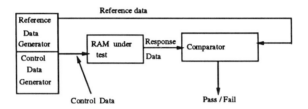

Deterministic RAM Test

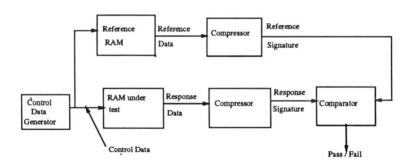

Pseudorandom RAM Test

Figure 5.45 Deterministic vs. pseudorandom testing

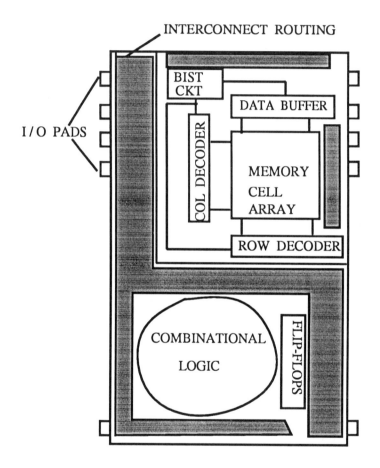

Figure 5.46 An embedded random-access memory

The memory is first initialized to a given state. In the following paragraphs, we describe what happens when the memory is initialized with all-0 or all-1.

Consider an arbitrarily chosen memory address. The binary representation of this address consists of, say, h 1s and $m-h$ 0s. The probability p_A of accessing this address is $p_A = p_a^h(1-p_a)^{m-h}$. The probability that a random vector writes to a particular address is, therefore, equal to $rp_c = p_c p_a^h(1-p_a)^{m-h}$. The probability that it writes a 1 to an arbitrarily chosen bit in a given address location is $w_0 = (1-p_d) * p_A * p_c$. The probability of reading from the address is $r = p_A * (1-p_c)$. Since the only possible operations when a memory location is accessed are read, write-0, and write-1, we must have the following:

$$p_A = w_0 + w_1 + r$$

We shall soon examine how the detection of different kinds of functional faults in the cell array, such as SAFs, CFs and PSFs, can be modeled using Markov chains. Such analysis enables us to compute the detection probability when pseudorandom patterns are applied to the memory.

5.9.2 Testability analysis for functional faults in embedded memories

Mazumder and Patel [99, 100, 103] have also proposed a complete design strategy for efficient and comprehensive random pattern testing of embedded RAMs in which neither are the address, data or read/write lines directly controllable nor are the output lines directly observable. The proposed test exploits an efficient testable design instead of using LFSRs, data registers, and multibit comparators for generating and applying test patterns and evaluating test responses. Test algorithms are accelerated by a factor of $0.5\sqrt{n}$.

The problems associated with the use of LFSRs are well-known. The major problem is aliasing which would cause faults to go undetected if the faulty signature happens to be equal to the fault-free one. Furthermore, random testing itself has the serious problem of deterministic memory initialization.

Empirical studies of the functional faults such as SAFs, CFs and PSFs have been performed by suitably representing the testing for such faults as Markov chains and then simulating these chains to derive detection test lengths. For detecting the common functional faults, the proposed technique needs only one second compared to about an hour needed by conventional random testing algorithms that perform sequential memory access, in a 1 Mb RAM.

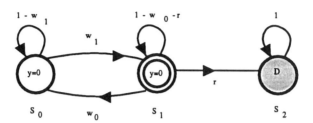

Figure 5.47 Markov model for a stuck-at-zero fault; courtesy [103]

5.9.3 Markov modeling for detecting functional faults

1. **Markov model for SAFs**: There are only two memory operations on a cell — read and write. Read operations are assumed to be fault-free. In a random test, these operations are applied at random to the cells and the fault coverage measured. Suppose that a memory cell is initially at state $y = 0$. Then, to detect a stuck-at-zero fault, we must first try to write a 1 on the cell and then read the contents of the cell. In a random test, these operations are not deterministic, but probabilistic in nature. So, if w_0, w_1, and r are the probabilities of writing a 0, writing a 1, and reading the cell, respectively, then the sequence of moves that will cause detection is depicted in the Markov chain of Figure 5.47.

In the Markov chain model, it is clearly noticed that detection will take place at state D. The probability of reaching different states, including D, is given by the transition matrix

$$M = \begin{bmatrix} 1 - w_1 & w_1 & 0 \\ w_0 & 1 - w_0 - r & r \\ 0 & 0 & 1 \end{bmatrix}$$

Initially, the memory is in state S_0 with a probability I_0 $(= p_0(S_0))$ or at state S_1 with a probability $1 - I_0$ $(= p_0(S_1))$. The cell cannot initially be at state S_2 which is an absorbing state. We shall use the following convention: the initial state is shown by a light circle, the state in which the fault is triggered is denoted by a ring-shaped double circle, and the absorbing state in which detection occurs is shown by a heavy circle. For a memory cell which is stuck-at-zero, the probability of the fault being triggered is w_1. After being triggered, the fault can be detected only if a read operation is performed; hence, the probability of detection is r (see

Figure 5.47). In our discussion, the symbol $p_j(S_i)$ will be used to denote the probability that the Markov chain is at state S_i after j steps.

For a given test length L, the *quality of detection* (Q_D) is defined as the probability $p_L(S_2)$, namely, the probability that the fault is detected after L test vectors are applied. The *escape probability* (e), is equal to $1 - Q_D = p_L(S_0) + p_L(S_1)$. These probabilities are computed recursively, yielding the following:

$$\begin{bmatrix} p_L(S_0) \\ p_L(S_1) \\ p_L(S_2) \end{bmatrix} = M^T \begin{bmatrix} p_{L-1}(S_0) \\ p_{L-1}(S_1) \\ p_{L-1}(S_2) \end{bmatrix}$$

$$= (M^T)^L \begin{bmatrix} I_0 \\ 1 - I_0 \\ 0 \end{bmatrix}$$

where M^T is the transpose of the transition probability matrix M, given above.

The Markov chain has been simulated using the above equation, with different values of I_0 and read and write probabilities, to arrive at estimates of the quality of detection Q_D ($= p_L(S_2)$). It has been observed that the quality of detection improves with an increase in the average number of test vectors per cell, L/n, which is commonly known as the *test length coefficient*. In all cases, however, a very high probability of detection (0.999) is obtained with less than 50 test vectors. Also, they noticed that for a large number of samples (cells), the fault is detected by applying only 10 vectors or less. The plot in Figure 5.48 shows these results. In this analysis, it is assumed that all the cells in the memory have uniform access probability, $p_a = 1/n$, where n is the number of cells. But in practice, due to the presence of combinational logic circuits very frequently, the access probability may not be uniform. If an address line is selected by a k-input AND or NOR gate, then, assuming uniform probability at all the k input lines, the probability of a 1 on the address line is $1/2^k$, and the probability of a 0 on the address line is $1 - 1/2^k$. These address line probabilities will be interchanged if the line is selected by a k-input OR or NAND gate. Thus the presence of combinational logic on the address path considerably modifies the access probability. Figure 5.50 illustrates how the test length coefficient changes with the signal probability of the address and data lines, where the signal probability has been varied over a wide range (from 0.1 to 0.9 in the case of the data line and from 0.25 to 0.75 in the case of address lines — beyond which the test length tends to increase rapidly).

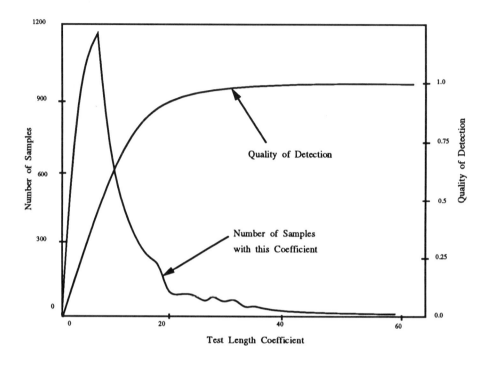

Figure 5.48 Quality of detection & number of samples vs. test length coefficient; courtesy [103]

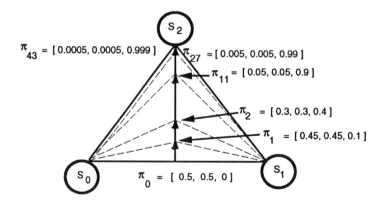

Figure 5.49 State probability diagram for stuck-at faults; courtesy [103]

BIST and DFT

Table 5.8 Test length coefficient for stuck-at faults; courtesy [103]

	$I_0 = 0$		$I_0 = 1$		$I_0 = 0.5$	
Q_D	$s-a-0$	$s-a-1$	$s-a-0$	$s-a-1$	$s-a-0$	$s-a-1$
0.1	2	1	1	2	1	1
0.2	3	1	1	3	1	1
0.3	4	1	1	4	2	2
0.4	5	2	2	5	3	3
0.5	7	2	2	7	4	4
0.6	8	3	3	8	6	6
0.7	10	4	4	10	8	8
0.8	13	7	7	13	10	10
0.9	18	11	11	18	15	15
0.99	33	27	27	33	31	31
0.999	49	43	43	49	47	47
0.9999	65	59	59	65	63	63
0.99999	81	75	75	81	79	79

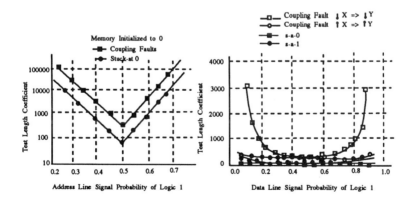

Figure 5.50 Signal probability vs. test length coefficient; courtesy [103]

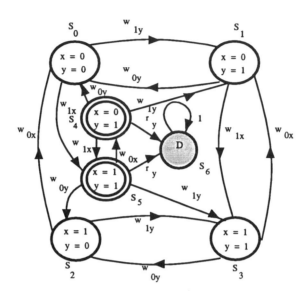

Figure 5.51 Markov model for a CFid $\langle \uparrow, 0 \rangle$; courtesy [103]

Figure 5.49 shows the state probability diagram for stuck-at faults. Table 5.8 displays the test length coefficient values for stuck-at faults.

We have also examined Markov models for coupling and pattern-sensitive faults. These models are briefly described below.

2. **Markov Models for 2-cell CFs**: Suppose a CF on two cells x and y is denoted by a doublet $\langle \Psi(x), \sigma_y \rangle$, where Ψ is a transition write operation on cell x, and $\sigma_y \in \{0, 1\}$ is the initial state of y. The four possible variants of this doublet are $\langle \uparrow, 0 \rangle$, $\langle \uparrow, 1 \rangle$, $\langle \downarrow, 0 \rangle$, and $\langle \downarrow, 1 \rangle$. The cells are said to be 1-way coupled if only one of these faults exist, otherwise, if a combination of two or more faults exist, they are 2-way coupled.

A separate Markov chain can be designed for each coupling fault. For example, the CFid $\langle \uparrow, 0 \rangle$ (i.e., the coupling fault observed when cell x undergoes an 'up'-transition, forcing y to change state if the initial state of y is 0), is depicted by the Markov chain shown in Figure 5.51.

In Figure 5.51, the coupling fault will clearly not be triggered at any of the four possible initial states of the cells x and y. The notations for various transition probabilities are self-explanatory; for instance, w_{0y} gives the probability of writing a 0 on cell y. The only way to trigger this coupling fault is to write a 1 on cell x when both x and y are in state 0, that is,

BIST and DFT

Table 5.9 Test length coefficient for coupling faults; courtesy [103]

$I_0 = 0$ Q_D	↓X ↓ ↓Y	↓X ↓ ↑Y	↑X ↓ ↓Y	↑X ↓ ↑Y	↑↓X ↓ ↓Y	↑↓X ↓ ↑Y	↑↓X ↓ $Y \to \overline{Y}$	↓X ↓ $Y \to \overline{Y}$	↑X ↓ $Y \to \overline{Y}$
0.1	8	6	6	2	5	2	2	5	2
0.2	12	10	10	4	8	4	4	7	4
0.3	16	14	14	8	11	7	5	9	5
0.4	21	19	19	12	14	10	7	11	7
0.5	27	25	25	18	18	14	9	14	10
0.6	34	32	32	25	23	18	12	17	13
0.7	44	41	41	35	29	25	15	22	17
0.8	57	55	55	48	38	33	19	28	23
0.9	79	77	77	70	53	48	27	38	33
0.99	154	152	152	145	102	98	53	72	67
0.999	228	226	226	220	151	147	79	106	102

Table 5.10 Test length coefficient for coupling faults; courtesy [103]

$I_0 = 0$ Q_D	↓X ↓ ↓Y	↓X ↓ ↑Y	↑X ↓ ↓Y	↑X ↓ ↑Y	↑↓X ↓ ↓Y	↑↓X ↓ ↑Y	↑↓X ↓ $Y \to \overline{Y}$	↓X ↓ $Y \to \overline{Y}$	↑X ↓ $Y \to \overline{Y}$
0.1	2	6	6	8	2	5	2	2	5
0.2	4	10	10	12	4	8	4	4	7
0.3	8	14	14	16	7	11	5	5	9
0.4	12	19	19	21	10	14	7	7	11
0.5	18	25	25	27	14	18	9	9	14
0.6	25	32	32	34	18	23	12	13	17
0.7	35	41	41	44	25	29	15	17	22
0.8	48	55	55	57	33	38	19	23	28
0.9	70	77	77	79	48	53	27	33	38
0.99	145	152	152	154	98	102	53	67	72
0.999	220	226	226	228	147	151	79	102	106

cause an 'up' transition on x and check the effect on y. So the path from S_0 through D via S_5 forms a checking sequence. A longer checking sequence can be obtained by writing a 0 on cell x from state S_5 (i.e., to take the chain to state S_4) and then reading y, because writing a 0 on x is assumed to have no further effect on y since this is a one-way coupling fault. The state at which detection takes place is S_6.

This model has been simulated and random test lengths corresponding to acceptable detection quality values have been recorded by the authors. For a detection quality of 99.9%, a test length coefficient of 225 is observed.

The authors have also constructed Markov chains for other kinds of CFs and have experimentally obtained test length coefficients that would give acceptable quality of detection for these faults. These are illustrated in Figures 5.52 and 5.53.

Test length coefficients for coupling faults have been presented in Tables 5.9, 5.10, and 5.11.

310   CHAPTER 5

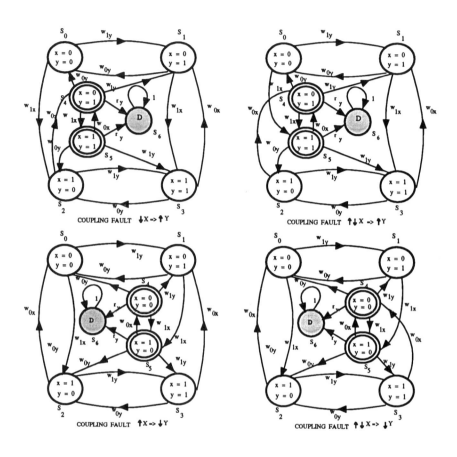

Figure 5.52   Markov model for various CFids; courtesy [103]

Table 5.11   Test length coefficient for coupling faults; courtesy [103]

$I_0 = 0$ Q_D	$\downarrow X$ $\Downarrow$ $\downarrow Y$	$\downarrow X$ $\Downarrow$ $\uparrow Y$	$\uparrow X$ $\Downarrow$ $\downarrow Y$	$\uparrow X$ $\Downarrow$ $\uparrow Y$	$\uparrow\downarrow X$ $\Downarrow$ $\downarrow Y$	$\uparrow\downarrow X$ $\Downarrow$ $\uparrow Y$	$\uparrow\downarrow X$ $\Downarrow$ $Y \rightarrow \overline{Y}$	$\downarrow X$ $\Downarrow$ $Y \rightarrow \overline{Y}$	$\uparrow X$ $\Downarrow$ $Y \rightarrow \overline{Y}$
0.1	5	5	5	5	3	3	2	3	3
0.2	9	9	9	9	6	6	4	4	5
0.3	13	13	13	13	9	9	5	7	7
0.4	18	18	18	18	12	12	7	9	9
0.5	24	24	24	24	16	16	9	12	12
0.6	31	31	31	31	21	21	12	15	15
0.7	40	40	40	40	27	27	15	20	20
0.8	54	54	54	54	36	36	19	26	26
0.9	76	76	76	76	50	50	27	36	36
0.99	151	151	151	151	100	100	53	70	70
0.999	225	225	225	225	149	149	79	104	104

BIST and DFT

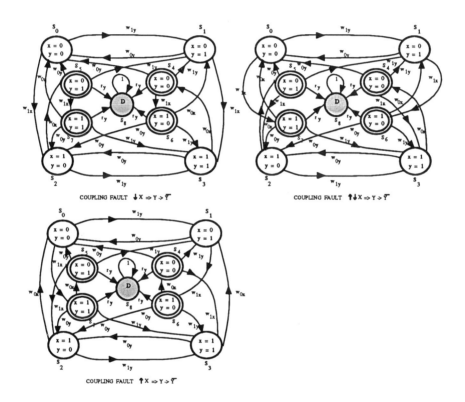

Figure 5.53 Markov model for various CFins; courtesy [103]

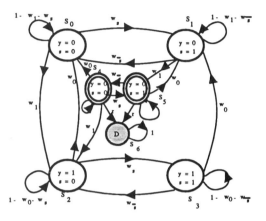

Figure 5.54 Markov model for an SNPSF $\langle \uparrow, S \rangle$; courtesy [103]

3. **Markov Model for NPSFs**: As expected, Markov models for NPSFs are somewhat more involved than those for 'simpler' faults like SAFs and CFs. We shall briefly discuss the model for a particular SNPSF here. In a Markov chain model, an SNPSF can be denoted by a doublet $\langle \Psi, S \rangle$, where S denotes a specific set of neighborhood patterns and Ψ denotes a faulty transition write operation on the base cell in presence of any pattern from that set. An ANPSF can be denoted by a triplet $\langle \sigma_x, \Psi(y), S \rangle$, which indicates that if the base cell is in state σ_x and a transition write Ψ is made on a neighborhood cell y, then the state of x changes if and only if the other cells store a pattern from a specific set of patterns S.

The Markov chain of Figure 5.54 shows the random test requirements for detecting the SNPSF $\langle \uparrow, S \rangle$, where the base cell x is prevented from making an operation $\uparrow$ in the presence of a specific set of patterns S. In the Markov chain, a state is labeled by a doublet $\langle y, S \rangle$, S being a binary value that denotes the presence ($S = 1$) or absence ($S = 0$) of a set S of specific patterns in the neighborhood.

This SNPSF can be detected by making the base cell y initially contain a 0 and setting up a neighborhood pattern p such that $p \in S$ (i.e., this can be represented by setting S to 1, as in state S_2 of the Markov chain of Figure 5.54), trying to write a 1 on the base cell y (state S_5), and then reading the value of y (state S_6). Another possibility is to remove the 'fault-causing' neighborhood pattern (i.e., set S to 0) and then read y, because the new pattern set up in the neighborhood will not have any effect on y. These possibilities can easily be seen by following around the arcs in Figure 5.54.

BIST and DFT

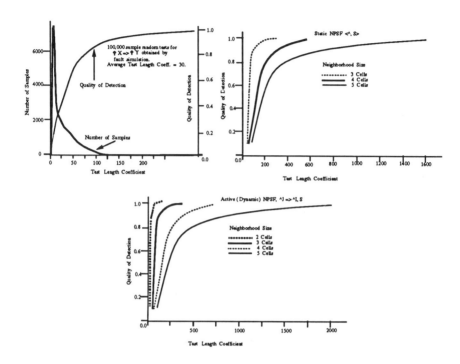

Figure 5.55 Quality of detection vs. test length coefficient; courtesy [103]

The above Markov chain is simulated for different neighborhood sizes. Varying neighborhood sizes would affect the values of w_{1S} and w_{0S} (i.e., the probabilities of installing a 'fault-causing' neighborhood pattern and of removing it, respectively). A test length coefficient of 1531 is seen to achieve a detection quality of 0.999 for this NPSF in a 5-cell neighborhood. These results are illustrated in Figure 5.55 and Tables 5.12 and 5.13.

Modeling ANPSFs is an even more involved process because of the potential need for more states, since each state is now encoded using 3 bits. Simulation results for estimating test length coefficients for one ANPSF [103] are shown in Table 5.13.

4. **Markov Model for data-line faults**: This category includes faults not only appearing on the data-lines, but also in the logic prior to the data lines. Suppose q is the probability that an incorrect word appears at a data-in line of the memory because of either a fault in the pre-logic or in the data-in line itself. When the memory is actively writing, the incorrect data will be written to some address with probability q. We now want

Table 5.12 Test length coefficients for static NPSF: $\langle \uparrow, S \rangle$; courtesy [103]

$I_0 = 0.5$	Neighborhood Size		
Q_D	3 Cells	4 Cells	5 Cells
0.1	7	13	25
0.2	13	26	51
0.3	20	40	81
0.4	28	57	115
0.5	37	77	155
0.6	49	101	204
0.7	63	132	268
0.8	84	176	358
0.9	120	250	511
0.999	239	499	1021
0.9999	356	748	1531

Table 5.13 Test length coefficients for active (dynamic) NPSF: $\langle \uparrow, S \rangle$; courtesy [103]

$I_0 = 0.5$	Neighborhood Size		
Q_D	3 Cells	4 Cells	5 Cells
0.1	6	12	22
0.2	14	27	53
0.3	23	44	88
0.4	33	65	129
0.5	45	88	176
0.6	59	118	235
0.7	78	155	310
0.8	104	208	416
0.9	149	299	598
0.999	299	600	1200
0.9999	449	901	1803

BIST and DFT

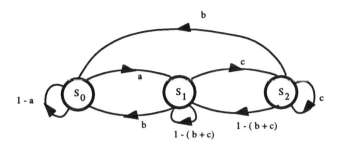

Figure 5.56 Markov model for a data-line fault or a pre-logic fault; courtesy [9]

to compute q', which is the probability that an incorrect word will be read from the data-out lines. This parameter q' is thereby the detection probability of the fault as measured at the data-out lines [9].

The Markov chain of Figure 5.56 shows the random test requirements for detecting such a fault. In Figure 5.56, the transition probability from state S_0 to state S_1 is the probability of writing an incorrect word into address A. For this transition to occur, the memory should be actively writing (probability p_c); the address selection lines should access address A (probability p_A), and the fault should appear at the data-in lines (probability q). The next step of the test is to read the faulty contents of the address A. This causes a transition from state S_1 to state S_2. The probabilities associated with these transitions are shown in the figure.

Bardell, et al. [9] have calculated the detection probability (the probability π_2 of being in state S_2) as $\pi_2 = (1 - p_c) * p_A * q$. The output detection probability q' is obtained by adding the contribution from all the addresses. This produces:

$$q' = \sum_{i=0}^{2^m - 1} \pi_2 = (1 - p_c)q$$

5. **Markov Model for address-line faults**: Suppose that there is a fault in the combinational logic driving the address decoder, and that the detection probability q of the fault at the decoder inputs is known. We also assume that the memory array and the internal logic of the decoder are fault-free. Hence, exactly two memory words A and B are affected every time the fault is present. When the fault is present, the decoder may write to B the word which should have been written to A. An output error is observed when the data read from B is different from the word stored in A.

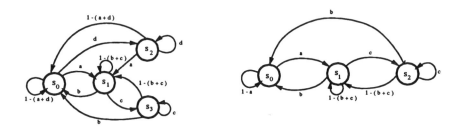

Figure 5.57 Markov model for the non-common and common addresses; courtesy [9]

Bardell, et al. have simplified the amount of computation required to understand various address logic faults, by assuming the *worst case* of such faults. The worst-case fault is called a *single-mapping* fault. When a single-mapping fault is present, all selected addresses are mapped to a single *common* address. The use of a single-mapping fault concept produces a lower bound on the detection probability for all kinds of address logic faults.

We now consider two Markov chains: one corresponding to the non-common address, and the other corresponding to the common address. These chains are shown in Figure 5.57.

Bardell, et al. have done an extensive analysis of the transition probabilities associated with these Markov chains and have obtained mathematical expressions for the output detection probability in steady state. These expressions are given in [9].

5.9.4 DFT architecture for Markov chain-based testability

The DFT architecture for deterministic parallel testing of parametric and pattern-sensitive faults (Figure 5.16, 5.18) proposed by Mazumder in [97] can be adapted for random testing in embedded applications. In such applications, the memory may be organized into single or multiple arrays inside the ULSI/WSI chip. Because of the routing complexity associated with other self-testing techniques, random pattern testing is likely to be more desirable in many applications where embedded registers and small amount of SRAMs will be scattered all over a chip.

BIST and DFT   317

Since the proposed test architecture compares memory cells simultaneously, the problem of observability will not exist and no LFSR will be needed nor will any predetermined pseudorandom sequences be necessary to test the memory. In order to ensure that the technique does not need any deterministic initialization, the word-line decoder is equipped with a latch which is set whenever a write operation is done into a cell on the word line. While reading the word line, the result of comparison activates the error decoder if the latch is set; otherwise, the decoder is disabled. This prevents incorrect comparisons due to initial arbitrary data in memory from corrupting the test results. The modified circuit with write-enable latches for word lines is shown in Figure 5.58. For each word line, there are two latches, L_j^E and L_j^O as shown in Figure 5.59. Each latch consists of inverters connected back to back. The LATCH ENABLE signal ODD is held high only if odd lines are selected in the write mode. Similarly, the signal EVEN is held high only if even lines are selected in the write mode. When the word line decoder selects a word line j, the word line j goes high; otherwise, it remains low. Thus the Write Sensing Latch, L_j^E is set to 1 and the pass transistor O_{E_j} turns on, holding the ERROR LATCH ENABLE to high, if both the word line j and the even bit lines are selected during a write operation. During a subsequent read operation, the latch L_j^E will be set to high, and the pass transistor will be selected if the EVEN bit lines and word line j are selected. The overall random testing technique is summarized in Figure 5.60.

As observed by Mazumder [103], the proposed technique has a test time of $O(\sqrt{n})$ as opposed to $O(n)$ for a sequential random test. It uses purely random input data (depending on the random logic level of the data-in line), unlike pseudorandom patterns as generated by an LFSR. Besides, the only overhead is a multibit 0/1 detector instead of an LFSR, register and comparator, as in sequential random test techniques. This technique does not require memory initialization and is based on a cheap form of self-comparison, eliminating the necessity of a gold unit. Finally, it can produce a high quality of detection (Q_D) for sufficiently long tests and has thereby good reliability.

5.10 SCAN BIST IMPLEMENTATION FOR EMBEDDED MEMORIES

Scan path-based testing is often used for embedded VLSI macrocells, and can also be implemented for embedded SRAMs. Nicolaidis, et al. [127] presented a new scan path technique for RAMs. They described several techniques for ad-

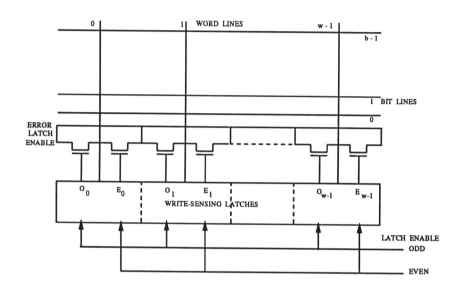

Figure 5.58 Modified circuit with write enable latch; courtesy [103]

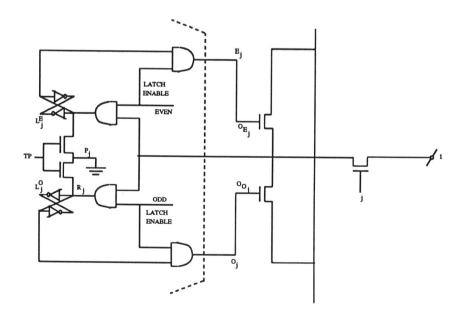

Figure 5.59 Design of write-sensing latch; courtesy [103]

BIST and DFT 319

1) Set the TEST mode to 1. From the tables of test length coefficients, find how many tests are needed for an "acceptable" quality of detection, Q_D. Let this number be N. Repeat steps 2 to 6 N times.

2) Select all the even bit lines or all the odd bit lines simultaneously.

3) Select a write word line arbitrarily.

4) Select 0 or 1 arbitrarily on the read/write line.

5) If the selected operation is write, set the flag on the selected word line in Write Sensing Latch. If the selected bit lines are odd, set the flag corresponding to the odd-half of the bit lines; otherwise, set the flag corresponding to the even-half.

6) If the selected operation is read, check if the corresponding latch flag is high. If high, the Error Detector is enabled and reports an error if the parallel comparator finds a mismatch. If the flag is low, Error Detector is disabled, and no mismatch is found.

Figure 5.60 Random-test algorithm for parallel DFT architecture

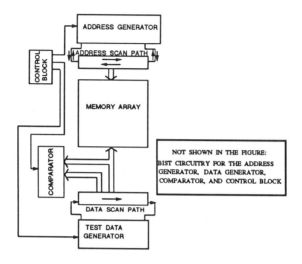

Figure 5.61 Scan path + comparator for output response verification; courtesy [127]

dress generation, test data generation, and output response verification related to scan path and BIST implementations for RAMs. Their basic BIST architecture employing scan paths consists of a reconfigurable test pattern generator, known as the TPG, for diverse macrocells that supplies the RAM *addresses*, a dedicated test data generator, and a signature analyzer. The scan path traverses through the circuit and various segments of it are used to provide input stimuli to various internal macrocells. Address input are injected through the scan path, while the RAM test data may be provided either with or without a separate scan path. Test responses of the embedded macrocells are captured in another scan path and fed to a signature analyzer. Additional circuits are provided to allow the test pattern generator, the control sequence generator and the signature analyzer to perform self-test.

Nicolaidis, et al. proposed the use of an up/down LFSR [124] during the address generation of the RAM (using the TPG), in order that a single-bit shift be needed to generate each new address pattern. This scheme requires the scan path portion of the RAM to be modified into a register that shifts both left and right. The various BIST implementations that they have proposed are shown in Figures 5.61, 5.62, and 5.63. The address generator implemented as an up/down LFSR and the sequence generator may both be tested by using the parity code, whereas the comparator may be tested by using a tree of double-rail checker cells [127].

BIST and DFT

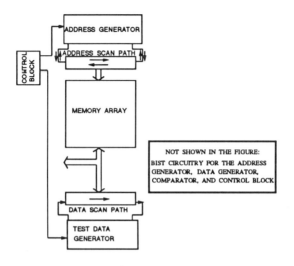

Figure 5.62 Single scan path for test data application and output response verification; courtesy [127]

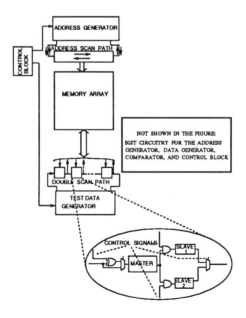

Figure 5.63 Modified double scan path; courtesy [127]

The main disadvantage of these BIST approaches is that though they generate memory addresses very efficiently (using single-bit shift per address pattern), the schemes in Figures 5.61 and 5.62 spend too much time shifting test data in and out. The scheme shown in figure 5.63 avoids this difficulty by allowing single-bit shifts for the data but ends up unnecessarily complicating the BIST design by causing a larger area overhead and a greater complexity of external response verification. All these schemes produce an area overhead of less than 7% for RAMs with more than 1024 words.

5.11 BOARD LEVEL SCAN-BASED DFT FOR EMBEDDED SRAMS

One approach for improving testability of embedded RAMs is to include a scan path within a RAM chip, as described in the previous section. This approach would cause a large increase in the cost of the RAM and would also impact the RAM's normal operation. In this section, a scan-based design for testability approach for embedded memories at the board level is described. An industry standard that allows board-level scan testing to be done easily is the IEEE-1149.1 (1990) standard, commonly known as boundary scan [61].

Boundary scan is the application of a scan path to the internal periphery of an IC to provide improved controllability and observability to the signal pins when the IC is mounted on a PCB and the pins are not physically accessible for being probed. Boundary scan, originally proposed by the Joint Test Action Group (JTAG) of the IEEE, is now IEEE standard P1149.1. The boundary scan method is found to accelerate self-testing of embedded SRAMs.

The boundary-scan standard [61] was initially conceived to test board interconnections without the need of additional probing. The standard requires the inclusion of boundary-scan functions on ICs. A *Boundary Scan Interconnect* (BSI) tester [79] utilizes a combination of module edge pins and the P1149.1 serial test bus to apply test stimuli and receive response test data. A *module under test* (MUT) is said to have a *full boundary scan implementation* if it has a latch on every signal pin. In that case, the test generation process for detection *and* location becomes quite straightforward — latches on devices that drive the net are *pseudoprimary* inputs and latches that receive data on a net are *pseudoprimary* outputs. By using a scan chain to scan data in and receive data out via these latches, functional and electrical faults can be detected and located. Memory devices that can be accessed from boundary scan latches and/or I/O of modules may be tested for interconnect failures via a scan chain. In this technique, an algorithmically generated set of input patterns is applied by the BSI tester. These patterns are designed to verify the interconnects for the address, data and control lines of the RAM array. One frequently used pattern set consists of a walking 1s data pattern. The general-purpose test program for testing interconnects can thereby be augmented with pattern and diagnostic data.

One boundary-scan controlled technique for testing embedded RAMs is as follows: boundary-scan instructions are used to scan in the address and data, set a memory strobe active to perform memory access, and set the memory strobe inactive to complete the cycle. Thus for each memory location access, multiple scans are performed and the technique is very scan-intensive and slow. For large memory arrays, the process would require millions of IEEE-1149.1 scan operations and take many hours to complete.

A better solution, proposed by Coleman and Thorpe [27] speeds up test execution time by two orders of magnitude. This technique employs IEEE-1149.1 controlled DFT (i.e., boundary-scan combined with DFT) with general purpose or ASIC components/macros. In this approach, the bus-interface unit is made boundary-scannable, and only the initial address and data patterns are scanned in from outside. After scanning in the 'seed' patterns, software (IEEE-1149.1) instructions are used to reconfigure the scan path of each bus-interface device into an LFSR or a binary counter. The same device may be reconfigured in various ways during the course of the test; for example, the data buffers are reconfigured into pseudorandom pattern generators (PRPG) during memory write and parallel signature analyzers (PSA) during memory read, using only some software codes. This approach drastically reduces the hardware complexity of scan testing and accelerates the testing process by allowing address and test patterns to be generated very fast (with single bit-shift per pattern generated).

The two techniques have been compared by Coleman and Thorpe [27] on a 256×8 bit memory array and associated bus interface and control logic. Explicit read/write operations are performed for this configuration using the IEEE-1149.1 EXTEST and Sample instructions, or alternatively, by employing the DFT capability controlled by boundary-scan methods. The latter is much faster, as shown in Table 5.14 [27].

Two methods of testing the RAM may be compared. The first method explicitly scans in the RAM array address, data and strobe signal; the second one executes IEEE-1149.1 controlled DFT, which generated the address, data and strobe

Table 5.14 Boundary-Scan versus Boundary-Scan + DFT; courtesy [27]

Mode	256 Accesses		1,000,000 Accesses	
	B.Scan	B.Scan+DFT	B.Scan	B.Scan+DFT
Time to apply	4.8 sec	0.011 sec	332.00 min	0.75 min
Scans	512	7	2,000,000	28,000
Patterns	512	512	2,000,000	2,000,000

signals automatically at the test clock rate of 6.25 MHz. The boundary-scan technique solves the problem of direct physical access, but is time-consuming. The second test clearly indicates the advantage of IEEE-1149.1 controlled DFT. It emulates the actual functional characteristics and timing speeds of memories being tested.

5.11.1 The test approach

Coleman and Thorpe have implemented the IEEE-1149.1-based DFT approach using Texas Instruments' 5-V and 3.3-V bus-interface and scan-support products, belonging to the Scope™ testability integrated circuit family [162]. Some of these bus-interface devices belong to a family of 8-bit parts, known as the Scope octals. These off-the-shelf devices offer the boundary-scan controlled DFT the functionality needed to test static memory. The Scope devices also supply the electrical signal conditioning and buffering typically designed around a microprocessor. Before describing the detailed implementation of this architecture in the following sections, we present a brief overview of the main hardware components and a block diagram to explain their inter-relationships.

For this IEEE-1149.1-oriented DFT approach for static RAMs, the hardware requirement is as follows:

1. **Test Access Port (TAP)** controller: This TAP controller is a 16-state machine, three of which deserve special mention in the context of DFT — *Update IR*, *Runtest/Idle*, and *Select DR*. These states play a role in the *test* mode during memory read and write operations. During the *Update IR* state, the control signals responsible for storing or retrieving data in test mode are activated. During the *Runtest/Idle* state, pseudorandom pattern generation takes place in case of *write* (i.e., generation of test patterns to be stored in the memory) and parallel signature analysis takes place in

BIST and DFT 325

case of *read* (i.e., analyzing the data retrieved from the memory). For read operation, the signature is read from the data register during *Select DR* state. For both read and write operations, the memory read or write operation for the next block of memory is also set up during this phase.

2. **Eight-bit Buffers and Transceivers (Octal Devices)**: These are special-purpose buffers equipped with boundary scan latches. These devices are capable of doing two things — generating pseudorandom patterns at their outputs during test writes and performing response compression at their inputs during test reads. There are separate buffers for address and data, ensuring complete isolation between address generation and data generation. The buses from the address buffers are unidirectional whereas those from the data buffers are bidirectional, allowing two-way flow of data (from memory during read followed by response compression, and to memory during pseudorandom pattern write). Each octal has two states — functional and idle.

3. **Boundary scan registers (BSR)**: Each octal device described above is provided with boundary scan registers which can be initialized by a DR scan. An octal is allowed access to its BSRs via a special instruction called *READBN*.

4. **Boundary control register (BCR)**: This register controls the loading of boundary scan instructions to different octal devices. It is activated via an IR scan that loads the devices with a special instruction called *SCANCN*.

5. **Instruction register (IR) and data register (DR)**: These store boundary-scan instructions and data respectively, for use in memory testing operations.

6. **Test access port PAL**: It is implemented with an octal device, called U_5, and generates the data and control signals for the TAP. It also receives a signal called *Memtest* which is activated and de-activated as the TAP keeps switching back and forth between its states.

The block diagram of Figure 5.64 displays the design of the test circuit. This block diagram is a simplified version of the one presented in [27]. Conceptually, this comprises all the basic functional components of the DFT design.

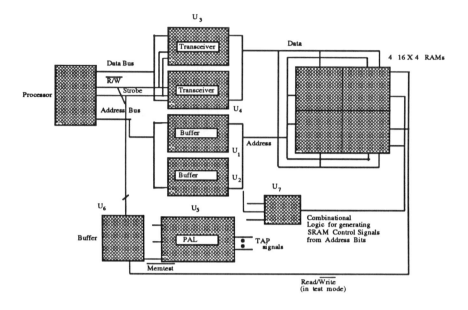

Figure 5.64 Bus-interface devices used in combined boundary-scan/DFT; courtesy [27]

5.11.2 Characteristics of the DFT architecture

The Scope 8-bit octal bus interface components used in [27] provide a flexible, generic DFT framework that supports several memory testing algorithms. The 8-bit BiCMOS technology (BCT) devices have circuits for pseudorandom pattern generation (PRPG), parallel signature analysis (PSA), concurrent PSA/PRPG, and Toggle/Sample (which is similar to PSA/PRPG with the added feature that the output pattern toggles at every clock in pattern generation mode (write), and the input data are sampled in signature analysis mode (read)). The 8-bit advanced BiCMOS technology (ABT) versions [162] provide, in addition, a count-up (for 256 different patterns) function at the outputs. The 18- and 20-bit Scope bus interface components generate 256-Kb and 1-Mb unique patterns, respectively, in each DFT execution and are also equipped with advanced capabilities, including the Count-up function of the ABT devices.

The example circuit uses the 8-bit BCT octal bus interface components [162]. The DFT functions for these components are described below.

Pseudorandom pattern generation (PRPG)

The scan sequence for a PRPG is outlined below (note: IR denotes instruction register and DR denotes data register):

(1) IR scan: the opcode for accessing the octal boundary-scan register is scanned from the instruction register;

(2) DR scan: the initial seed (16 bits) for pseudorandom-pattern generation is scanned into the boundary-scan register;

(3) IR scan: the opcode for the scanning sequence ($SCANCN$) is scanned from the instruction register;

(4) DR scan: the boundary-control register is accessed to scan a special string that encodes the PRPG mode;

(5) two IR scans: the first one causes the test access port (TAP) to enter the Run-Test/Idle state after which the device starts generating pseudorandom patterns at its outputs.

The 16-bit PRPG function using the above scan sequence can be used to generate input patterns for the data bus when performing memory write operations. This function would not normally be used for generating address patterns because it would take more than 256 clock cycles to ensure that all 256 memory addresses are generated.

Parallel signature analysis (PSA)

The PSA function causes data appearing on the eight functional data inputs to be compressed into a 16-bit signature. The scan sequence for a PSA is described briefly below:

(1) IR scan: the opcode for accessing the octal boundary-scan register is scanned from the instruction register;

(2) DR scan: the initial seed (16 bits) is scanned;

(3) IR scan: the opcode for the scanning sequence ($SCANCN$) is scanned from the instruction register;

(4) DR scan: the boundary-control register is accessed to scan a special string that encodes the PSA mode;

(5) two IR scans: after the first scan is completed and the TAP enters the Run-Test/Idle state, 8-bit PRPG patterns are generated and the inputs are compressed;

(6) DR scan: the signature produced is scanned into the boundary register.

A combined PRPG/PSA can be implemented using similar scan operations as the above two, if the 16-bit seed value is split into two 8-bit seeds, one for the PRPG and one for the PSA and the appropriate code is loaded for the combined PRPG/PSA mode. After the TAP enters the Run-Test/Idle state, the octal's outputs begin generating PRPG patterns and the inputs are compressed. Thus the 8-bit PRPG function can be used to produce memory addresses. However, an 8-bit PSA function might lead to more aliasing problems and is thereby not recommended for sampling memory data.

Memory read and write operations

Memory read and write operations during the above testing are implemented using boundary-scan operations. The following steps execute the DFT memory write operation:

(1) Load all Scope octals with an instruction for accessing the octal boundary-scan registers for each device while the octals remain functional. This involves an IR scan.

(2) Initialize the boundary-scan registers of each octal device. This involves a DR scan. A boundary-scan register called U_1 is set up with the desired RAM memory address and registers U_2, U_3, and U_4 are loaded with PRPG seed values. The *Memtest* signal (low active) and the *read/write* signal are asserted (*write* in this case) in a register U_5.

(3) An IR scan causes U_2, U_3, and U_4 to be loaded with an instruction which allows access to the boundary control register. U_1 and U_5 are loaded with the Bypass instruction.

(4) A DR scan loads the boundary control registers U_2, U_3 and U_4 with the PSA/PRPG code.

BIST and DFT 329

(5) An IR scan loads U_2, U_3 and U_4 with the instruction $RUNT$ that initiates the running of the test and loads U_1 and U_5 with the $EXTEST$ instruction. The $RUNT$ instruction is a signal that causes the TAP to enter the *Runtest/Idle* state in which the octals U_2, U_3, and U_4 begin generating PRPG patterns. The $EXTEST$ instruction allows the values already loaded on U_5 to be asserted when the TAP enters the *Update-IR* state.

(6) An IR scan puts the octals into their functional mode and disables the *Memtest* signal of U_5, preventing the test signal to be generated again when the *Runtest/Idle* state is re-entered by the TAP.

(7) A DR scan sets up the memory write operation for the next block of memory.

Steps (1) & (2) above are similar to memory write, except that U_3 and U_4 are loaded with a PSA seed value instead of a PRPG seed value. Steps (3) & (4) are similar to memory write, except that in Step (4), a special PSA code is also loaded. Step (5) is similar to memory write, except that U_3 and U_4 begin compressing a signature instead of generating PRPG patterns after the TAP enters the *Runtest/Idle* state. Steps (6) & (7) are similar to the corresponding steps in memory write, but at Step (7), the signature is read from octals U_3 and U_4 *before* the memory read operation is set up for the next block of memory.

5.12 ADVANTAGES OF THIS APPROACH

The above approach requires the IEEE-1149.1 hardware to be included *only* in the bus-interface unit and not in the RAM under test itself. A problem with boundary-scan is that it cannot be embedded easily within a wide range of devices, particularly those that have a high transistor count, such as RAMs. Embedded boundary-scan would double, or even triple, the size of a RAM device, and is thereby highly impractical. The technique of performing RAM testing using the the bus-interface unit would thereby provide the improved test access to the pins of the embedded RAM devices under test without requiring that the RAM chips or boards have a boundary-scan implementation. Furthermore, several megabits of embedded RAM are typically used at the board level, causing millions of data patterns to be required for testing each word and each bit. Reconfiguring the scan path of the bus-interface unit would achieve test acceleration at the board-level by at least two orders of magnitude. Furthermore, this approach is very flexible and generic and can be used with any device under test connected to the system buses, and not just RAM. This is due to the fact that the bus-interface unit that has direct access to the device under test can be used both for test generation and signature analysis. This approach does not require the device under test to have any BIST circuitry or embedded scan path, and can thereby be used with the vast majority of commercial devices which do not have any self-testing capability. Also, the simultaneous PSA/PRPG modes of testing provide the possibility for parallel testing with only slight modification of the basic test structure, but with considerable modification of the software test instructions.

5.13 CONCLUSION

BIST and DFT techniques of various kinds are of increasing importance these days because a large percentage of memory devices used are embedded SRAMs for which the address and data buses and control lines cannot easily be controlled and observed. The different BIST/DFT architectures described try to achieve several objectives: a high fault coverage, low area overhead, low power dissipation, negligible impact on the access time of the RAM, high speed of testing (for example, using parallel comparison and error detection). In some cases, the testable architecture is designed to preserve the original RAM contents so as not to require that these contents be saved in backup storage prior to testing.

BIST and DFT

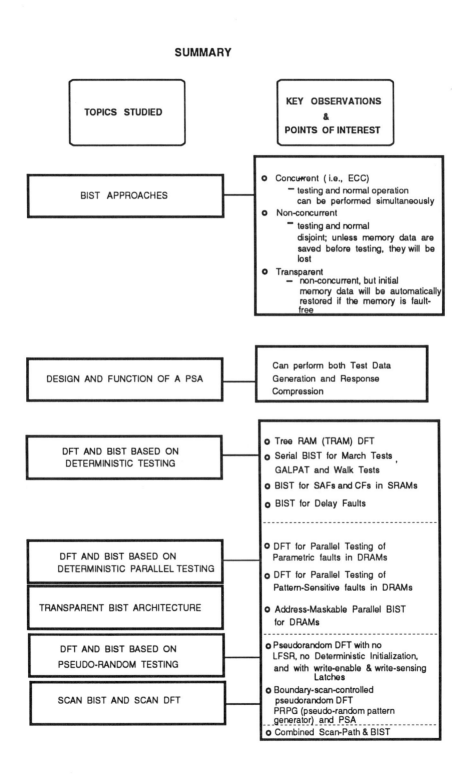

5.14 PROBLEMS

5.14.1 Worked-out Example

1. In this chapter, we have described DFT for parametric and pattern-sensitive faults in DRAMs using modified bit-line decoder and parallel comparator. Design a BIST approach to perform parallel testing on a DRAM array logically tessellated into square tiles consisting of 4 cells.

   (**Sample Answer Sketch:**) As with the DFT technique for parallel and parametric testing, in the BIST mode, the normal 1-out-of-4 decoder is modified to select multiple bit lines. In the test mode, all the bit lines in an even or odd group are selected in parallel to execute a read or write operation. As before, in a write operation, the same data is written on all the even or all the odd cells accessed simultaneously.

   The modified decoder circuit is illustrated in Figure 5.65 [96]. Its design may be used for any non-reconvergent decoder topology by adding a two-input OR gate at each bit-line decoder output, so that each decoder output drives one input of the corresponding OR gate and an external control signal drives the other input. In normal mode, the external control signal of each OR gate is grounded, and the decoder outputs are passed directly onto the corresponding OR gate outputs. In test mode, the bit lines are selected by the external control signals and not by the decoder outputs. To illustrate this technique, let us assume that the bit lines decoders are made of NOR plane of a two-level PLA logic. The bit-line decoder has been modified as shown in Figure 5.65.

   In Figure 5.65, lines L_1 and L_2 are added to select the odd and even bit lines in parallel. Transistors Q_0 through Q_{b-1} are used to select the bit lines by L_1 and L_2. For normal operation mode, TEST = 0 and both L_1 and L_2 are 1. Thus the bit lines are selected only by the decoder inputs a_0 through a_k, where $k = \lg b - 1$. In test mode, all the address inputs are set to 0, and TEST = 1. Transistor Q_b pulls down the product line 0 to zero while every other product line is pulled down to 0 by one or more cross-point decoding transistors. If $L_1 = 0$ and $L_2 = 1$, all the even bit lines will be selected in parallel, and likewise, if $L_1 = 1$ and $L_2 = 0$, all the odd bit lines will be selected in parallel.

   The design of the parallel comparator and error detector is described earlier in this chapter. Let us now examine the parallel BIST algorithms that can be implemented with this test structure. As shown in Figure 5.30, the memory is tessellated into tiles consisting of four types of cells 1, 2, 3 and 4. The rationale behind dividing the memory plane into four types of cells

BIST and DFT

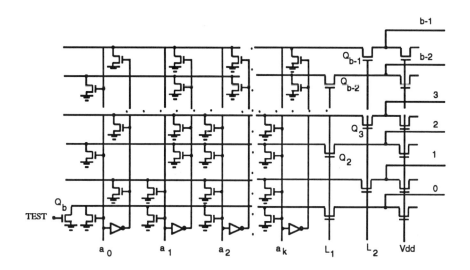

Figure 5.65 Modified bit-line decoder

is to classify the eight neighbors of a base cell as logical three neighbors (namely, bit-line, word-line and diagonal). For practical applications, the leakage/coupling effect between the base cell and its bit-line, word-line and diagonal neighbors is maximum when identical cells in the corresponding neighborhood are identical. This is especially true in high-density, single-cell DRAM, where the RAM cell is present at the cross-points of bit-lines and word-lines.

The tetromino tiles shown in Figure 5.66 can tessellate the memory plane in four different ways depending on the type of cell at the top right corner. These four types of tessellations are called 1, 2, 3 and 4, respectively. The BIST algorithm consists of tessellating the memory and performing a particular set of operations on the cells of the tile. In other words, all cells of one type on a word line are read/written in parallel.

Suppose we define a 'unit' operation by the symbol k_y, where $k \in \{0, 1, 2, 3\}$ and $y \in \{0, 1\}$, to consist of: reading in parallel all the cells numbered k, writing 1 in parallel in all the cells numbered k, and then reading in parallel each of the four types of cells in the memory. This operation is repeated in order to detect all 9-cell neighborhood pattern-sensitive faults. For example, if $k = 0$ and $y = 1$, we obtain the following sequence of operations.

Repeat the following steps for all odd word lines (columns):

1. Read in parallel all type 0 cells in a word line (column).

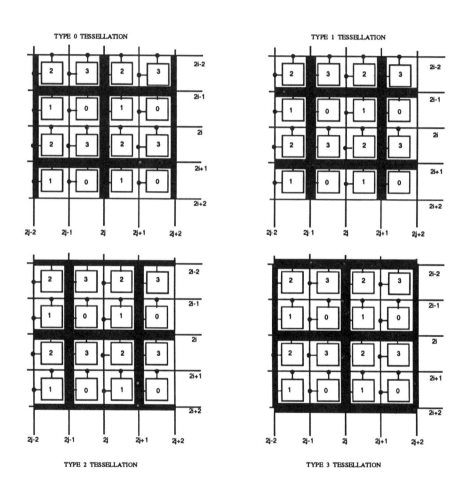

Figure 5.66 Four types of tessellations

2. Write in parallel value 1 in type 0 cells in a word line (column).
3. Read in parallel all type 0 cells in a word line (column).
4. Read in parallel all type 1 cells in a word line (column).
5. Read in parallel all type 2 cells in a word line (column).
6. Read in parallel all type 3 cells in a word line (column).

Clearly, a $b \times w$ RAM takes a total of $5w/2$ read and $w/2$ write operations to perform k_y operations on all the type k cells in the memory. The test would thereby consist of an initialization sequence followed by a sequence of operations of type k_y such that every edge of a 4-dimensional hypercube is traversed, where each vertex denotes a 4-tuple of values applied to the 4 types of cells. As we have seen before, the most efficient way of doing this is via an Eulerian cycle which requires a total of $195w$ operations — $35w$ reads and $160w$ writes. It can be seen quite easily that this algorithm would automatically test the decoder for stuck-at and bridging faults. The test for the parallel comparator is described earlier in this chapter. The modified bit-line decoder is tested as follows:

(a) Initialize all the cells on the first word line to 0.
(b) For all bit lines (in ascending order of address) in the selected first word line: read the cell at the cross-point and find out if it is 0, then write a 1.
(c) For all bit lines (in descending order of address) in the selected first word line: read the cell at the cross-point and find out if it is 1, then write a 0.

This simple algorithm has a complexity of $4b$.

This testable scheme is highly suitable for BIST implementation. It may be noted that in order that a scheme be used for BIST implementation, it should

(a) use the testing algorithms that give high fault coverage,
(b) have minimal additional hardware,
(c) have minimal performance degradation,
(d) use simple, testable additional logic,
(e) match with existing design constraints.

The proposed scheme uses only $2b + 13$ extra transistors in each memory subarray to compare the data and to set the ERROR latch. It needs only $b+1$ transistors for modifying the decoder. These additional transistors can

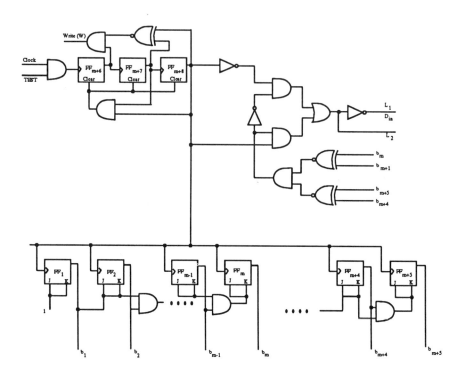

Figure 5.67 Test generator circuit

be easily fitted within the pitch width limitations for high density memories. Also, we have shown that these additional hardware can be easily tested. Now we shall discuss the implementation details of the proposed algorithm.

Intuitively, the proposed algorithm is generated by making an interlaced scanning along both the bit lines and the word lines. The bit lines are scanned concurrently and the word lines are scanned sequentially. Thus, to select w word lines a $m = \log w - 1$ bit synchronous binary counter is used. The least significant bit in the word line address is toggled alternately to select the odd or even word lines. The test generation circuit is shown in Figure 5.67.

In the test mode, L_1, L_2, Data In (D_{in}) and Write (W) signals are generated as shown. The corresponding waveforms which are applied to the RAM are shown in Figure 5.68. The circuit simplicity and its ease of testability provide an elegant BIST solution for RAM applications. For a 256 Kb RAM organized into four identical square subarrays, the tech-

BIST and DFT

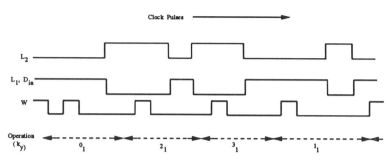

(Note: The whole test consists of a sequence of 66 operations, of which only four have been shown.)

Figure 5.68 Waveforms at the RAM control input

Table 5.15 Comparison of Different PSF test algorithms

RAM Size	256 Kb	1 Mb	4 Mb	16 Mb
No. of Partitions (p)	4	8	8	16
Hayes' Test $(3k+2)2^k n$	28.5 s	114.1 s	456.3 s	1825.3 s
Suk & Reddy's Test $(k+5)2^{k-1}n$	8.4 s	33.6 s	134.2 s	536.9 s
Proposed Test $195(n/pe)^{0.5})$	9.2 ms	12.8 ms	25.6 ms	36.4 ms

nique needs only 19 flip-flops as opposed to 427 flip-flops used for 8 Kb RAM in [65]. Moreover, the test algorithm in [65] does not detect dynamic pattern-sensitive faults. Kinoshita and Saluja in [71] describe a BIST for detecting NPSFs using programmable ROM microcodes. Their testing overhead is very large compared to the BIST described here, also, their test algorithms have $O(n)$ complexity as opposed to $O(\sqrt{n})$ complexity described here. Nicolaidis in [124] also proposes a built-in test algorithm for detecting the 2-coupling faults described in [83], by using an LFSR with Berger code encoding; however, the hardware overhead for his scheme is much higher than the one described here.

Table 5.15 compares the running times of the different PSF test algorithms.

5.14.2 Exercises

1. Compare the various schemes by which scan-path testing can be combined with embedded memory BIST. Make a comparison of these schemes on the basis of (i) circuit complexity, (ii) area overhead, (iii) impact on memory access time, and (iv) speed of testing. For a hint, see [127].

2. Describe test structure built out of boundary-scannable bus-interface devices (such as Texas Instruments' Scope Octals [162]) that can provide DFT functions for DRAMs (i.e., address, data, and control signals generation) at the board level, and design memory tests using these ICs. Suggest efficient software techniques to implement march and checkerboard tests using such test structure. Make a detailed comparison of the testing complexity.

3. Design two tables, one comparing various BIST/DFT approaches for DRAMs, and another doing the same for SRAMs. Compare these approaches on the basis of faults covered, area overhead, testing speeds, layout characteristics and power consumption. Which techniques do you think hold out promise for the twenty-first century ?

4. Describe various trends in state-of-the-art BIST circuits used in commercial RAMs. How much has the percentage of their use grown over the last ten years ? What is the current market status for such RAMs, and what has the impact of BIST been on the price (in dollars) of memory devices ?. How does the cost of implementing BIST in RAMs compare with that of testing RAMs with an external tester ?

5. Design Markov chains for the pseudorandom pattern testability of the following types of 3-coupling faults:

   - a cell k is prevented from undergoing a transition because of a fixed pair of values in cells i and j;
   - a cell k undergoes a spurious transition because of a fixed pair of values in cells i and j.

   Also plot the quality of detection versus the test length coefficient for each of these faults.

6
CONCLUSION

The book deals with issues germane to testing and testable designs of commercial static and dynamic RAMs (SRAMs and DRAMs). It describes, in Chapter 1, the technological advances that have brought the DRAM of yester-years to the modern DRAM of the multimegabit era. In this twenty-year long journey through the history of commercial SRAMs and DRAMs, we have discovered the use of these devices in myriad applications; nowadays, RAMs are probably the most ubiquitous electronic devices in the market. The advances in RAM technology have resulted in faster, cheaper and smarter SRAMs and DRAMs on the one hand and have made the problems of testing and fault-tolerance of these devices more challenging on the other. New processing techniques are often associated with new defect mechanisms, which, in turn, require the use of more sophisticated testing techniques. Since RAMs are used in a wide range of applications, some of them being extremely failure-critical (for example, computers that perform real-time control of spacecraft and those that monitor chemical reactions with potentially hazardous consequences), the proper functioning of RAMs is of paramount importance. In a computer, a RAM may store a large volume of data and also a lot of user programs at any given time. Any failure in the RAM may, therefore, have catastrophic results.

Chapters 2,3, and 4 describe the testing problem and its various distinct characteristics. Chapters 2 and 3 deal with electrical and functional testing, respectively. In electrical testing, circuit parameters like voltages, currents and timing delays are measured at the input-output interface, and then compared with their expected values. In functional testing, the test patterns consist of voltage levels (logic-0 and logic-1) applied to various cells in the memory. In this approach, only voltage levels (and not actual voltage values) of cells and interconnects are compared with the expected values. In functional testing, the test patterns themselves may be derived from either layout-oriented calculations (such as the Inductive Fault Analysis (IFA) approach), or layout-independent criteria (such as tests for stuck faults derived without layout considerations), both of which have been examined closely. In Chapter 3, we have described test algorithms not only for standard RAMs but also for specialized RAM devices — such as FIFOs, double-buffered memories (DBMs) and pointer-addressed memories (PAMs).

Since new manufacturing processes are constantly associated with new defect mechanisms only some of which produce observable faulty behavior, a finer examination of the circuit layout is necessary to test for defects in a more comprehensive manner. A defect may not produce an immediate failure of the chip but may cause unreliable future operation; hence, repairing or replacing such a chip may be advisable. Also, observing defects with a scanning electron microscope (SEM) zooming into the layout provides the manufacturer valuable feedback regarding processing flaws. The number of defects counted from an SEM photomicrograph can be used to calculate the yield of a process and a low yield should warrant a close scrutiny of the masking steps by the manufacturer.

Hence, defect modeling at the layout level is very important nowadays. This approach to testing is described as being technology and layout-related, or parametric. In this approach, the test patterns are derived from layout-oriented fault modeling criteria; moreover, the actual values of circuit parameters (for example, the quiescent power supply current I_{DDQ}) are measured during the test. We have described various schemes for technology- and layout-related testing in Chapter 4. Chapter 5 deals with design for testability (DFT) and built-in self-testing (BIST) approaches that describe special hardware for applying various tests on the memory devices. In particular, a BIST approach allows a test program to be run autonomously within the chip, with no external control or test data. Built-in self-testing is of great use in eliminating the high cost of external testing and allowing embedded memories to be tested efficiently. For memory devices that are embedded within larger circuits, such as microprocessors, ASICs, and embedded controllers (such as the ABS of automobiles), or those that are used in space, oceanic and avionic applications, the address, data and control lines are very difficult to control and observe externally. We have described DFT and BIST approaches for various kinds of memories, such as SRAMs, DRAMs and content-addressable memories (CAMs) in Chapter 5.

This book is useful not only to manufacturers of (stand-alone) SRAM and DRAM chips, but also to designers of high-density VLSI chips and systems such as microprocessors built using deep submicron processing technology. Testing of embedded RAMs in such gargantuan systems is a very challenging issue,

Conclusion

because of peculiar processing and circuit-related defects on the one hand, and the difficulty of accessing the internal nodes on the other. Nowadays, embedded memories are often provided with redundant rows and columns to make them repairable. For such memories, it is very important to use efficient testing schemes to diagnose layout defects. Such defects can then be repaired using laser personalization or built-in self-repair circuitry. This book gives a compendium of useful testing algorithms and test hardware for such memories. The appendices provide the reader with a good understanding of the trends in processing technologies, circuit parameters, integration levels, and commercial markets for RAMs.

A
GLOSSARY

AC parametric test (Chapter 2) An electrical test to ensure that the output changes for a device occur within a time that satisfies the timing specifications for the device.

Access time test (Chapter 2) An electrical test that verifies the memory access time, using either an external tester, a characterization test, or a production test.

Address maskable BIST (Chapter 5) Replacing normal mode column decoding by a special test mode decoding that triggers multiple column decoder outputs simultaneously, using circuitry that *match* the data with the expected value, and thereby detects errors.

Address transition detection (ATD) (Chapter 1) A mechanism for producing a pulse when a transition occurs at the address inputs.

Aliasing (Chapter 5) A situation in which a faulty and a fault-free memory word produce identical signatures.

Arbitration test (Chapter 2) A two-step test for dual-port RAMs consisting of contention circuitry testing followed by testing of a set of semaphore flags.

Bathtub curve (Chapter 2) A 'bathtub'-shaped curve showing the failure rate of a memory device as a function of time after fabrication.

Battery RAM (Chapter 1) A RAM that can be operated by a battery: it requires low operating power and has built-in circuitry to protect stored data.

Bit-line decoder (Chapter 5) Column decoder.

Bit-line-to-word-line crosstalk (Chapter 5) Unwanted capacitive coupling between the bit and word lines.

Bit-line voltage imbalance (Chapters 4,5) An erroneous read operation produced by precharge voltage level degradation due to leakage current.

Boundary scan (IEEE Std. 1149.1) (Chapter 5) Applying a scan path to the outer periphery of a device to control and observe the input/output pads.

Built-in self-test (Chapter 5) The incorporation of circuitry within a memory device that can perform tests autonomously.

Butterfly (Chapter 3) A functional test for stuck-at faults in cells and for coupling faults between a cell and its rookwise neighboring cells located within a fixed distance.

Cache tag RAM (Chapter 1) A CAM which stores a tag that represents a few bits of stored data; it is used for the associative match operation in cache memories to determine hits or misses.

Characterization test (Chapter 2) A test in which a set of functional tests are applied repetitively, and the exact values of various DC or AC parameters are measured at each stage, resulting in a pass/fail decision; the test is repeated for different pairs of chosen variables, for example, V_{CC} and some other parameter.

Checkerboard test (Chapter 3,4) A functional test in which the memory array is divided into two groups of cell to form a checkerboard pattern, with 1s being written in cells of one group and 0s being written in the other group.

Concurrent testing (Chapter 5) Also known as error-correction it is a BIST scheme that allows testing to be performed during normal operation, without the requirement of a separate *test mode*.

Contact test (Chapter 2) A test consisting of drawing current out of a device and measuring the voltage at the input pin.

Content addressable memory (CAM) (Chapters 1,4,5) A RAM in which retrieval takes place by providing a part of the data as input; also described as an *associative memory*, the circuit searches for a match when data are provided at the input, and reports the address at which the match is found.

Glossary

Coupling fault (CF) (Chapter 3) A fault that causes the state (i.e., contents) or a write transition in one cell to affect the state or a transition in another – based on the type of coupling, we have *inversion coupling* (CFin), *idempotent coupling* (CFid), *state coupling* (SCF), *bridging* (BF), k-coupling, and so on.

DC parametric test (Chapter 2) An electrical test to ensure that the DC voltage and current parameters at the input and output pins have proper values.

Defect (Chapters 3,4) A physical variation in the layout of a device caused by a process-related error.

Defect model (Chapter 4) A set of defects in the layout of a system regarded as most likely or important.

Defective hashing (Chapter 3) A fault simulation technique in which the addresses of the faulty circuit are recorded in a hash table for efficient retrieval.

Delay time testing (Chapter 4) Measuring the access time of a RAM using test patterns generated from layout-based considerations.

Divide-and-conquer approach (Chapter 3) A functional test approach which recursively partitions the memory into k equal parts, and treats all the cells in one part as base cells, and each of the other parts as groups of cells that are jointly coupled to the base cells.

Double-buffered memory (DBM) (Chapter 3) A dual-port SRAM each cell of which consists of two latches – a master and a slave, with conditional buffering between them.

Dual-port RAM (Chapters 1,2,4) A RAM circuit with two independent address, data and control interfaces for accessing the memory locations for reading or writing.

Dynamic RAM (DRAM) (Chapters 1-5) A random-access memory which stores a bit of data as a tiny quantity of charge on a capacitor (which is a few fF).

Electrical testing (Chapter 2) Creating a fault model describing the electrical behavior observed at the input-output interface of a good and a faulty memory device, and designing a test that verifies this electrical interface behavior in terms of voltage and current values, and timing delays.

Embedded RAM (Chapter 5) A RAM which forms a part of a bigger circuit, such as a microprocessor; for such RAMs, the address, data and control lines are not directly controllable and observable from outside.

Eulerian sequence (cycle) (Chapters 3,5) A cycle that travels each arc of a graph exactly once, and comes back to the starting vertex.

Fast-page mode (Chapter 1) A timing mode to enable fast operation of RAMs, it is a technique in which external signals are presented to the input port of the address buffer after the row address has been latched; this allows the address buffer port to remain open during the read operation, thereby causing rapid access.

Fault (Chapter 3) An externally observable difference between a good and a bad memory system.

Fault model (Chapter 3) A set of faults in a memory system regarded as most likely or important.

Fault simulation (Chapter 3) Simulating the effect of test patterns on a memory with and without errors, and comparing their results.

First-in first-out (FIFO) memory (Chapter 1,3) A memory device where the data is either stored and then retrieved in a first-in first-out manner, or the data simply passes through; this arrangement is useful for communication buffering type of application.

Functional testing (Chapter 3) Creating a fault model describing only the logical behavior of a good and a faulty memory, and designing a test that verifies this logical behavior in terms of only voltage levels of output pins (level-0 and level-1).

GALCOL (Chapter 3) *GAL*loping pattern with *COL*umn read, a functional test for stuck-at and coupling faults.

GALPAT (Chapter 3) *GAL*loping *PAT*tern of 0s and 1s, a functional test for stuck-at and coupling faults.

GALROW (Chapter 3) *GAL*loping pattern with *ROW* read, a functional test for stuck-at and coupling faults.

Glossary 347

Ground transient current monitor (GTCM) (Chapter 4) A current monitor used in I_{DD} testing.

Half-V_{CC} sensing (Chapter 1) Reducing the precharge level to approximately half of V_{CC} to improve access time and alleviate the electromigration problem and voltage drop across resistors by reducing the peak currents during sensing and bit-line precharge.

Hamiltonian cycle (Chapters 3,5) A cycle that visits every vertex in a graph exactly once and come back to the starting vertex.

High electron mobility transistor (HEMT) (Chapter 4) Also known as a MODFET (modulation doped FET) or TEGFET (two dimensional electron gas FET), they are somewhat similar to JFETs or MESFETs and are formed by growing heterostructures of GaAs and AlGaAs epitaxially using the MBE (molecular beam epitaxy) technique; electrons in a HEMT have more mobility than those in a MOSFET or a JFET device.

Homing sequence (Chapter 3) An input sequence that takes a finite-state machine to a final state that can be determined uniquely from the machine's response to the input sequence, regardless of the machine's initial state.

I_{DD} testing (Chapter 4) The dynamic power supply current drawn by a memory circuit.

I_{DDQ} testing (Chapter 4) A test technique based on monitoring the quiescent power supply current (I_{DDQ}) to detect the presence of any defect that may cause an abnormally high value of this current.

Inductive Fault Analysis (IFA) (Chapter 3) A layout-based fault modeling and testing approach comprising: layout defect generation, defect placement, extraction of schematic and electrical parameters, and evaluation of the results of testing the defective memory.

Interrupt test (Chapter 2) A test for a dual-port RAM performed by simultaneously writing data into the interrupt location and monitoring the $\overline{INT}$ output of the opposite port.

Leakage test (Chapter 2) An electrical test that verifies that the current drawn by a device lies within specifications.

Linear feedback shift register (LFSR) (Chapter 5) A set of flip-flops connected as a shift register, and XOR gates which allow the output of certain cells to be fed back to the input cell.

March test (Chapters 3,4,5) A test which applies a finite number of read and write operations to each memory cell before proceeding to the next cell, and traverses the memory only along a certain predetermined address sequence or its reverse.

Mealy machine (or Mealy automaton) (Chapter 3) A model of a finite-state machine in which transition arcs are associated with output symbols; faulty and fault-free memory devices can be represented as Mealy machines, whose states encode the memory contents and transitions represent transition writes under fault-free ocnditions.

Modified bit-line decoder (Chapter 5) A decoder with some additional control logic to bypass the address selection mechanism and select the output by brute-force.

Moving inversions (MOVI) (Chapters 1,2,3) A test for memory access time.

Multiple input shift register (MISR) (Chapter 5) A shift register with multiple inputs that can be loaded in parallel.

Multiplexed addressing (Chapter 1) Time-multiplexing row and column addresses on the same pins.

Multi-port RAM (Chapter 4) A RAM circuit with multiple read and write interfaces (address, data and controls) for accessing the memory locations for reading or writing.

Non-concurrent testing (Chapter 5) A BIST scheme in which testing and normal operation are disjoint activities – that is, there is a *normal* mode of operation and a *test* mode during which normal operation is suspended.

Open (Chapter 4) A missing conductor (such as first or second metal) in a layout.

Output drive current test (Chapter 2) An electrical test that verifies that the output voltage level is maintained for a specified output driving current.

Glossary

Output short current test (Chapter 2) An electrical test to verify that the output current drive capability is sustained at high and low output voltages.

Parallel comparator and error detector (Chapter 5) A circuit that can compare multiple bit-line voltage levels in parallel and detect errors if any.

Parallel signature analyzer (PSA) (Chapter 5) A multiple-input shift register (MISR) that can compute the compressed version of the response data produced when a read operation is performed on the memory; it has two modes – a 'compress' mode, when the signature is being computed, and a 'scan' mode, when it is being scanned out; in some designs, a PSA can also be used for providing input patterns to the memory, such PSAs may have an additional 'parallel load' mode.

Pattern-sensitive fault (PSF) (Chapter 3) A fault that causes the contents or a write transition of a cell to be dependent on a pattern of data stored in the other cells; if the other cells are confined to a neighborhood instead of being spread all over the memory, we have a **neighborhood PSF** (NPSF).

Phase generator (Chapter 1) A circuit that comes into play at the beginning of a DRAM memory cycle when $\overline{RAS}$ goes low; this circuit inverts and adjusts the control signal, and drives the word address decoder, initiating word decoding.

Planar cell (Chapter 1) A RAM cell that utilizes a two-dimensional storage capacitor.

Pointer-addressed memory (PAM) (Chapter 3) A FIFO, with a shift register to control the accessing sequence.

Power consumption test (Chapter 2) A test that determines the worst case power consumption under static and dynamic conditions.

Power supply voltage transition test (Chapter 5) A test to verify memory read and write operations at minimum and maximum supply voltages.

Preset distinguishing sequence (Chapter 3) An input sequence to a finite-state machine such that the output sequence produced, in response, by the machine is different for each initial state.

Production test (Chapter 2) A test that verifies whether some chosen parameters are in agreement with device specifications, when the device is operated under nominal conditons.

Propagation delay test (Chapter 2) A test to measure the delay of the change in output in response to a change in input; frequently performed by *pulse testing*.

Pseudo-random testing (Chapter 5) A test in which read and write operations are determined pseudo-randomly during the test.

Pseudo-random pattern generator (PRPG) (Chapter 5) A device that can provide pseudo-random test patterns to a memory.

Pseudostatic DRAM (PSRAM) (Chapter 1) A single-transistor-cell, synchronous DRAM provided with built-in refresh control and capable of being used in SRAM sockets.

Refresh (Chapter 1) An operation to replenish the charge in a DRAM storage capacitor after a (destructive) read operation; it is done with the help of a sense amplifier and a bidirectional bus that propagates an amplified value back to the bit-lines.

Refresh line stuck-at fault (Chapter 4) A DRAM fault that causes data leakage.

Restore generator (Chapter 1) A circuit that comes into play after a memory cycle, when the *RAS* timing is terminated; it generates a *restore pulse* to reset all delayed drivers in the timing chain, and pull all word lines to ground.

Retention fault, or sleeping sickness (Chapter 4) A fault affecting DRAMs that causes a DRAM cell to lose charge by leakage from storage capacitances; this typically occurs when the DRAM cell is left unaccessed for some time, between 100 μs and 100 ms.

Rise and fall time tests (Chapter 2) Measuring the time interval between two 'edges' of a pair of voltage waveforms to determine whether the rise (or fall) time of the signals has a proper value (i.e., agrees with the device specifications).

Row/Column PSF (Chapter 3) A pattern-sensitive fault model which assumes that the contents of a cell are influenced by all the cells located in the same row or column.

Glossary

Running time test (Chapter 2) A test that determines the fastest running speed of a memory device with respect to read/write operations.

Scaling (down) (Chapter 1) Designing RAM devices using a process with a smaller minimum feature width.

Schmoo plot (Chapter 2) A plot showing the regions of correct and incorrect operation of a chip with respect to a pair of parameter values.

Scrambled pattern-sensitive fault (Chapter 3) A pattern-sensitive fault (PSF) model that can be used to produce test patterns for a memory for which the logical and topological addresses are not identical, in terms of cell neighborhood and proximity.

Self-timed RAMs (Chapter 1) A RAM with inputs clocked into the memory on the edge of the system clock.

Semaphore (Chapter 2) A latch accessible to any port of a multi-port RAM, but can be released only to one port at a time.

Sense amplifier (Chapter 1) A device that latches up bit-line signals during read operation by amplifying the difference in voltage between bit and $\overline{bit}$ lines; in a **shared** sense amplifier, the right (left) bit-line pair can be selectively disconnected, in a **distributed** sense and restore amplifier, the bit-line length is effectively halved and the latches in each side of the amplifier independently sense and restore the bit-line pair on their side.

Sense amplifier recovery fault (Chapter 2) A fault causing the sense amplifier to respond slowly when a bit is read from the memory, especially after the sense amplifier has read a long string of bits of the opposite value.

Serial BIST (Chapter 5) A BIST technique in which the test data are applied in a bit-serial fashion with shifting.

Setup, hold and release time tests (Chapter 2) Electrical tests that measure propagation delays of input signals with respect to clock signals and vice versa.

Short (Chapter 4) An extra, unwanted conductor (such as metal) between two points in a layout.

Signature (Chapter 5) The result of compressing (using a signature analyzer, such as a PSA) the data read from a memory word.

Single-ended write (Chapters 4,5) Employing a single I/O line to write into the bit lines in a DRAM; this may potentially cause errors because writing on one half bit line will be controlled by the I/O line driver, and writing on the other half by the sense amplifier; consequently, the 0 level may differ in the two halves.

Sliding Diagonal (Chapter 3) A functional test in which a diagonal of base cells is written with some values, which are then shifted out diagonally one step at a time north-east and south-west; this test is designed to detect coupling faults between diagonal and non-diagonal cells.

SVCTEST (Chapter 3) A functional test for V-coupling faults.

Soft defect detection (SDD) (Chapter 4) A technique providing a complete data retention test for a CMOS SRAM array; it consists of two parts – an *open circuit* test that checks for connectivity of the P-type load transistors, and a *cell array* test that carefully monitors the standby array current to detect abnormally high leakage.

Stacked capacitor (Chapter 1) A capacitor stacked over the bit line above the surface of the silicon.

Standby current test (Chapter 2) Measuring currents at the ports of a dual-port RAM.

Static column mode (Chapter 1) A fast-access timing mode in DRAMs which eliminates the long precharge time during a page-mode operation; it basically simplifies the functions of the CAS signal by an internal read operation of the column circuitry which is triggered by ATD (address-transition detection) instead of the $\overline{CAS}$ signal.

Static data loss (Chapter 4) Loss of data caused by leakage currents in SRAM cells.

Static RAM (SRAM) (Chapters 1-5) A random-access memory which stores a bit of data in a latch formed by a pair of cross-coupled inverters.

Stuck-at fault (SF or SAF) (Chapter 3) A fault that causes a memory cell to have a fixed voltage level (either 0 or 1) that cannot be changed in any manner; based on this value, we have a stuck-at-zero (SA0) and a stuck-at-one (SA1) fault.

Glossary

Surrounding gate transistor (SGT) (Chapter 1) A cell with a gate electrode surrounding the sidewalls of a pillar silicon island; in this cell, the source, gate and drain are vertically arranged, and the side-walls of the pillar silicon island form the channel region.

Symmetric PSF (SPSF) (Chapter 5) A pattern-sensitive fault model that partitions the cells in a 9-cell neighborhood into four logical groups – the base cell, bit-line neighbors, word-line neighbors, and diagonal neighbors; each group of cells is assigned one value during the test.

Synchronizing sequence (Chapter 3) An input sequence to a finite-state machine that forces the machine to a fixed final state regardless of the initial state.

Synchronous RAMs (Chapter 1) Self-timed RAMs.

Technology and layout-related (or parametric) testing (Chapter 4) Creating a fault model from a layout-based defect model of a memory device, and designing functional and electrical tests based on such a model; the term 'parametric' refers to *process* parameters, and not AC or DC parameters.

Tessellation (Chapter 5) Any manner of tiling a region of the plane using a set of identical, interlocked shapes.

Test length coefficient (Chapter 5) The average number of operations per cell to detect a certain fault with a given escape probability.

Threshold test (Chapter 2) An electrical test that determines the maximum and minimum output voltages required to cause the output voltage to switch from high to low, and from low to high, respectively.

Timing chain (Chapter 1) A number of clock phases generated sequentially in a driver chain with proper delays in dynamic DRAM circuits.

Transition fault (TF) (Chapter 3) A fault that causes a cell to fail to undergo a low-to-high or a high-to-low voltage transition when a transition write operation is performed on it; this is different from the stuck fault because the cell can interact with neighboring cells and can still undergo a spurious transition.

Transmission-line effect (Chapter 4) Delay through the polysilicon and diffusion lines that are used for the word or bit lines in DRAMs, causing a weak signal to be delivered to cells located far away from the sense amplifiers.

Transparent testing (Chapter 5) A non-concurrent BIST scheme in which the memory contents on-line remain unchanged if the memory is fault-free, after the test is completed.

Tree RAM DFT (Chapter 5) A DFT scheme in which the memory is partitioned into subarrays that are tested in parallel through a tree-type switching (decoder) network.

Trench capacitor (Chapter 1,5) A capacitor cell folded vertically into the surface of the silicon in the form of a trench, to obtain greater area.

Unrestricted PSF (UPSF) (Chapter 3) A pattern-sensitive fault model which assumes that a cell's contents can be influenced by all the other cells in the memory.

Variable hold time (VHT) (Chapter 4) A layout defect causing the hold time to become variable.

V_{DD} transient current monitor (VTCM) (Chapter 4) A current monitor used in I_{DD} testing.

Virtually static DRAM (VSRAM) (Chapter 1) A DRAM with the refresh totally transparent to the user; it consists of a PSRAM combined with intermittently-generated refresh signals with a built-in refresh timing circuitry.

Voltage bump test (Chapter 2) An electrical production test in which the supply voltage V_{CC} is fluctuated and the results produced by read operations observed.

Walking 1/0 (Chapter 3) A functional test for stuck-at and coupling faults.

Word-line decoder (Chapter 5) Row decoder.

Write recovery fault (Chapters 1,5) A write operation at a wrong address during an address change.

B
COMMERCIAL RAM DATA

A large number of chip manufacturers have been involved in RAM design. Tables showing their commercial products are illustrated below. Blank entries indicate data unavailable or not ascertained at the time of writing the book.

Table B.1 DRAMs - Manufacturer: Texas Instruments

Capacity	1 Mb	4 Mb	16 Mb	64 Mb
Year	1986	1986	1990	1991
Process	CMOS	CMOS	CMOS	CMOS
Design Rule (μ)	1.0	1.0	0.6	0.4
Technology	LDD Tr	2P2M	3P2M	
Chip Area (mm^2)	50.3	100	144	268.1
Cell Area (μ^2)	21.25	8.9	4.8	2.0
Storage Capacitor	Trench	Trench	Trench	Trench
Access Time (ns)	120-150		60	40
Cycle Time (ns)	230			
Power Diss. (Active) (mW)	357		450	
Power Diss. (Standby) (mW)	11			
Voltage (V)	5	5	3/4/5	3.3

Table B.2 DRAMs - Manufacturer: Toshiba

Capacity	256 Kb	1 Mb	4 Mb	16 Mb	64 Mb
Year	1983	1985 (2)	1986,87	1988	1991
Process	NMOS	NMOS, CMOS	CMOS	CMOS	CMOS
Design Rule (μ)	2.0	1.2	1.0, 0.9	0.5	0.4
Technology	2P	LDD Tr	2P2M	3P3M	
Chip Area (mm^2)	46.0	54.2	137, 111.2	136.9	196.4
Cell Area (μ^2)	77	29.16	17.4, 13.7	4.8	1.53
Storage Capacitor	Planar	Planar	Trench	Stacked	Trench
Access Time (ns)	34-94	100-120		70	33
Cycle Time (ns)	260	190			
Power Diss. (Active) (mW)	170	330		600	
Power Diss. (Standby) (mW)	15	11			
Voltage (V)	5	5	5	4/5	3.3

Table B.3 DRAMs - Manufacturer: NEC

Capacity	256 Kb	1 Mb	4 Mb	64 Mb
Year	1980	1984	1986	1990
Process	NMOS	NMOS	CMOS	CMOS
Design Rule (μ)	1.3	1.0	0.8	
Technology	2P	Conv. Tr	2P1M	
Chip Area (mm^2)	34.0	48.6	99.2	
Cell Area (μ^2)	67	20.4	10.6	
Storage Capacitor	Planar	Trench	Trench	Stacked
Access Time (ns)	90	100-150		
Cycle Time (ns)		200		
Power Diss. (Active) (mW)	250	550		
Power Diss. (Standby) (mW)	10	27.5		
Voltage (V)	5	5	4/5	3.3

Table B.4 DRAMs - Manufacturer: IBM

Capacity	4 Mb	16 Mb
Year	1987	1990
Process	CMOS	CMOS
Design Rule (μ)	0.8	0.5
Technology	1P2M	2P2M
Chip Area (mm^2)	78	140.9
Cell Area (μ^2)	10.6	4.13
Storage Capacitor	Trench	Trench
Access Time (ns)		50
Voltage (V)		3.3/5

Table B.5 DRAMs - Manufacturer: Mitsubishi

Capacity	256 Kb	1 Mb	4 Mb	16 Mb	64 Mb
Year	1983	1985	1987	1989	1991
Process	NMOS	CMOS	CMOS	CMOS	CMOS
Design Rule (μ)	2.0	1.0	0.8	0.5	0.4
Technology	2P	Conv. Tr	3P1M		
Chip Area (mm^2)	47.55	64.9	72.3	134	233.8
Cell Area (μ^2)	98	29.92	10.9	4.8	1.7
Storage Capacitor	Planar	Planar	Trench	Stacked	Stacked
Access Time (ns)	25-100	100-150		60	45
Cycle Time (ns)	210	190			
Power Diss. (Active) (mW)	250	412		450	
Power Diss. (Standby) (mW)	12	22			
Voltage (V)	5	5		3.3	3.3

Table B.6   DRAMs - Manufacturer: Hitachi

Capacity	1 Mb	4 Mb	16 Mb	64 Mb
Year	1985	1987	1988	1990
Process	CMOS	CMOS	CMOS	CMOS
Design Rule (μ)	1.3	0.8	0.6	0.3
Technology	LDD Tr	2P2M	2M	
Chip Area (mm^2)	64.0	110.8	141.9	199.5
Cell Area (μ^2)	35.7	14.7	4.2	1.28
Storage Capacitor	Planar	Stacked	Stacked	Stacked
Access Time (ns)	100-150		60	50
Cycle Time (ns)	190			
Power Diss. (Active) (mW)	385		420	
Power Diss. (Standby) (mW)	11			
Voltage (V)	5	5	3.3/5	1.5/3.3

Table B.7   DRAMs - Manufacturer: Matsushita

Capacity	1 Mb	4 Mb	16 Mb	64 Mb
Year	1985	1987	1988	1991
Process	NMOS	CMOS	CMOS	CMOS
Design Rule (μ)	1.2	0.8	0.5	0.4
Technology	LDD Tr	2P2M	2P2M	
Chip Area (mm^2)	55.5	67.1	93.8	234.4
Cell Area (μ^2)	35.1	8.0	3.3	2.0
Storage Capacitor	Planar	Trench	Trench	Stacked
Access Time (ns)	100-120		65	50
Cycle Time (ns)	190			
Power Diss. (Active) (mW)	522		450	
Power Diss. (Standby) (mW)	35.8			
Voltage (V)	5	5	3/4/5	3.3

Table B.8 DRAMs - Manufacturer: Fujitsu

Capacity	256 Kb	1 Mb	4 Mb	64 Mb
Year	1984	1985 (2)	1987	1991
Process	NMOS	NMOS (both)	CMOS	CMOS
Design Rule (μ)	2.5	1.4, 1.0	0.7	0.4
Chip Area (mm^2)	34.13	54.7, 60.3	63.7	224.7
Cell Area (μ^2)	68	26.46, 29.24	7.5	1.8
Storage Capacitor	Planar	Stacked, Planar	Stacked	Stacked
Access Time (ns)	15-80	100-150		40
Cycle Time (ns)	200	210,230		
Power Diss. (Active) (mW)	300	550		
Power Diss. (Standby) (mW)	15	24.8		
Voltage (V)	5	5		3.3

Table B.9 DRAM - Manufacturer: Siemens

Capacity	4 Mb
Year	1988
Process	CMOS
Design Rule (μ)	0.75
Chip Area (mm^2)	91.3
Cell Area (μ^2)	10.6
Storage Capacitor	Trench
Access Time (ns)	80
Cycle Time (ns)	160
Fast Page Mode Cycle Time (ns)	45
Power Diss. (Max) (mW)	495
Power Diss. (Standby) (mW)	5.5

Table B.10 DRAMs - Manufacturer: NTT

Capacity	256 Kb	1 Mb	16 Mb
Year	1980	1984	1987
Process	Mo-Poly	CMOS	CMOS
Design Rule (μ)	1.0	0.8	0.7
Chip Area (mm^2)	18		147.7
Cell Area (μ^2)		20	4.9
Storage Capacitor	Planar	Trench	n/a
Access Time (ns)			80
Power Diss. (Active) (mW)			500
Voltage (V)			3.3

Table B.11 DRAM - Manufacturer: AT&T

Capacity	1 Mb
Year	1985
Process	CMOS
Design Rule (μ)	1.3
Chip Area (mm^2)	47.3
Cell Area (μ^2)	24.12
Storage Capacitor	Trench
Voltage (V)	5

Table B.12 DRAM - Manufacturer: Intel

Capacity	1 Mb
Year	1986
Process	CMOS
Design Rule (μ)	
Chip Area (mm^2)	
Cell Area (μ^2)	
Storage Capacitor	Planar
Voltage (V)	5 V (with N-well bias 2.5 V above V_{CC})

Table B.13 DRAM - Manufacturer: Samsung

Capacity	16 Mb
Year	1989
Process	CMOS
Design Rule (μ)	0.55
Technology	3P3M
Chip Area (mm^2)	134.7
Cell Area (μ^2)	4.4
Storage Capacitor	Stacked
Access Time (ns)	60
Power Diss. (Active) (mW)	500
Power Diss. (Standby) (mW)	
Voltage (V)	4/5

Table B.14 DRAM - Manufacturer: Motorola

Capacity	256 Kb
Year	1983
Process	NMOS
Design Rule (μ)	2.0
Technology	1P/2P
Chip Area (mm^2)	46
Cell Area (μ^2)	84
Storage Capacitor	Planar
Access Time (ns)	≤ 90
Cycle Time (ns)	250
Power Diss. (Active) (mW)	225
Power Diss. (Standby) (mW)	15
Voltage (V)	5

Table B.15 SRAMs - Manufacturer: Sony

Capacity	1 Mb	4 Mb
Year	1987	1989
Process	Mix-MOS (P-well CMOS + Poly load resistor)	R-load
Design Rule (μ)	1.0	0.5
Technology	2P2M	2P2M
Chip Area (mm^2)	109	130
Cell Area (μ^2)	74	21.2
Access Time (ns)	35	25
Voltage (V)	5	3.3

Table B.16 SRAMs - Manufacturer: Toshiba

Capacity	1 Mb	4 Mb
Year	1987, 1989	1990
Process	Mix-MOS	P-load
Design Rule (μ)	0.6, 0.8	0.5
Technology	2P2M	3P2M
Chip Area (mm^2)	105, 107	136
Cell Area (μ^2)	53, 50	20.3
Access Time (ns)	25, 8	23
Voltage (V)	5	3.3/5

Table B.17 SRAM - Manufacturer: Philips

Capacity	1 Mb
Year	1987, 1988
Process	Mix-MOS
Design Rule (μ)	0.7
Technology	1P2M
Chip Area (mm^2)	93
Cell Area (μ^2)	60
Access Time (ns)	25
Voltage (V)	5

Commercial RAM data

Table B.18 SRAMs - Manufacturer: Mitsubishi

Capacity	1 Mb	4 Mb
Year	1987-90	
Process	Mix-MOS, CMOS	R-load, P-load
Design Rule (μ)	0.6-0.8	0.6
Technology	3P1M, 3P2M	4P2M
Chip Area (mm^2)	82-88	150, 134
Cell Area (μ^2)	39-44	18.6, 19.5
Access Time (ns)	34, 14, 7	20
Voltage (V)	5	3.3, 3

Table B.19 SRAMs - Manufacturer: Hitachi

Capacity	1 Mb	4 Mb
Year	1987-89	1990
Process	Mix-MOS, CMOS	P-load
Design Rule (μ)	0.5-0.8	0.6
Technology	3P1M, 2P2M	4P2M
Chip Area (mm^2)	79, 94, 55	122
Cell Area (μ^2)	45, 44, 21	17
Access Time (ns)	42, 15, 9	23
Voltage (V)	5	3/5

Table B.20 SRAMs - Manufacturer: Fujitsu

Capacity	1 Mb	4 Mb
Year	1987, 1988	1990
Process	Mix-MOS	BiCMOS
Design Rule (μ)	0.8, 0.7	0.5
Technology	3P1M, 3P2M	4P2M
Chip Area (mm^2)	94, 90	152
Cell Area (μ^2)	41	18.5
Access Time (ns)	44, 18	10
Voltage (V)	5	3.4/5

Table B.21 SRAMs - Manufacturer: NEC

Capacity	1 Mb	4 Mb
Year	1988, 1990	1990
Process	NMOS	R-load
Design Rule (μ)	0.8	0.55
Technology		3P2M
Chip Area (mm^2)	81, 113	143
Cell Area (μ^2)	41, 45	19
Access Time (ns)	35, 5	15
Voltage (V)	5	4/5

C
MARKET FOR RAMS

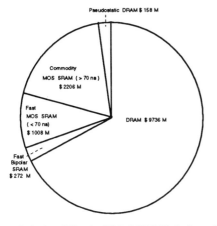

Semiconductor RAM market 1989 (total: $12.03 billion). (Source: Dataquest.)

THE DRAM MARKET

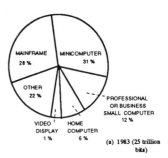

(a) 1983 (25 trillion bits)

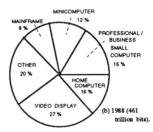

(b) 1988 (461 trillion bits).

Source: Texas Instruments

REFERENCES

[1] Abadir, M.S. and Reghbati, H.K., "Functional Testing of Semiconductor Random-Access Memories," *Computing Surveys of ACM*, vol. 15, no. 3, 1983, pp. 174-198.

[2] Abe, M., et al., "Recent Advances in Ultrahigh-Speed HEMT LSI Technology," *IEEE Trans. on Electron Devices*, vol. 36, no. 10, October 1989, pp. 2021-2031.

[3] Abe, M., et al., "Ultra-High-Speed HEMT Integrated Circuits," *Semiconductors and Semimetals*, vol. 24, Academic Press, San Diego, 1987, pp. 249-278.

[4] Abe, M. and Mimura, T., "Ultra-High-Speed HEMT LSI Technology for Supercomputer," *IEEE J. Solid-State Circuits*, vol. 26, no. 10, October 1991, pp. 1337-1344.

[5] Abe, M. and Yokoyama, N., "Semiconductor Heterostructure Devices," *Gordon and Breach Science Publishers*, New York, 1989.

[6] Anami, K., et al., "Design Consideration of a Static Memory Cell," *IEEE J. Solid-State Circuits*, vol. SC-18, no. 4, August 1983, pp. 414-417.

[7] Ando, M., et al., "A 0.1 μA Standby Current Bouncing-Noise-Immune 1 Mb SRAM," *Symp. on VLSI Technology*, 1988, pp. 49-50.

[8] Anholt, R.E. and Swirhun, S.E., "Experimental Investigation of the Temperature Dependence of GaAs FET Equivalent Circuits," *IEEE Trans. on Electron Devices*, vol. 39, no. 9, 1992, pp. 2029-2036.

[9] Bardell, P.H., et al., "Built-In Testing for VLSI : Pseudo-Random Techniques," *John-Wiley & Sons*, New York, 1987.

[10] Barnes, J.J., et al., "Circuit Techniques for a 25 ns 16K × 1 SRAM Using Address-Transition Detection," *IEEE J. Solid-State Circuits*, vol. SC-19, no. 4, 1984, p. 455-461.

[11] Benowitz, N., et al., "An Advanced Fault Isolation System for Digital Logic," *IEEE Trans. on Computers*, vol. C-24, May 1975, pp. 489-497.

[12] Berge, C., "Graphs and Hypergraphs," *New York: American Elsevier Pub. Co.: North-Holland Pub. Co.*, 1976.

[13] Bondy, J. and Murty, U.S.R., "Graph Theory with Applications," *New York : American Elsevier Pub. Co.*, 1976.

[14] Breuer, M.A. and Friedman, A.D., "Diagnosis and Reliable Design of Digital Systems," *Computer Science Press*, Potomac, MD, 1976.

[15] Brzozowski, J.A. and Jürgensen, H., "A Model for Sequential Machine Testing," *Research Report CS-88-12, Department of Computer Science, University of Waterloo*, April 1988.

[16] Brzozowski, J.A. and Cockburn, B.F., "Detection of Coupling Faults in RAMs," *Journal of Electronic Testing: Theory and Applications*, vol. 1, no. 2, May 1990, pp. 151-162.

[17] Canali, C., et al., "Trap-Related Effects in AlGaAs/GaAs HEMTs," *IEE Proceedings - Part G*, vol. 138, no. 1, February 1991, pp. 104-108.

[18] Chakraborty, K. and Mazumder, P., "Technology and Layout-Related Testing of Static Random-Access Memories," *Journal of Electronic Testing: Theory and Applications, Special Issue on Memory Testing*, vol. 5, no. 4, November 1994, pp. 347-365.

[19] Chappel, B., et al., "Stability and SER Analysis of Static RAM Cells," *IEEE J. Solid-State Circuits*, vol. SC-20, no. 1, February 1985, pp. 383-390.

[20] Chatterjee, P.K., et al., "High-Density Dynamic MOS Memory Devices," *IEEE J. Solid-State Circuits*, vol. SC-14, April 1979, pp. 486-497.

[21] Chuang, C.T. and Lu, P.F., "On the Scaling Property of Trench Isolation Capacitance for Advanced High-Performance ECL Circuits," *IEEE Intl. Electron Devices Meeting Dig. Tech. Papers*, 1989, pp. 799-802.

[22] Cocking, J., "RAM Test Patterns and Test Strategy," *Dig. Papers, 1975 Semiconductor Test Symposium*, Cherry Hill, NJ, October 14-16, 1975 (IEEE Computer Society, 1975), pp. 1-8.

[23] Cockburn, B.F. and Brzozowski, J.A., "Near-Optimal Tests for Classes of Write-Triggered Coupling Faults in RAMs," *Journal of Electronic Testing: Theory and Applications*, vol. 3, no. 3, August 1992, pp. 251-264.

REFERENCES

[24] Cockburn, B.F., "Deterministic Tests for Detecting Single V-Coupling Faults in RAMs," *Journal of Electronic Testing: Theory and Applications*, vol. 5, no. 3, 1994, pp. 91-113.

[25] Cockburn, B.F. and Sat, Y.-F. Nicole, "A Transparent Built-in Self-Test Scheme for Detecting Single V-Coupling Faults in RAMs," *Records of the 1994 IEEE International Workshop on Memory Technology, Design, and Testing*, pp. 119-124.

[26] Cockburn, B.F. and Shen, L., "An Optimal Test for a Realistic DRAM Fault Model," *Sixth Workshop on New Directions for Testing*, Montreal, Canada, May 20-22, 1992.

[27] Coleman, Jim, and Thorpe, Richard, "Boundary Scan Speeds Static Memory Tests," *Electronic Design*, February 1993, pp. 61-72.

[28] Cunningham, B.T., et al., "Fault Characterization and Delay Fault Testing of GaAs Logic Circuits," *Proc. IEEE Intl. Test Conference*, 1986, pp. 836-842.

[29] "Cypress Semiconductor CMOS/BiCMOS Databook," February 1, 1989.

[30] Daehn, W. and Gross, J., "A Test Generator IC for Testing Large CMOS-RAMs," *Proc. IEEE Intl. Test Conference*, 1986, pp. 18-24.

[31] De Jong, P. and Van de Goor, A.J., Comments on "Test Pattern Generation for API Faults in RAM," *IEEE Trans. on Computers*, vol. C-37, no. 11, 1988, pp. 1426-1428.

[32] De Jonge, J.H. and Smeulders, A.J., "Moving Inversions Test Pattern is Thorough, Yet Speedy," *Computer Design*, May 1976, pp. 169-173.

[33] Dekker, R., "Fault Modeling and Self-Test of Static Random Access Memories," *TUD report 1-68340-28(1987)25. Department of Electrical Engineering, Delft University of Technology*: Delft, The Netherlands, 1987.

[34] Dekker, R., et al., "Fault Modeling and Test Algorithm Development for Static Random Access Memories," *Proc. IEEE Intl. Test Conference*, 1988, pp. 343-352.

[35] Dekker, R., et al., "Realistic Built-in Self-Test for Static RAMs," *IEEE Design and Test of Computers*, vol. 6, no. 1, February 1989, pp. 26-34.

[36] Dekker, R., et al., "A Realistic Fault Modeling and Test Algorithms for Static Random Access Memories," *IEEE Trans. on Computers*, vol C-9, no. 6, 1990, pp. 567-572.

[37] Dumas, J.M., et al., "Investigation of Interelectrode Metallic Paths Affecting the Operation of IC MESFETs," *Proc. of the Gallium Arsenide IC Symposium*, 1987, pp. 15-18.

[38] Eaton, S.S., et al., "A Ferroelectric Nonvolatile Memory," *IEEE Intl. Solid-State Circuits Conf. Dig. Tech. Papers*, February 1988, pp. 130- .

[39] Edwards, L., "Low Cost Alternative to Hamming Codes Corrects Memory Errors," *Computer Design*, 1981, pp. 143-148.

[40] Eklund, R., et al., "A 0.5 μm BiCMOS Technology for Logic and 4 Mb-Class SRAMs," *IEEE Intl. Electron Devices Meeting Dig. Tech. Papers*, 1989, pp. 425-428.

[41] Erslab, P. and Lanteri, J.P., "GaAs FET MMIC Switch Reliability," *Proc. of the Gallium Arsenide IC Symposium*, 1988, pp. 57-60.

[42] Franch, R.L., et al., "A Large V_{DS} Data Retention Test Pattern for DRAMs," *IEEE J. Solid-State Circuits*, vol. 27, no. 8, 1992, pp. 1214-1217.

[43] Franklin, M., et al., "Row/Column Pattern-Sensitive Fault Detection in RAMs via Built-in- Self-Test," *IEEE Intl. Symp. on FTC*, 1989, pp. 36-43.

[44] Franklin, M. and Saluja, K.K., "Built-in Self-Testing of Random-Access Memories," *Computer*, vol. 23, no. 10, October 1990, pp. 45-56.

[45] Franklin, M., et al., "A Built-In Self-Test Algorithm for Row/Column Pattern-Sensitive Faults in RAMs," *IEEE J. Solid-State Circuits*, vol. 25, no.2, April 1990, pp. 514-523.

[46] Franklin, M. and Saluja K.K., "An Algorithm to Test RAMs for Physical Neighborhood Pattern-Sensitive Faults," *Proc. IEEE Intl. Test Conference*, 1991, pp. 675-684.

[47] Fuentes, A., et al., "Random Testing versus Deterministic Testing of RAMs," *Proc. 16th IEEE Intl. Conf. Fault-Tolerant Comput. Symp. (FTCS-16)*, July 1986, pp. 266-271.

[48] Furnweger, C., "Dynamic Interfacing, Spare Cells Raise 16K Static RAMs Speed and Yield," *Electronics*, Mar. 24, 1982, pp. 121-124.

[49] Giles, G. and Hunter, C., "A Methodology for Testing Content-Addressable Memories," *Proc. IEEE Intl. Test Conference*, 1985, pp. 471-474.

REFERENCES

[50] Grosspietsch, K.E., et al., "The Concept of a Fault-Tolerant and Easily Testable Associative Memory," *Proc. 16th IEEE Intl. Conf. Fault-Tolerant Comput. Symp. (FTCS-16)*, July 1986, pp. 34-39.

[51] Gubbels, W., et al., "A 40ns/100pF Low-Power Full-CMOS 256K (32K × 8) SRAM," *IEEE J. Solid-State Circuits*, vol. SC-22, no. 5, October 1987, pp. 741-747.

[52] Hayden, J., et al., "A High-Performance Sub-Half Micron CMOS Technology for Fast SRAMs," *IEEE Intl. Electron Devices Meeting Dig. Tech. Papers*, 1989, pp. 417-420.

[53] Hayes, J.P., "Detection of Pattern-Sensitive Faults in Random-Access Memories," *IEEE Trans. on Computers*, vol. C-24, no. 2, 1975, pp. 150-157.

[54] Hayes, J.P., "Testing Memories for Single-Cell Pattern-Sensitive Faults in Semiconductor Random-Access Memories," *IEEE Trans. on Computers*, vol. C-29, no. 3, pp. 249-254.

[55] Healy, J.T., "Automatic Testing and Evaluation of Digital Integrated Circuits," *Reston Publishing Company, Inc., a Prentice-Hall Company*, Reston, VA, 1981.

[56] Hisamoto, D., et al., "A Fully Depleted Lean-channel Transistor (DELTA) – A Novel Vertical Ultra-Thin SOI MOSFET," *IEEE Intl. Electron Devices Meeting Dig. Tech. Papers*, pp. 833-836.

[57] "Hitachi IC Memory Products Databook," *1987/88, DL113R2*.

[58] Hori, T., "1/4 μm LATID (LArge-Tilt-angle Implanted Drain) Technology for 3.3 V Operation," *IEEE Intl. Electron Devices Meeting Dig. Tech. Papers*, 1989, pp. 777-784.

[59] Horiguchi, M., et al., "An Experimental Large-Capacity Semiconductor File Memory Using 16-Levels/Cell Storage," *IEEE J. Solid State Circuits*, vol. 23, no. 1, 1988, pp. 27-33.

[60] Huisman, L.M., "Simulation of Embedded Memories by Defective Hashing," *IBM J. Res. Develop.*, vol. 34, no. 2/3, March/May 1990, pp. 289-298.

[61] IEEE Standard 1149.1-1990, "IEEE Standard Test Access Port and Boundary Scan Architecture," *IEEE Standards Board*, 345 East 47th Street, New York, NY 10017, May 1989.

[62] Ikoma, T., "Very High Speed Integrated Circuits: Heterostructure," *Academic Press, Inc.*, San Diego, CA, 1990.

[63] Inoue, J., et al., "Parallel Testing Technology for VLSI Memories," *Proc. IEEE Intl. Test Conference*, 1987, pp. 1066-1071.

[64] Isomura, S., et al., "A 36 kb/2 ns RAM with 1 kG/100 ps Logic Gate Array," *IEEE Intl. Solid-State Circuits Conf. Dig. Tech. Papers*, February 1989, pp. 26-27.

[65] Jain, S.K. and Stroud, C.E., "Built-in Self Testing of Embedded Memories," *IEEE Design and Test of Computers*, October 1986, pp. 27-37.

[66] Jarwala, N. and Pradhan, D., "An Easily Testable Architecture for Multimegabit RAMs," *Proc. IEEE Intl. Test Conference*, 1987, pp. 750-758.

[67] Johnson, C.L., "ATE Considerations for Board and System Test," *Application Note : Micro Control Company*, ©1980 Micro Control Company, 1987.

[68] Kang, S.W. and Lee, J.Y.M., "Low-Temperature Degradation Studies of AlGaAs/GaAs Modulation-Doped Field Effect Transistors," *Solid-State Electronics*, vol. 34, no. 12, 1991, pp. 1415-1419.

[69] Kantz, D., et al., "A 256 Kb DRAM with Descrambled Redundancy Test Capability," *IEEE J. Solid-State Circuits*, vol. SC-19, October 1984, pp. 596-602.

[70] Kinoshita A., et al., "A Study of Delay Time on Bit Lines in Megabit SRAMs," *IEICE Trans. Electron Devices*, vol. E75-C, no. 11, November 1992, pp. 1383-1386.

[71] Kinoshita, K. and Saluja, K.K., "Built-in Testing of Memory using On-chip Compact Testing Scheme," *Proc, IEEE Intl. Test Conference*, 1984, pp. 271-281.

[72] Kishore, J.K., et al., "Testing of Semiconductor Random Access Memories, a Survey" : *ISRO Satellite Center*, Bangalore, India, 1985.

[73] Knaizuk Jr., J. and Hartmann, C.R.P., "An Algorithm for Testing Random-Access Memories," *IEEE Trans. on Computers*, vol. C-26, no. 4, 1977, pp. 414-416.

[74] Knaizuk Jr., J. and Hartmann, C.R.P., "An Optimal Algorithm for Testing Stuck-at-Faults in Random-Access Memories," *IEEE Trans. on Computers*, vol. C-26, no. 11, 1977, pp. 1141-1144.

REFERENCES

[75] Kobayashi, N., et al., "A Fully Operational 1 Kb HEMT Static RAM," *IEEE Trans. on Electron Devices*, vol. ED-33, no. 5, May 1986, pp. 548-553.

[76] Kohavi, Z., "Switching and Finite Automata Theory, 2nd Ed.," *McGraw-Hill Publishing Co.*, 1986.

[77] Komatsu, T., et al., "A 35-ns 128K × 8 CMOS SRAM," *IEEE J. Solid-State Circuits*, vol. SC-22, no. 5, October 1987, pp. 721-726.

[78] Kuo, C., et al., "Soft-Defect Detection (SDD) Technique for a High-Reliability CMOS SRAM," *IEEE J. Solid-State Circuits*, vol. 25. no. 1, February 1990, pp. 61-67.

[79] Lefebvre, M.F., "Test Generation : A Boundary Scan Implementation for Module Interconnect Testing," *Proc. IEEE Intl. Test Conference*, 1991, pp. 88-95.

[80] Lo, T.C. and Guidry, M.R., "An Integrated Test Concept for Switched-Capacitor Dynamic MOS RAMs," *IEEE J. Solid-State Circuits*, vol. SC-12, December 1977, pp. 693-703.

[81] Lu, N.C. and Chao, H.H., "Half-VDD Bit-Line Sensing Scheme in CMOS DRAMs," *IEEE J. Solid-State Circuits*, vol. SC-19, no. 4, 1984, pp. 451-454.

[82] Lu, N.C., et al., "A substrate-plate trench-capacitor (SPT) memory cell for dynamic RAMs," IEEE J. Solid-State Circuits, vol. SC-21, October 1986, pp. 627-634.

[83] Marinescu, M., "Simple and Efficient Algorithms for Functional RAM Testing," *Proc. IEEE Intl. Test Conference*, 1982, pp. 236-239.

[84] Maly, W. (1985), "Modeling of Lithography-Related Yield Losses for CAD of VLSI Circuits," *IEEE Trans. on CAD*, CAD-4, no. 3, 1985, pp. 166-177.

[85] Maly, W., et al., "Yield Diagnosis through Interpretation of Tester Data," *Proc. IEEE Intl. Test Conference*, 1987, pp. 10-20.

[86] Maly, W. and Naik, S., "Process Monitoring-Oriented Testing," *Proc. IEEE Intl. Test Conference*, 1989, pp. 527-532.

[87] Maly, W., et al., "VLSI Yield Prediction and Estimation : a Unified Framework," *IEEE Trans. on Computer-Aided Design of Integrated Circuits*, vol. CAD-5, no. 1, January 1986, pp. 114-130.

[88] Matsuda, Y., et al., "A New Array Architecture for Parallel Testing in VLSI Memories," *Proc. IEEE Intl. Test Conference*, 1989, pp. 322-326.

[89] Maurer, R.H., et al., "Failure Analysis of Aged GaAs HEMTs," *29th Annual Proceedings, Reliability Physics* 1991, pp. 214-223.

[90] Max Cortner, J., "Digital Test Engineering," *John Wiley and Sons*, NY, 1987.

[91] Mazumder, P., and Patel, J.H., "Testable RAM Design," *SRC Corporate Research*, 1986 Annual Report.

[92] Mazumder, P. and Patel, J.H., "A Novel Fault-Tolerant Design of Testable Dynamic Random Access Memory," *Proc. 1987 IEEE Intl. Conference on Comp. Design (ICCD '87): VLSI in Computers & Processors*, October 1987, pp. 306-309.

[93] Mazumder, P., et al., "Design and Algorithms for Parallel Testing of Random-Access and Content-Addressable Memories," *Proc. Design Automation Conference (DAC'87)*, June 1987, pp. 688-694 (nominated for the Best Paper Award).

[94] Mazumder, P., and Patel, J.H., "Methodologies for Testing Embedded Content-Addressable Memories," *Proc. 17th IEEE Intl. Conf. Fault-Tolerant Comput. Symp. (FTCS-17)*, July 1987, Pittsburgh, Pennsylvania, pp. 270-275.

[95] Mazumder, P., "Testing and Fault-Tolerance Aspects of High-Density VLSI Memory," Ph.D. dissertation, Univ. of Illinois, Dept. Elec. Comput. Eng., August 1987.

[96] Mazumder, P., and Patel, J.H., "An Efficient Built-in Self-Testing for Random-Access Memory," *Proc. IEEE Intl. Test Conference*, September 1987, pp. 1072-1077.

[97] Mazumder, P., "Parallel Testing of Parametric Faults in a Three-Dimensional Random-Access Memory," *IEEE J. Solid-State Circuits*, vol. 23, no. 4, August 1988, pp. 933-941.

[98] Mazumder, P. and Patel, J.H., "Parallel Testing of Parametric Faults in DRAM," in *Advanced Research in VLSI: Design and Applications of Very Large Scale Systems*, Leighton and Allen (editors), MIT Press, 1988. (Presented at the 5-th Massachusetts Institute of Technology Conference on VLSI).

[99] Mazumder, P., et al., "Methodologies for Testing Embedded Content-Addressable Memories," *IEEE Trans. on Computer-Aided Design of Integrated Circuits and Systems*, January 1988, pp. 11-20.

[100] Mazumder, P., "An Efficient Design of Embedded Memories for Random Pattern Testability," *Proc. IEEE Intl. Conf. on Wafer Scale Integration*, January 1989, San Francisco, pp. 230-237.

[101] Mazumder, P., and Patel, J.H., "Parallel Testing for Pattern-Sensitive Faults in Semiconductor Random-Access Memories," *IEEE Trans. on Computers*, vol. 38, no. 3, March 1989, pp. 394-407.

[102] Mazumder, P., and Patel, J.H., "An Efficient Built-in Self-Testing Algorithm for Random-Access Memory," *IEEE Trans. on Industrial Electronics* (Special Issue on Testing), vol. 36, no. 3, May 1989, pp. 246-253.

[103] Mazumder, P. and Patel, J.H., "An Efficient Design of Embedded Memories and their Testability Analysis using Markov Chains," *Journal of Electronic Testing : Theory and Applications*, vol. 3, 1992, pp. 235-250.

[104] Mazumder, P., and Patel, J.H., "An Efficient Design of Embedded Memories and their Testability Analysis using Markov Chains," *Proc. Intl. Conf. Wafer Scale Integration*, January 1992, pp. 389-400.

[105] McAdams, H., et al., "A 1 Mb CMOS Dynamic RAM with Design-For-Test Functions," *IEEE J. Solid-State Circuits*, vol. SC-21, October 1986, pp. 635-641.

[106] McAnney, W.H., et al., "Random Testing for Stuck-At Storage Cells in an Embedded Memory," *Proc. IEEE Intl. Test Conference*, 1984, pp. 157-166.

[107] Meershoek, R.F.M., "Functional and I_{DDQ} Testing on a Static RAM," *TUD report 1-68340-28(1990)04. Department of Electrical Engineering, Delft University of Technology*: Delft, The Netherlands, 1990.

[108] Meershoek, R., et al., "Functional and I_{DDQ} Testing on a Static RAM," *Proc. IEEE Intl. Test Conference*, 1990, pp. 929-937.

[109] "Micro Control Company Test Application Series : An Introduction to Digital Testing," ©*1980 Micro Control Company*, 00004, 1980.

[110] "Micro Control Company Test Application Series: Monitored Burn-In, an Overview," ©*1980 Micro Control Company*, 00046, 1980.

[111] "Micro Control Company Test Application Series: Testing Memory Chips," ©*1980 Micro Control Company*, 00046, 1980.

[112] Mohan, S. and Mazumder, P., "Fault Modeling and Testing of GaAs Static Random Access Memories," *Proc. IEEE Intl. Test Conference*, 1991, pp. 665-674.

[113] Mohan, S. and Mazumder, P., "Analytical and simulation studies of failure modes in SRAMs using High electron mobility transistors," *IEEE Trans. on Computer-Aided Design of Circuits and Systems*, vol. 12, no. 12, December 1993, pp. 1885-1896.

[114] Morooka, Y., et al., "An Address Maskable Parallel Testing for Ultra High Density DRAMs," *Proc. IEEE Intl. Test Conference*, 1991, pp. 556-563.

[115] "MOS Memory Data Book : Commercial and Military Specifications," *Texas Instruments*, 1984.

[116] "Motorola Memory Databook," 1989.

[117] Mundy, J.L., et al., "Low-Cost Associative Memory," *IEEE J. Solid-State Circuits*, vol. SC-7, 1972, pp. 364-369.

[118] Murakami, S., et al., "A 21 mW 4 Mb CMOS SRAM for Battery Operation," *IEEE Intl. Solid-State Circuits Conf. Dig. Tech. Papers*, February 1991, pp. 46-47.

[119] Nadeau-Dostie, B., et al., "Serial Interfacing for Embedded Memories," *IEEE Design and Test of Computers*, vol. 7, no. 2, 1990, pp. 52-63.

[120] Naik, S., et al., "Failure Analysis of High Density CMOS SRAMs Using Realistic Defect Modeling and I_{DDQ} Testing," *IEEE Design and Test of Computers*, vol. 10, June 1993, pp. 13-23.

[121] Nair, R., et al., "Efficient Algorithms for Testing Semiconductor Random Access Memories," *IEEE Trans. on Computers*, vol. C-27, no. 6, 1978, pp. 572-576.

[122] Nair, R., "Comments on 'An Optimal Algorithm for Testing Stuck-at Faults in Random-Access Memories'," *IEEE Trans. on Computers*, vol. C-28, no. 3, 1979, pp. 258-261.

[123] Naono, N. and Asano, M., "What's Happening to the DRAM Market?," *Semiconductor International*, June 1993, pp. 154-157.

[124] Nicolaidis, M., "An Efficient Built-In Self-Test Scheme for Functional Test of Embedded RAMs," *Proc. 15th IEEE Intl. Conf. on Fault-Tolerant Comput. Symp., (FTCS-15)*, July 1985, pp. 118-123.

REFERENCES

[125] Nicolaidis, M., "Transparent BIST for RAMs," *Proc. IEEE Intl. Test Conference*, 1992, pp.598-607.

[126] Nicolaidis, M. and Kebichi, O., "Transparent BIST for RAMs versus Standard BIST," *IEEE Design and Test of Computers*, 1994.

[127] Nicolaidis, M., et al., "Trade-offs in Scan Path and BIST Implementations of RAMs," *Journal of Electronic Testing: Theory and Applications*, vol. 5, no. 2, May 1994, pp. 273-283.

[128] Noble, W.P., et al., "Parasitic Leakage in DRAM Trench Storage Capacitor Vertical Gated Diodes," *IEEE Intl. Electron Devices Meeting Dig. Tech. Papers*, no. 14.5, 1987, pp. 340-343.

[129] Notomi, S., et al., "A High-Speed 1K × 4-bit Static RAM using 0.5μm-gate HEMT," *Proc. of the Gallium Arsenide IC Symposium*, 1986, pp. 177-180.

[130] Oberle, H.D., et al., "Enhanced Fault Modeling for DRAM Test and Analysis," *Digest of the 1991 IEEE VLSI Test Symp.*, Atlantic City, NJ, April 15-17, 1991, (IEEE Comp. Soc., Washington, 1991), pp. 149-154.

[131] Oberle, H.D. and Muhmenthaler, P., "Test Pattern Development and Evaluation for DRAMs with Fault Simulator RAMSIM," *Proc. IEEE Intl. Test Conference*, 1991, pp. 548-555.

[132] O'Conner, K., "A Prototype 2K × 8b Pipelined Static RAM," *Proc. International Solid-State Circuits Conference*, 1989, pp. 66-67 (omitted).

[133] Papachristou, C.A. and Sahgal, N.B., "An Improved Method for Detecting Functional Faults in Random-Access Memories," *IEEE Trans. on Computers*, vol. C-34, no. 2, 1985, pp. 110-116.

[134] Pavio, J.S. and Rhine, D., "GaAs MMIC Evaluation of Via Fracturing," *Proc. of the Gallium Arsenide IC Symposium*, 1986, pp. 305-307.

[135] Pinkham, R., et al., "Video RAM Excels at Fast Graphics," *Electronic Design*, August 18, 1983, pp. 161-172.

[136] Prince, B., "Semiconductor Memories - A Handbook of Design, Manufacture, and Application, 2nd Ed.," *John Wiley & Sons.*, NY, 1991.

[137] Raposa, M.J., "Dual-Port Static RAM Testing," *Proc. IEEE Intl. Test Conference*, 1988, pp. 362-368.

[138] Ritter, H.C. and Muller, B., "Built-in Test Procesor for Self-Testing Repairable Random Access Memories," *Proc. IEEE Intl. Test Conference*, 1987, pp. 1078-1084.

[139] Roesch, W.J. and Peters, M.F., "Depletion Mode GaAs Reliability," *Proc. of the Gallium Arsenide IC Symposium*, 1987, pp. 27-30.

[140] Saluja, K.K. and Kinoshita, K., "Test Pattern Generation for API Faults in RAM," *IEEE Trans. on Computers*, vol. C-34 (3), 1985, pp. 284-287.

[141] Sarkany, E.F. and Hart, W.S., "Minimal Set of Patterns to Test RAM Components," *Proc. IEEE Intl. Test Conference*, 1987, pp. 759-764.

[142] Sasaki, K., et al., "A 23 ns 4 Mb CMOS SRAM with 0.5 μA Standby Current," *IEEE Intl. Solid-State Circuits Conf. Dig. Tech. Papers*, February 1990, pp. 130-131.

[143] Sawada, K., et al., "A 30-μA Data-Retention Pseudostatic RAM with Virtually Static RAM Mode," *IEEE J. Solid-State Circuits*, vol. 23, no.1, February 1988, pp. 12-19.

[144] Scholz, H.N., et al., "A Method for Delay Fault Self-Testing of Macrocells," *Proc. IEEE Intl. Test Conference*, 1993, pp. 253-261.

[145] Seevinck, E., et al., "Static-Noise Margin Analysis of MOS SRAM Cells," *IEEE J. Solid-State Circuits*, vol. SC-22, no. 5, pp. 748-754, October 1987, pp. 748-754.

[146] *Semiconductor International*, June 1993, p. 18.

[147] Shah, A.H., et al., "A 4 Mb DRAM with Trench-Transistor Cell," *IEEE J. Solid-State Circuits*, vol. SC-21, October 1986, pp. 618-627.

[148] Shen, J.P., et al., "Inductive Fault Analysis of CMOS Integrated Circuits," *IEEE Design & Test of Computers*, 1985, pp. 13-26.

[149] Shen, Lin and Cockburn, B.F., "An Optimal March Test for Locating Faults in DRAMs," *record of the 1993 IEEE Intl. Workshop on Memory Testing*, San Jose, CA, August 9-10, 1993.

[150] Shinohara, H., et al., "A 45ns 256K CMOS Static RAM with a Tri-Level Word Line," *IEEE J. Solid-State Circuits*, vol. SC-20, no. 5, October 1985, pp. 929-934.

[151] Soares, R., editor, "GaAs MESFET circuit design," *Artech House*, Boston, 1988.

[152] Soden, J.M., et al., "CMOS IC Stuck-Open Fault Electrical Effects and Design Considerations," *IEEE Intl. Test Conference*, 1989, pp. 423-430.

[153] Spence, R. and Soin, R.S., "Tolerance Design of Electronic Circuits," *Addison-Wesley Publishing Company*, Wokingham, England, 1988.

[154] Sridhar, T., "Analysis and Simulation of Parallel Signature Analyzers," *Proc. IEEE Intl. Test Conference*, 1983, pp. 656-661.

[155] Sridhar, T., "A New Parallel Test Approach for Large Memories," *Proc. IEEE Intl. Test Conference*, 1985, pp. 462-470.

[156] Stein, M.L., "An Efficient Method of Sampling for Statistical Circuit Design," *IEEE Trans. on Computer-Aided Design of Integrated Circuits*, vol. CAD-5, no. 1, January 1986, pp. 23-29.

[157] Su, S. and Makki, R.Z., "Testing of Static Random-Access Memories by Monitoring Dynamic Power Supply Current," *Journal of Electronic Testing: Theory and Applications*, 1992, pp. 265-278.

[158] Suk, D.S. and Reddy, S.M., "Test Procedures for a Class of Pattern-Sensitive Faults in Semiconductor Random-Access Memories," *IEEE Trans. on Computers*, vol. C-29, no. 6, 1980, pp. 419-429.

[159] Suk, D.S. and Reddy, S.M., "A March Test for Functional Faults in Semiconductor Random-Access Memories," *IEEE Trans. on Computers*, vol. C-30, no. 12, 1981, pp. 982-985.

[160] Sunouchi, K., et al., "Double LDD Concave (DLC) Structure for Sub-Half Micron MOSFETs," *IEEE Intl. Electron Devices Meeting Dig. Tech. Papers*, 1988, pp. 226-229.

[161] Takato, H., et al., "High-Performance CMOS Surrounding Gate Transistor (SGT) for Ultra-High Density LSIs," *IEEE Intl. Electron Devices Meeting Dig. Tech. Papers*, 1988, pp. 223-225.

[162] Texas Instruments, "Boundary-Scan Logic IEEE Std. 1149.1 (JTAG): 5 V and 3.3 V Bus-Interface and Scan Support Products," *Data Book, Advanced System Logic Products, Texas Instruments*, ©1994.

[163] Thatte, S.M. and Abraham, J.A., "Testing of Semiconductor Random-Access Memories," *Proc. 7th Annual Intl. Conf. on Fault Tolerant Computing*, 1977, pp. 81-87.

[164] Treuer, R., "Built-In Self-Diagnosis for Repairable Embedded RAMs," *Ph.D. thesis*, 1993, not yet published.

[165] Van de Goor, A.J., "Testing Semiconductor Memories - Theory and Practice," *John Wiley & Sons*, UK, 1991.

[166] Van de Goor, A.J., et al., "Locating Bridging Faults in Memory Arrays," *Proc. IEEE Intl. Test Conference*, 1991, pp. 685-694.

[167] Van de Goor, A.J., et al., "Functional Memory Array Testing," *Proc. of COMPEURO90 Conference*, IEEE Tel-Aviv, 1990, pp. 408-415.

[168] Van de Goor, A.J. and Zorian, Y., "Effective March Algorithms for Testing Single-Order Addressed Memories," *Journal of Electronic Testing, Theory and Applications*, 1994.

[169] Van Sas, J., et al., "Test Algorithms for Double-Buffered Random-Access and Pointer-Addressed Memories," *IEEE Design and Test of Computers*, 1993.

[170] Vida-Torku, E.K., et al., "Test Generation for VLSI Chips with Embedded Memories," *IBM J. Res. Develop.*, vol. 34, no. 2/3, March/May 1990, pp. 276-288.

[171] Wada, T., et al., "A 34 ns 1 Mb CMOS SRAM Using Triple Polysilicon," *IEEE J. Solid-State Circuits*, vol. SC-22, no. 5, October 1987, pp. 727-732.

[172] Walker, H. and Director, S., "VLASIC : A Catastrophic Fault Yield Simulator for Integrated Circuits," *IEEE Trans. CAD Circuits and Systems*, vol. CAD-5, no. 4, October 1986, pp. 541-546.

[173] Wang, L.K., et al., "Characteristics of CMOS Devices Fabricated Using High Quality Thin PECVD Gate Oxide," *IEEE Intl. Electron Devices Meeting Dig. Tech. Papers*, 1989, pp. 463-466.

[174] Wang, N., "Digital MOS Integrated Circuits : Design for Applications," *Prentice Hall*, NJ 07632, 1989.

[175] Weste, N. and Eshraghian, K., "Principles of CMOS VLSI Design, a Systems Perspective," *Addison-Wesley Publishing Company*, 2nd Ed., 1993.

[176] Winegarden, S. and Pannell, D., "Paragons for Memory Test," *Proc. IEEE Intl. Test Conference*, 1981, pp. 44-48.

[177] Yamamoto, S., et al., " A 256K CMOS SRAM with Variable Impedance Data-Line Loads," *IEEE J. of Solid-State Circuits*, vol. SC-20, no. 5, October 1985, pp. 924-928.

REFERENCES

[178] Yananaka, T., et al., "A 25 μm^2, New Poly-Si PMOS Load (PPL) SRAM Cell Having Excellent Soft Error Immunity," *IEEE Intl. Electron Devices Meeting Dig. Tech. Papers*, 1988, pp. 48-51.

[179] Yaney, D.S., et al., "A Meta-Stable Leakage Phenomenon in DRAM Charge Storage - Variable Hold Time," *IEEE Intl. Electron Devices Meeting Dig. Tech. Papers*, no. 14.4, 1987, pp. 336-339.

[180] Yang, P., et al., "An Integrated and Efficient Approach for MOS VLSI Statistical Circuit Design," *IEEE Trans. on Computer-Aided Design of Integrated Circuits*, vol. CAD-5, no. 1, 1986, pp. 5-14.

[181] Yeager, H.R. and Dutton, R.W., "Users Guide to the Stanford HEMT SPICE model," *Stanford Electronics Laboratories*, 1987.

[182] Yeager, H.R. and Dutton, R.W., "Circuit Simulation Models for the High Electron Mobility Transistor," *IEEE Trans. on Electron Devices*, vol. ED 33, no. 5, May 1986, pp. 682-692.

[183] You, Y. and Hayes, J.P., "A Self-Testing Dynamic RAM Chip," *IEEE J. Solid-State Circuits*, vol. SC-20, February 1985, pp. 428-435.

[184] Yuzuriha, K., et al., "A New Process Technology for a Polysilicon Load Resistor Cell," *Symp. VLSI Circuits Dig. Tech. Papers*, 1989, pp. 162-163.

INDEX

AC parametric test, 58
  characterization, 63
  propagation delay, 63
  rise and fall times, 60
  running time, 64
  sense amp recovery, 64
  set-up, hold and release times, 60
  write recovery, 64
Address-maskable BIST, 278
Algorithm A (Nair
  et al.), 107
Algorithm B (Nair
  et al.), 107
Alpha-particle, 16, 22
ATD (address transition detection, 29
ATD (address transition
  detection), 31
Base cell, 91, 95
Bathtub curve, 49
BIST (built-in self-testing), 222
Bit-to-word-line crosstalk, 258
Butterfly, 95
Cell design, 4
CF (coupling fault), 86, 201
  k-coupling, 106
    multiple, 107
    restricted, 107
  BF (bridging fault), 87
  CFid (idempotent), 90
  CFin (inversion), 90
  DCF (dynamic), 86
  SCF (state), 86–87, 90

Column address select timing, 10
Complete test, 92
Concurrent BIST, 223
Dark current, 211, 255
Data retention test, 165
DBM (double-buffered memory), 136
DC parametric test, 52
  contact, 53
  leakage, 55
  output drive current, 57
  output short current, 57
  power consumption, 55
  threshold, 56
  voltage bumping, 57
Decoder, 4
Defective hashing, 151
Delay fault BIST, 246
Delay testing, 169
Deterministic comparison, 228
Dielectric pinhole defects, 255
Double LDD concave transistor, 36
DRAM, 3
  battery back-up, 21
  bit density, 19
  dual-port, 23
  embedded, 21
  fast, 22
  isolated storage plate, 16
  march tests, 109
  multi-megabit, 14
  multilevel storage, 22

PSRAM (Pseudostatic DRAM), 20
refresh, 12, 22
silicon files, 21
stacked capacitor, 14
trench capacitor, 14, 255
triple-well, 16
video DRAM (VDRAM), 22
VSRAM (Virtually Static DRAM), 20
Dual-port SRAM testing, 65
  circuit-dependent tests, 67
  contention, 69
  semaphore, 69
ECC (error-correcting code), 223
Electrical testing, 46
  characterization, 47
  production, 48
Eulerian cycle, 84, 97, 116, 272
Fast-page mode, 26
Fault detection, 90
Fault location, 90
Fault model
  electrical, 46
  functional, 86
  parametric, 159
  technology- and layout-related, 163
Fault simulation, 147
Ferroelectric materials, 16
Field-inversion current, 211, 255
Folded bit-line architecture, 14
Functional fault model
  AF (address decoder fault), 90
  CF (coupling fault), 86
  PSF (pattern-sensitive fault), 111
  SAF (stuck-at fault), 86
  TF (transition fault), 86, 90
Functional fault models, 85
Functional testing, 76
  non-march test, 92

fault modeling, 85
layout-independent, 76, 78
layout-oriented, 76, 132
march test, 99
Mealy automaton, 79
NPSF detection and location, 126
transparent, 146
GALCOL, 95
GALPAT ('gal'loping pattern), 92
GALROW, 95
Gate leakage current, 255
Ground bounce, 31
Half-V_{CC} bit-line sensing, 27
Hamiltonian sequence, 116, 273
HEMT, 183
IDD testing, 159, 179
IDDQ testing, 159, 172
IFA (Inductive Fault Analysis), 133, 140, 176
IFA-13 ($13N$ Test), 135
IFA-9 ($9N$ test), 134
Irredundant test, 92
Isolation leakage, 205
LFSR (linear-feedback shift register), 227
March test, 99
  Suk's second algorithm (March B), 106
  ATS ('A'lgorithmic 'T'est 'S'equence), 102
  March X, 106
  March Y, 106
  Marinescu's algorithm (March C-), 106
  MATS, 103
  MATS+, 103
  MATS++, 106
  Suk's algorithm (March A), 106
Markov chain, 308
MISR (multiple-input shift register), 229

Index

Modified bit-line decoder for BIST, 252
MOVI (Moving Inversions), 109
Multi-megabit RAMs, 33
Multiplexed addressing, 9
Mutual comparison, 229
Neighborhood, 91
Non-march test, 92
   k-coupling faults, 97–98
   butterfly, 95
   checkerboard, 92
   divide-and-conquer approach, 96
   GALCOL, 95
   GALPAT ('gal'loping pattern), 92
   GALROW, 95
   sliding diagonal, 95
   walking 1/0, 95
Nonconcurrent BIST, 223
NPSF (neighborhood PSF), 113
   physical(or scrambled), 142
   ANPSF(active), 113
   APNPSF (active/passive), 124
   PNPSF (passive), 113
   SNPSF(static), 113
   symmetric, 269
Open, 160, 179
PAM (pointer-addressed memory), 136
Parallel BIST for NPSFs, 266
Parallel comparator faults, 277
Parallel comparator for BIST, 252
Parallel parametric testing, 261
Parametric fault model
   bit-line voltage imbalance, 207
   sleeping sickness, 204
   bit-line voltage imbalance, 214, 255
   bit-line/word-line coupling, 209
   refresh line stuck-at, 207
   retention, 200, 204
   stuck-open fault, 200
Parametric fault, 159
Power dissipation reduction, 31
   pulsed word line, 32
   bit-line equalization, 32
   latched-column, 32
Power supply voltage transition test, 264
PSA (Parallel Signature Analyzer), 231
Pseudorandom BIST, 300
PSF (pattern-sensitive fault), 111
   NPSF (neighborhood), 112
   row/column, 112
   UPSF (unrestricted), 112
RAM (random-access memory), 3
   bipolar, 162
   DRAM (dynamic), 3
   GaAs (Gallium Arsenide), 182
   MOS, 165
   SRAM (static), 3
Read operation, 12
Read, 194
   destructive, 277
Row address select timing, 10
SAF (stuck-at fault), 86, 88
   SA0 (stuck-at zero), 88
   SA1 (stuck-at one), 88
SDD (soft-defect detection), 165
Sense amplifier, 7, 12, 27
Serial BIST, 237
SGalpat, 238
SGT (surrounding gate transistor), 35
Short, 87, 160, 181, 203
Signature, 229
Sliding diagonal, 95
SMarch, 238
SMarchdec, 238
SOI (silicon-on-insulator), 35
SRAM, 3
   BiCMOS, 20
   bipolar, 20

BRAM (battery RAM), 24
CAM (content-addressable
  memories), 25, 282
CMOS/ECL-compatible, 20
dual-port, 24, 65
FIFO (first-in
  first-out) memories, 25, 110
GaAs, 20
STRAM (self-timed or
  synchronous), 30
Stuck-open faults, 135
Submicron transistors, 34
Subthreshold leakage, 205
SVCTEST, 107
SWalk, 238
Test length coefficient, 313
Test-LTCid, 106
Test-LTCin, 106

Test-UCin, 106
Test-US, 103
Test-UT, 105
Testing, 37, 46, 76
TF (transition fault), 86
Tiling method, 117
Topological checkerboard, 205
Transmission line effects, 258
Transparent BIST, 223, 294
Tree RAM BIST, 234
Two-group method, 119
VHT (variable hold time), 206
Walking 1/0, 95
Weak-inversion current, 208, 255
Write operation, 11, 85, 86, 195, 275
Write-sensing latch, 317